The Maintenance and Management of Roadways and Bridges

American Association of State
Highway and Transportation Officials
444 North Capitol Street, NW, Suite 249
Washington, DC 20001

tel 202-624-5800
fax 202-624-5806
web www.aashto.org

© Copyright 1999, by the American Association of State Highway and Transportation Officials. *All Rights Reserved*. Printed in the United States of America. This book, or parts thereof, may not be reproduced in any form without permission of the publishers.

ISBN: 1-56051-135-4

Published by the
American Association of State Highway and Transportation Officials
444 North Capitol Street, N.W.
Suite 249
Washington, D.C. 20001
(202) 624-5800
www.aashto.org

EXECUTIVE COMMITTEE
2000

Voting Members

Officers:
    President: Thomas R. Warne, Utah
    Vice President: E. Dean Carlson, Kansas
    Secretary-Treasurer: Clyde Pyers, Maryland

Regional Representatives:
    Region I: James Sullivan, Connecticut
                   William Ankner, Rhode Island
    Region II: Kam K. Movassaghi, Louisiana
                   David McCoy, North Carolina
    Region III: James C. Codell, III, Kentucky
                   Cristine Klika, Indiana
    Region IV: Sid Morrison, Washington
                   Pete Rahn, New Mexico

Nonvoting Members

Immediate Past President: Dan Flowers, Arkansas
AASHTO Executive Director: John C. Horsley

HIGHWAY SUBCOMMITTEE ON MAINTENANCE

Tom Norton, Colorado, Chairman
Ray Bass, Alabama, Vice Chairman
James B. Sorenson, Federal Highway Administration, Secretary

ALABAMA, Francis Allred, Ray Bass, Mike Harper, John E. Lorentson
ALASKA, Frank Richards
ARIZONA, James Dorre
ARKANSAS, Jim Barnett, Ralph Hall
CALIFORNIA, Randell Iwasaki
COLORADO, Tom Norton, Ed Fink, William Reisbeck
CONNECTICUT, Louis R. Malerba, Michael D. Turano
DELAWARE, Charles Lightfoot
DISTRICT OF COLUMBIA, Luke Dipompo, Gary Burch
FLORIDA, Sharon Holmes
GEORGIA, Larry Seabrook, Steve Henry, Van Frazier
HAWAII, Charles S. Yonamine
IDAHO, Dave Jones
ILLINOIS, Jack M. Hook, Joe Hill
INDIANA, Timothy D. Bertram
IOWA, Neil Volmer, Leland D. Smithson, John Selmer
KANSAS, Jaci Vogel, Dean M. Testa
KENTUCKY, William A. Crace, Joe K. Deaton
LOUISIANA, William H. Temple, Gill Gautreau, Karl Finch
MAINE, Jerome A. Casey, Brian W. Pickard, Marc H. Guimont
MARYLAND, Russell A. Yurek
MASSACHUSETTS, Gordon A. Broz
MICHIGAN, Calvin Roberts
MINNESOTA, Rodney Pletan, Jim Lilly, Mark Wikelius
MISSISSIPPI, John Vance, Mark C. McConnell
MISSOURI, Clif Jett
MONTANA, John Blacker
NEBRASKA, Paul Cammack
NEVADA, Rod Johnson, Frank G. Taylor
NEW HAMPSHIRE, Stephen W. Gray, Edward Welch
NEW JERSEY, Rod Roberson
NEW MEXICO, Dennis Ortiz, Ernest D. Archuleta
NEW YORK, Edward G. Fahrenkopf, Clifford A. Thomas
NORTH CAROLINA, David A. Allsbrook, Jr., W. S. (Steve) Varnedoe, Lacy D. Love
NORTH DAKOTA, Jerry Horner
OHIO, Don R. Conaway, Keith C. Swearingen
OKLAHOMA, John Fuller, Kevin S. Bloss
OREGON, Ken Stoneman, Doug Tindall
PENNSYLVANIA, James S. Moretz, Robert M. Peda, Donald F. Wise
PUERTO RICO, Wilfredo Jirau
RHODE ISLAND, Thomas Jackvony, Jr.

SOUTH CAROLINA, Ron Wertz
SOUTH DAKOTA, Norm Humphrey, Michael Durick
TENNESSEE, Gerald Gregory, Rod Boehm, C. Alan Pinson
TEXAS, Joe S. Graff, Zane Webb
U.S. DOT, James B. Sorenson (FHWA), Don Steinke
UTAH, Dan Julio
VERMONT, David C. Dill, Paul E. Corti
VIRGINIA, Andrew V. Bailey II, Erle W. Potter
WASHINGTON, John Conrad, Ken Kirkland
WEST VIRGINIA, Julian Ware, Joseph Deneault
WISCONSIN, David Vieth, Paulette Hanna, Thomas Lorfeld
WYOMING, Gene Roccabruna
ALBERTA, Ted Harrison
BRITISH COLUMBIA, Rodney Chapman
MARIANA ISLANDS, John C. Pangelinan
NEW BRUNSWICK, Kenneth Connell
NOVA SCOTIA, Martin Delaney
ONTARIO, Ray Girard
SASKATCHEWAN, Gordon King
NEW JERSEY TURNPIKE AUTHORITY, Dave Wingerter
U.S. FOREST SERVICE, John W. Bell

National Cooperative Highway Research Program Advisory Panel SP20-07, Task 64
Leland D. Smithson, Chairman
Andrew V. Bailey II
Barkley H. Berry
John Conrad
Gary Hoffman
Jim D. Jackson
Robert W. Moseley
John O'Doherty
Amy Steiner, AASHTO Liaison
Ken Kobetsky, AASHTO Liaison
Andrew Mergenmeier, FHWA Liaison
Frank Lisle, TRB Liaison
Dr. Amir Hanna, PE, NCHRP Staff

PREFACE

This manual is based on the information contained in the *AASHTO Maintenance Manual - 1987*, the *AASHTO Manual for Bridge Maintenance - 1987*, the *Guide for Bridge Maintenance Management - 1980*, materials contributed by the various state departments of transportation, materials contributed by the Federal Highway Administration, recent highway transportation literature on maintenance, and field visits and interviews conducted by Kenneth A. Brewer as the principal author under National Cooperative Highway Research Program (NCHRP) Project SP20-07, Task 64. This document was prepared under the guidance of the NCHRP SP20-07, Task 64, Advisory Panel chaired by Mr. Leland D. Smithson of the Iowa DOT.

Intended for persons early in their career in roadway and bridge maintenance, the *Maintenance Manual* will assist them in understanding the various processes, methods, and materials that are applied to maintain the bridge and highway system effectively. Also intended to be a dynamic document, the manual will need to be updated as research and applications change the relevant content sections. Historically, highway maintenance has encompassed personnel management, materials selection, equipment use, and application of methods to react to problems in bridges and roadways, budgeting, and planning. The field is now evolving into a new engineering science of its own, resulting from a synergistic combination of traditional civil engineering, industrial engineering, business management, materials science, industrial psychology, and environmental science. This manual is envisioned as the first step in documenting that transition; this first-generation text will lead and guide the professional development of the engineering managers of roadway and bridge maintenance in the beginning of the 21st century.

This manual, however, will need to be used in conjunction with your own agency's maintenance operation policy manual, maintenance standards publication, and organizational operation guides. The effective engineering manager of maintenance in the 21st century will have to supplement the education and experience brought to the maintenance position with self-study, continuing professional education, and on-the-job training in management principles and concepts.

Some illustrations and examples in this manual show a particular product, brand, trademark, agency, or company. AASHTO does not endorse any manufactured product, licensed process, or other proprietary item shown in this manual. All such illustrations are provided to make the accompanying text more understandable, and no reference to such items in this manual should be construed as an endorsement.

Table of Contents

1.0 INTEGRATED MAINTENANCE MANAGEMENT
1.1 INTRODUCTION 1-1
   1.1.1 Evolution of Maintenance Management Concepts and Systems 1-1
   1.1.2 Development of Integrated Management Systems 1-3
   1.1.3 Reactive vs. Preventive Maintenance 1-5
1.2 MANAGEMENT AND ADMINISTRATION 1-8
   1.2.1 Functional Aspects 1-8
      1.2.1.1 Organization for Maintenance 1-13
      1.2.1.2 Planning and Budgeting for Maintenance 1-17
      1.2.1.3 Scheduling Maintenance 1-18
      1.2.1.4 Data Systems to Support Maintenance 1-20
      1.2.1.5 Incident Management and Emergency Response in Maintenance 1-20
      1.2.1.6 Work Zone Safety 1-21
   1.2.2 Human Resources 1-22
      1.2.2.1 Motivation 1-22
      1.2.2.2 Training 1-23
      1.2.2.3 Personnel Safety 1-23
      1.2.2.4 Recruitment and Development 1-24
   1.2.3 Technical Resources 1-24
      1.2.3.1 Equipment for Maintenance 1-24
      1.2.3.2 Materials for Maintenance 1-24
      1.2.3.3 Models Available for Management of Resources 1-25
1.3 1991 ISTEA AND IMPACTS ON MAINTENANCE 1-25
1.4 INTEGRATED SYSTEMS 1-25
   1.4.1 Advantages and Benefits 1-25
   1.4.2 Limitations and Costs 1-26
1.5 DEVELOPING ISSUES 1-26
   1.5.1 Sustainable Development 1-26
   1.5.2 Designing for Maintainability 1-26
   1.5.3 Environmental Sensitivity 1-26
   1.5.4 Application of TQM Principles 1-27
   1.5.5 Litigation Involving Maintenance 1-27
   1.5.6 Rational and Probabilistic Models to Predict Maintenance Needs and Effects ... 1-27
1.6 REFERENCES 1-28

2.0 ROADWAY MAINTENANCE AND MANAGEMENT
2.1 ROADWAY MAINTENANCE 2-1
   2.1.1 Introduction 2-1
      2.1.1.1 Preventive Roadway Maintenance Philosophy 2-1
      2.1.1.2 Interaction With Bridge Maintenance and the Management of
            Other Functions 2-1
      2.1.1.3 Interaction With the Design of Facilities 2-2
      2.1.1.4 Role of Roadway Maintenance in the Safe, Efficient Operation of Facilities 2-2

    2.1.1.5 Value of Roadway Maintenance to Agency 2-3
    2.1.1.6 Value of Roadway Maintenance to the User Community 2-4
  2.1.2 Maintenance of Roadway Surfaces 2-4
    2.1.2.1 Planning and Budgeting 2-5
    2.1.2.2 Scheduling 2-6
    2.1.2.3 Maintenance Needs by Type of Surface 2-7
    2.1.2.4 Materials for Roadway Surface Maintenance 2-19
    2.1.2.5 Methods for Roadway Surface Maintenance 2-24
  2.1.3 Shoulder Maintenance 2-35
    2.1.3.1 Impact of Shoulder Type on Maintenance 2-36
    2.1.3.2 Scheduling Shoulder Maintenance 2-38
  2.1.4 Roadway Drainage Maintenance 2-39
    2.1.4.1 Maintenance of Roadway Surface Drainage 2-40
    2.1.4.2 Maintenance of Subsurface Drainage 2-43
    2.1.4.3 Other Considerations 2-45
  2.1.5 Roadside Maintenance 2-46
  2.1.6 Maintenance of Safety Features 2-54
  2.1.7 Building Facilities Related to Roadway Maintenance 2-56
  2.1.8 Special Pathways 2-59
  2.1.9 Snow and Ice Control 2-59
  2.1.10 Traffic Control Devices and Roadway System Instrumentation 2-67
  2.1.11 Roadway, Tunnel, and Bridge Illumination 2-74
  2.1.12 Roadways on Bridges and in Tunnels 2-80
  2.1.13 Environmental Aspects of Roadway Maintenance 2-84
  2.1.14 Maintenance Work Zone Traffic Control 2-88
  2.1.15 General Maintenance Worker Safety Notes 2-92
  2.1.16 Developing Issues 2-92
  2.1.17 REFERENCES 2-95

2.2 ROADWAY MANAGEMENT 2-102
  2.2.1 Introduction 2-102
  2.2.2 Interaction With Other Highway System Management Systems 2-102
  2.2.3 Roadway Inspection and Condition Inventory Process 2-103
  2.2.4 Environmental and Nonroadway Issues 2-104
  2.2.5 Developing Issues 2-104
  2.2.6 REFERENCES 2-106

3.0 BRIDGE MAINTENANCE AND MANAGEMENT

3.1 BRIDGE MAINTENANCE 3-1
  3.1.1 Introduction 3-1
    3.1.1.1 Load-Carrying Capacity 3-1
    3.1.1.2 Structural Systems 3-3
    3.1.1.3 Bridge Maintenance Concepts 3-11
  3.1.2 Traveled Surface 3-12
    3.1.2.1 Concrete Surfaces 3-13
    3.1.2.2 Asphalt Surfaces 3-15
    3.1.2.3 Stone and Brick Surfaces 3-16
    3.1.2.4 Steel Grate Surfaces 3-17
    3.1.2.5 Wood Surfaces 3-17
  3.1.3 Structural Decks 3-17
    3.1.3.1 Introduction 3-17
    3.1.3.2 Preventive Maintenance of Concrete Decks 3-18
    3.1.3.3 Concrete Deck Sealing 3-19
    3.1.3.4 Concrete Deck Patching 3-19

3.1.3.5 Epoxy Deck Patching ... 3-21
3.1.3.6 Asphaltic Concrete Patching ... 3-21
3.1.3.7 Emergency Full-Depth Patching ... 3-21
3.1.3.8 Crack Sealing ... 3-22
3.1.3.9 Overlays to Bridge Decks ... 3-22
3.1.3.10 Concrete Deck Replacement ... 3-24
3.1.3.11 Timber Bridge Decks ... 3-24
3.1.3.12 Steel Grid Decks ... 3-25
3.1.3.13 Maintaining Deck Joints ... 3-26
3.1.3.14 Deck Drainage Systems ... 3-33
3.1.3.15 Curbs, Sidewalks, and Railings ... 3-34
3.1.4 Superstructure ... 3-37
3.1.4.1 Jacking and Supporting the Superstructure ... 3-37
3.1.4.2 Bearing Maintenance and Repairs ... 3-40
3.1.4.3 Beam Repair ... 3-46
3.1.4.4 Truss Repair ... 3-54
3.1.5 Substructure ... 3-58
3.1.5.1 Problems Associated With Substructure Caps ... 3-59
3.1.5.2 Maintenance ... 3-59
3.1.5.3 Preventive Maintenance ... 3-59
3.1.5.4 Repair Process ... 3-60
3.1.5.5 Repairing Substructures Above Water ... 3-61
3.1.5.6 Underwater Repair of Substructures ... 3-65
3.1.5.7 Pile and Bent Repair ... 3-70
3.1.6 Watercourse and Embankments ... 3-80
3.1.6.1 Removing Debris From Channels ... 3-80
3.1.6.2 Scour Protection and Repair ... 3-83
3.1.7 Protective Systems ... 3-90
3.1.7.1 Protecting the Substructure ... 3-90
3.1.7.2 Spot Painting to Protect the Superstructure ... 3-91
3.1.8 Environmental Aspects ... 3-98
3.1.8.1 Lead-Based Paint Removal ... 3-98
3.1.8.2 Other Environmental Concerns ... 3-99
3.1.8.3 Hauling and Disposal Regulations ... 3-101
3.1.8.4 Hazardous Wastes ... 3-102
3.1.8.5 Historic Structures ... 3-102
3.1.8.6 Noise Control ... 3-102
3.1.8.7 Creosote-Treated Timber ... 3-102
3.1.8.8 Worker Safety ... 3-103
3.1.8.9 Work Site Safety Review ... 3-105
3.1.8.10 Toxic Materials ... 3-106
3.1.8.11 Confined Spaces ... 3-106
3.1.8.12 Fall Protection, Rigging, Scaffolding, and Hoisting ... 3-109
3.1.8.13 Safety Review ... 3-120
3.1.9 Movable Bridges ... 3-121
3.1.9.1 Types of Movable Bridges ... 3-121
3.1.9.2 Structural Maintenance ... 3-122
3.1.9.3 Machinery and Equipment Maintenance ... 3-122
3.1.9.4 Inspection and Maintenance of Specific Parts and Components ... 3-123
3.1.9.5 Inspection Items ... 3-126
3.1.10 Maintenance Work Zone Traffic Control ... 3-128
3.1.10.1 Control for Work Zones ... 3-128
3.1.10.2 Temporary Structures ... 3-133

- 3.1.11 Developing Issues 3-134
  - 3.1.11.1 Nonmetallic Repair of Steel Bridge Components 3-134
  - 3.1.11.2 Nonmetallic Bridges 3-134
- 3.1.12 REFERENCES 3-135

3.2 BRIDGE MANAGEMENT AND BRIDGE MAINTENANCE MANAGEMENT 3-136
- 3.2.1 Introduction 3-136
  - 3.2.1.1 Bridge Maintenance Management Process 3-136
  - 3.2.1.2 Federal Bridge Management System Requirements 3-138
  - 3.2.1.3 PONTIS 3-138
  - 3.2.1.4 System-Level vs. Project-Level Management Decisions 3-138
  - 3.2.1.5 System Interfaces 3-139
- 3.2.2 Planning and Scheduling Bridge Maintenance 3-140
  - 3.2.2.1 Long-Range Planning and Scheduling 3-140
  - 3.2.2.2 Work Orders 3-140
  - 3.2.2.3 Job Layout 3-140
  - 3.2.2.4 Short-Term Scheduling Procedures 3-142
  - 3.2.2.5 Graphical Methods of Scheduling 3-143
- 3.2.3 Performing the Work 3-143
  - 3.2.3.1 Job Execution 3-143
  - 3.2.3.2 Personnel Resources 3-144
  - 3.2.3.3 Equipment 3-144
  - 3.2.3.4 Materials 3-145
- 3.2.4 Reporting Bridge Maintenance Accomplishments 3-145
  - 3.2.4.1 Why Report? 3-145
  - 3.2.4.2 Report Requirements 3-145
  - 3.2.4.3 Reporting Procedures 3-146
- 3.2.5 Contract Maintenance for Bridges 3-146
  - 3.2.5.1 Lump Sum Contracts 3-147
  - 3.2.5.2 Unit Price Contracts 3-148
  - 3.2.5.3 Cost Reimbursement Contracts 3-148
  - 3.2.5.4 Negotiated Contracts 3-149
  - 3.2.5.5 Cost Comparison of Contracting vs. Performing Work In-house 3-150
- 3.2.6 Quality Control and Quality Assurance of Bridge Maintenance Operations ... 3-150
  - 3.2.6.1 QC at the Work Site 3-150
  - 3.2.6.2 Technical Site Review 3-151
  - 3.2.6.3 Traffic Control Site Review 3-151
  - 3.2.6.4 Site Review of Tools and Equipment Use 3-151
  - 3.2.6.5 Site Review of Rigging and Climbing Practices 3-151
  - 3.2.6.6 Budget Monitoring 3-152
  - 3.2.6.7 Schedule Monitoring 3-152
  - 3.2.6.8 Methods of Quality Assurance 3-152
- 3.2.7 Bridge Inspection 3-153
  - 3.2.7.1 Inspection Process 3-153
  - 3.2.7.2 NBI Program 3-153
  - 3.2.7.3 Performing the Inspection 3-153
  - 3.2.7.4 Inspection Reports 3-156
  - 3.2.7.5 Inspection Performed by Maintenance Supervisors 3-156
  - 3.2.7.6 Resource Estimating 3-156
  - 3.2.7.7 Identifying Fracture-Critical Bridges 3-157
  - 3.2.7.8 Testing Existing Bridge Components 3-159
  - 3.2.7.9 Tests for Steel Members 3-161
  - 3.2.7.10 Tests for Timber Members 3-162
  - 3.2.7.11 Load Tests 3-163

    3.2.8 Preliminary Site Visit to Plan a Bridge Repair 3-163
        3.2.8.1 Basic Information Required 3-163
        3.2.8.2 Channel Damage 3-164
        3.2.8.3 Approach Damage 3-164
        3.2.8.4 Deck Damage 3-165
        3.2.8.5 Superstructure Damage 3-166
        3.2.8.6 Substructure Damage 3-167
        3.2.8.7 Emergency Damage 3-167
    3.2.9 REFERENCES 3-168

4.0 EQUIPMENT SYSTEMS
4.1 INTRODUCTION 4-1
4.2 EQUIPMENT APPLICATION 4-2
4.3 MAINTENANCE OF EQUIPMENT 4-4
4.4 ACQUISITION AND REPLACEMENT 4-5
4.5 REFERENCES 4-8

5.0 MAINTENANCE RESEARCH AND DEVELOPMENT
5.1 ADMINISTRATION 5-1
5.2 METHODS 5-2
5.3 MATERIALS 5-2
5.4 HUMAN RESOURCES AND PERSONNEL 5-3
5.5 EQUIPMENT 5-3
5.6 INTERNATIONAL ASPECTS AND TECHNOLOGY TRANSFER 5-3
5.7 REFERENCES 5-4

6.0 TORT LITIGATION
6.1 TORT CLAIMS 6-1
6.2 REFERENCES 6-3

7.0 INDEX 7-1

1.0 INTEGRATED MAINTENANCE MANAGEMENT
1.1 INTRODUCTION
When highway departments were first formed in the early 20th century, they were seen as construction organizations. Design was a minor responsibility of the organizations. Agency leadership quickly realized, however, that constructing a road properly required a technically adequate and implementable design. As a fledgling road network developed, network maintenance became a recognized activity important to the long-term success and stability of the highway system. During World War II, very little investment was made in the highway system, enabling the United States to devote its resources to the war effort. After World War II, the deteriorated highway system was rapidly expanded and rehabilitated. This increased level of investment to expand and rebuild the highway network created an awareness of the need to begin a rational management process of highway maintenance.

1.1.1 Evolution of Maintenance Management Concepts and Systems
One of the first thorough, systematic studies of highway maintenance attempting to identify the degree to which the engineering management of maintenance was an important component of successfully maintaining a highway network was the Iowa State Highway Maintenance Study [1]. The 12-month study (August 1959 to August 1960) focused on three areas of activity:
(1) Maintenance of the structural or physical highway itself.
(2) Provision of services, such as snow removal and detour upkeep.
(3) Management of the above activities.

A detailed supplement to the study report documented each type of activity conducted by the Iowa State Highway Commission within the study areas during the study periods [2]. Taken together, these two documents were the factual database showing the need for the systematic management of maintenance and the need for research specifically addressed to improving maintenance for the next 15 years.

A significant milestone that focused attention on maintenance management concepts and systems was the 1968 Maintenance Management Workshop [3]. This workshop included an overview of what the various states had accomplished in terms of managing maintenance up to 1968 and what were seen as the next decade's challenges. Building on this workshop, maintenance engineers began to apply statistical quality control to highway maintenance as early as 1971 [4].

As statistical quality control entered the field of maintenance management, so did business management principles. Many departments began to adopt a "management by objectives" approach. Target goals for efficiency and productivity were established with a "bottom-line" analysis for fiscal control and budgetary review of programs. Positive effects of this approach were examination for areas of unnecessary duplication, identification of contract maintenance opportunities, and regular managerial review of programs. This approach also led to research on the development of maintenance levels of service and guidelines for moving to a "level of service" approach to managing maintenance [5]. The downside of this approach included some agencies suffering from micromanagement, lower levels of management in some agencies developing low morale from loss of decisionmaking authority, and severe cost-reduction measures in some agencies generating employment fluctuations. Application of business management principles to maintenance has evolved to the use of strategic planning as a model for maintenance management. One state, for example, has developed a maintenance performance plan with three separate areas of strategic emphasis [6]:

(1) Developing an effective work force.
(2) Optimizing resources.
(3) Practicing sound stewardship of the environment.

This state's adoption of a strategic planning model of management resulted in reorganization of the maintenance management function, decentralizing much of the decisionmaking to engineering managers of major areas of the state. Four functional areas of activity now group personnel and resources:

(1) Bridge maintenance and operations.
(2) Maintenance (roadway) operations.
(3) Maintenance (roadway) programs.
(4) Maintenance (roadway) services.

Management of highway maintenance typically uses information from several different database areas [6]:

- The maintenance management system (MMS).
- Payroll reports.
- Equipment records.
- Accounting reports.
- The transportation inventory database (a record of the infrastructure).
- Bridge inspection processes and reports.
- The pavement management system (PMS).

Maintenance managers typically have an MMS available that is intended to control and manage the flow of this information so that each level of maintenance manager can make decisions supported by a rational database. The flow and information management process in a typical MMS [6] is objectively oriented to provide specific numeric values to everything that happens with an associated cost. In this sense, the process is a classical engineering economy systems analysis. Many MMSs that operate in this fashion have their organizational roots in the principles and processes published in National Cooperative Highway Research Program (NCHRP) Report 131 on performance budgeting [7]. While the process is not typically integrated across all elements of a transportation department's activities that have a bearing on the need for or conduct of highway maintenance, it does incorporate the critical information elements needed to manage the maintenance function internally.

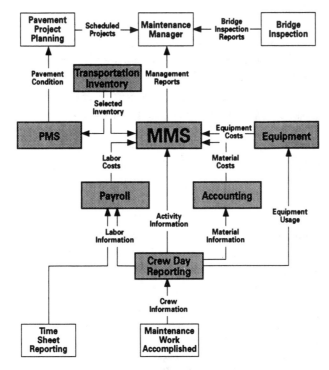

Reference: Maintenance Management System Study for the Iowa DOT *by Wilbur Smith Associates, June 1993.*

Maintenance Management System Information Flow

1.1.2 Development of Integrated Management Systems

Efforts have begun to develop integrated management systems for maintenance. It is not currently known if an integrated management system for maintenance should be a variation of the typical MMS shown above or a central processing host to all of the various functional information bases within a department (of which the MMS is just one). The information above reveals how input databases from an expansion of functional activities continue to broaden the sensitivity of the maintenance manager's decisions. Most departments of transportation (DOTs) are traditionally top-down in their management control. Pressures are mounting, however, to adopt bottom-up planning and the principles of "total quality management" (TQM) to the extent that TQM can be realistically applied to a public sector service function. Computer hardware and software are capable of processing and organizing information in quantities, in levels of inquiry, at speeds, and in input/output forms that were unthinkable when MMSs were initiated in the early 1970s.

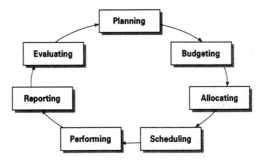

Reference: Maintenance Management System Study for the Iowa DOT *by Wilbur Smith Associates, June 1993.*

Maintenance Management Activity Flow

Chapter 1 Integrated Maintenance Management

A comprehensive attempt to determine what an integrated management system that encompasses maintenance should be has recently been reported in NCHRP Report 363 [8]. That effort identified three areas of characteristics or developments needed to produce integrated management systems for maintenance. The study suggested that a very important change would be to ensure that all data collection and maintenance activity reports take place at the level where the information can actually be used to improve or change the way business is done.

A concept scheme was developed that can guide an agency attempting to create an integrated management system. It was recognized that whatever degree of integration an agency chooses to pursue, the next generation of MMSs will be more integrated across maintenance and other functional databases within a department and among the various activities within maintenance. Because business management is a much larger market for management technology (both hardware and software) than are the U.S. DOTs, transportation agencies can "borrow" or transfer technology from business advances in information processing, telecommunication, interactive display, and the study of information's role on the decisionmaking process without ever moving to an integrated management system. However, for those agencies wishing to move to an integrated management system, a "hub-and-spoke" concept has been proposed to organize the information flow (see figure below) [8].

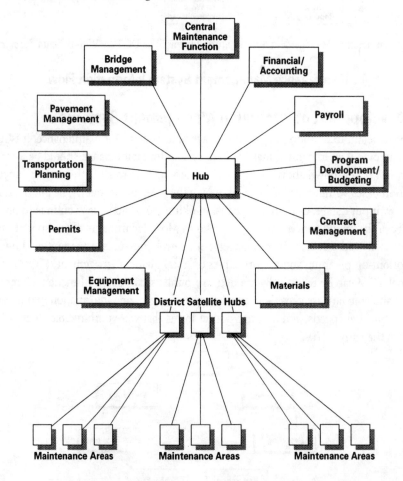

Hub-and-Spoke Maintenance Management Structure

1.1.3 Reactive vs. Preventive Maintenance

Roadway and bridge maintenance began with an implied or directly stated intention of "fixing" what was wrong with any part of a highway facility. A side-by-side attitude that developed in maintenance is best expressed by the adage, "If it isn't broken, don't fix it!" Thus, by the very nature of the intent that maintenance was to "respond" to "problems," the focus of maintenance planning has evolved to be reactive, that is, creating programs in response to problem conditions that are generally evident. As rational engineering economic analysis was applied to the development of all kinds of highway projects from the 1950s forward, the benefit-cost analysis was organized in a way that assumed that the maintenance costs of all alternative projects were direct annual costs. Unfortunately, this caused the planning and budgeting process to reinforce a reactive approach to highway maintenance. The traditional economic analysis of alternatives in the planning and budgeting process places anything that has a life longer than 1 year outside the realm of maintenance and into the realm of capital construction, for which an annualized capital cost must be calculated. For over 20 years, maintenance engineering managers and other individuals have been moving away from this reactive focus toward a preventive maintenance approach. The push to change the focus of maintenance from reactive to preventive measures was accelerated by Strategic Highway Research Program (SHRP) projects in pavement maintenance. The project statement for these projects directed that the research efforts should seek methods of establishing more effective preventive maintenance than damage repair in preserving and maintaining the highway infrastructure. The SHRP projects primarily focused on extending pavement service life with surface treatments. However, many other activities still need to be approached from a preventive maintenance view. Eventually, an integrated management system will need to examine each maintenance activity, considering the merit of moving from a reactive ("fix it when it's broken") maintenance approach to a preventive ("a stitch in time saves nine") maintenance approach with anticipated costs and long-term infrastructure consequences.

To initiate or extend a preventive maintenance philosophy to any maintenance activity requires an assessment of why an activity is done when. To complete the transition from a reactive maintenance philosophy for an activity requires that some definitive threshold be established at which the planned maintenance activity is programmed, scheduled, performed, and subjected to review for quality, productivity, efficiency, and cost-effectiveness. An activity must be reviewed to determine if it is being performed "on demand" or "on schedule." When an activity to preserve some highway characteristic is performed on demand, it should be evaluated to see if planning, programming, and scheduling the activity as part of a preventive maintenance policy can optimize the maintenance effort in a highway system life-cycle cost analysis.

EXAMPLE

Surface sealing of asphalt pavements

Some agencies seal an asphalt surface every fifth year of its life from the initial year of construction until the pavement condition suggests it is ready for replacement. What is the reason for such an "on-schedule" (time-interval) surface sealing policy? Perhaps budgetary limits suggest that only about one-fifth of a network needing surface sealing can be accomplished in 1 year. Perhaps the agency has an accumulated "expert experience" that once every 5 years is sufficient to rejuvenate the surface and sufficiently extend its life to the desired replacement interval. Perhaps some roadway surface sections in which varying frequencies of surface sealing were applied have been monitored for pavement durability, and an overall pattern of optimum cost-effectiveness of pavement life extension has been noted at about 5-year intervals. Perhaps the cycle of 5-year intervals has been in use for the past 20 years; it is the historical pattern of budgeting and planning, and everyone seems satisfied with it.

Seal coat over new pavement

When the fundamental reasons for performing surface sealing are examined, a preventive maintenance policy can begin to be formulated. Is sealing done to limit permeation of water into the pavement structure? Is it to change the surface skid resistance? Is it intended to add a small fraction of pavement strength? Is it to provide a top-wearing surface so the top structural layer will not be eroded by tire wear? Is it to provide a clean surface for traffic stripes? Is it to provide sharp delineation of traffic lanes from the shoulder? Is it to forestall the need to replace the structural pavement as long as possible? Once the agency's various reasons for surface sealing have been identified, the importance of each reason needs to be estimated. This suggests that observation records on what maintenance is done will have to be kept on inventory sections of roads. These records will allow an estimate to be made of the effectiveness of a surface treatment in preserving the desired roadway condition. Knowing what rehabilitation has to be done to restore the surface to a desired condition if a maintenance treatment fails, what the rehabilitation costs are, and what the cost of the treatment is facilitates development of a preventive maintenance policy based on the cost-effectiveness of maintaining the highway system's integrity. SHRP Project H-101, Pavement Maintenance Effectiveness, was directed toward creating such a policy. Although some test section data are still being collected, the project has yielded two documents to assist in defining surface treatments in a preventive maintenance context. SHRP Manual 3033, *Rating Preventive Maintenance Treatments*, aids in evaluating the quality of crack sealing, chip sealing, and slurry sealing for asphalt pavements as preventive maintenance treatments [9]. SHRP Manual 3034, *Specifications for Preventive Maintenance Treatments*, describes specifications for climatic regions of the United States to initiate preventive maintenance policies for crack sealing, chip sealing, slurry sealing, joint sealing, and undersealing [10]. These documents are not the final answers in preventive maintenance for surface treatments, but they do significantly accelerate the pace at which agencies are beginning to share information to produce analyses for preventive maintenance policies [11].

Examining an activity, such as surface sealing, from a preventive maintenance approach can, in some cases, radically change the maintenance planning thinking. For example, an agency may refuse to do chip sealing because of complaints regarding windshield damage caused by loose cover material. When a cost-effectiveness analysis from a preventive maintenance approach is applied, an agency may opt not to do chip seals. Considering the political constraints that the complaints impose, slurry seals or thin overlays may yield the most cost-effective treatment. Furthermore, if a chip seal is equally effective at a lower cost than slurry seal or thin overlay, then a value can be placed on the "complaints" and traditional engineering economic analyses can assist in making a decision to continue or change a maintenance planning policy. A preventive maintenance approach may have a different effect too. For example, the preceding figure shows a newly overlaid asphalt pavement that has had a surface chip seal applied before opening the resurfaced roadway to traffic. In this case, the "maintenance treatment" has been applied immediately at the completion of the "pavement construction." The preventive maintenance approach encourages rethinking the relationship of maintenance to construction and operation.

EXAMPLE

Roadside mowing

Mowing is not a maintenance activity usually thought of in the context of preventive maintenance. It is discussed here to illustrate the impact of looking at any maintenance activity as "on demand" versus "on schedule" to see if the reasons for doing an activity lend it to a preventive maintenance approach.

Early fall mowing to limit snow drifting

When the reasons for doing roadside mowing are identified, the possible benefits of mowing can be related to the cost, and the quality standards can be examined to see if adjustments can be made consistent with achieving the desired benefits. If the roadside is mowed to create a "mowed lawn" appearance, then it most likely has to be done on a time schedule. On the other hand, if mowing is done to aid in maintaining a natural rustic (wildflower) look, then mowing will only be required according to the conditions necessary to ensure retention of the native species and control noxious weeds. If mowing is done to limit snow drifting in winter, then a seasonal weather condition (frost that kills plants) makes the mowing a part of the winter preventive maintenance policy for winter maintenance. This list of thoughts to determine the root cause for roadside mowing can be expanded. Each reason can then provide a decision whether the reason contributes to preventive maintenance by maintaining the infrastructure condition, traffic service, or safety, etc., which can provide a way of estimating any benefit to the frequency of mowing and the quality standards used to evaluate mowing. Then a cost-effectiveness basis can be provided from a preventive maintenance approach, or the mowing program can be identified as an activity that can stand on its own in alternative budget planning.

In this way, any maintenance activity can be examined for its role, if it has one, in preventive maintenance. Every way in which an activity can be associated with preventive maintenance increases the degree to which an integrated maintenance management system can be built for forward-thinking planning and maintenance management.

1.2 MANAGEMENT AND ADMINISTRATION
1.2.1 Functional Aspects

One objective of the American Associaion of State Highway and Transportation Officials (AASHTO) is to encourage a pattern of uniform systems for highway maintenance management within the various states. This objective has taken on increased significance as a result of the 1991 Integrated Surface Transportation Efficiency Act. Within any given state, maintaining the highway network to preserve the capital investment in the network and to provide the services needed by highway users is a very complex management and administrative responsibility. Maintenance engineers and managers are expected to understand a wide variety of functional aspects of managing complex organizations and to assemble a staff capable of applying sufficient detailed knowledge in each functional area.

Highway maintenance encompasses a program to preserve and repair a system of roadways with its elements to its designed or accepted configuration and to an accepted quality of roadway performance. Elements of the highway network to be maintained include travel way surfaces, shoulders, roadsides, drainage facilities, bridges, tunnels, signs, markings, lighting fixtures, and truck weighing and inspection facilities. The program includes such traffic services as lighting and signal operation, snow and ice removal, operation of roadside rest areas, operation of movable span bridges, and roadside beautification.

Highway maintenance programs are developed to reduce the negative effects (real or perceived) of weather, vegetation growth, deterioration of roadway elements, traffic wear on surfaces, vandalism, and other types of damage. Deterioration includes the effects of aging, material failures, obsolete designs, and construction imperfections. The maintenance organization conducting this program is judged by performance criteria internal to the organization; by the confidence of those providing the funds for maintenance that resources are being wisely allocated and used; by the perception of the traveling public as to how well the program efforts are addressing maintenance issues important to the public; and finally, by peer review and judgment.

The maintenance program may be conducted completely in-house; conducted with private contracts for selected activities; or in extreme cases, conducted completely through contracts with outside vendors. Most states operate by contracting selected activities to private vendors while retaining all activities deemed to be best provided by in-house personnel, ensuring proper quality control and quality assurance. Thus, the maintenance program will necessarily include maintenance and repair of certain buildings, stockpiles, and equipment essential to performing all of the functional aspects of the maintenance program retained in-house. Maintenance of these buildings, stockpiles, and equipment is not directly related to maintaining the highway but is a part of the highway maintenance program.

While the boundary between "maintenance" and "construction" varies according to local policies, local contract administration, and the legal guidelines for administering funds available for maintenance and construction, maintenance is generally limited to activities that do not increase the capability, strength, or capacity of an element of a facility. At times a maintenance and repair activity improves a roadway element, but those times are usually limited to replacing a destroyed element with an improved version developed since the original installation. For example, an old design guardrail end section may be hit and destroyed; in the repair process, a newer design unit is installed that upgrades and improves the safety capability of the guardrail end section as a maintenance activity. However, a contract to replace all of the old guardrail end sections across an entire region would be a "construction and rehabilitation" contract excluded from the functional maintenance program. The AASHTO-adopted guidelines for the distinction between maintenance and construction are included here, along with the AASHTO-adopted definition of highway construction, to further clarify the functional aspect of the highway maintenance program.

SUMMARY TABLE
Distinction Between Construction and Maintenance

Construction		Maintenance	
Construction and Reconstruction	**Betterment**	**Physical Maintenance**	**Traffic Services**
Traveled Way All operations on new location involving considerable reconstruction to modern standards for 500' or more. Widening sufficient to change geometric type, as from a two-lane to a three- or four-, from a three- to a four-, or from a four- to a six-lane highway, with all lanes not less than 11'. Addition of 500' or more of frontage road in any 1 mile.	Placing new loose material on road surfaces to substantially increase thickness of surfacing beyond that originally built for 500 or more continuous feet. Improvement of surface to higher type for 500' or more. Resurfacing of hard surfaces with bituminous material 1/4" thick or more for a length of 500 continuous feet or more. Replacement of existing pavement with higher standard for 500' or more. Widening with no change in number of lanes. Addition of less than 500' of frontage road in any 1 mile.	Scarifying, reshaping, applying dust palliatives, and restoring material losses; patching, mudjacking, joint filing, crack sealing, surface treating, etc. Resurfacing of hard surfaces with bituminous material less than 3/4" thick. Replacement of traveled way in kind for less that 500 continuous feet. Replacement of unsuitable base materials in patching operations.	Removal of snow and ice, and related operations such as sanding, chemical applications, etc. Restoring pavement stripes and markings, and replacing raised pavement markers.
Shoulders and Side Road Approaches All work incidental to above. Original surfacing for 500 or more continuous feet.	All work incidental to above. Resurfacing, stabilizing, or widening of shoulders and side road approaches for a length of 500 continuous feet of more.	All work incidental to above. Restoring material losses. Replacement of shoulder in kind. Reseeding and resodding.	Same as above.
Roadsides All work incidental to above.	Widening the roadbed. Substantial slope flattening or landscape treatment.	Restoration of erosion controls. Removing slides, reshaping drainage channels and side slopes, mowing and tree trimming. Replacing topsoil, sod, shrubs, etc. Chemical spraying.	Erection of snow fences. Opening of inlets clogged with snow and ice. Removal of litter.
Intersections All work incidental to above. Complete reconstruction of intersections, including changes in type of intersection as from plain to a grade separation and ramps.	Nominal channelization of intersections. Addition of auxiliary lanes.	As covered by other items in this column.	As covered by other items in this column.
Alignment Changes New location. Appreciable alignment changes for 500' or more with or without change in highway type.	Minor changes in alignment with profile, such as easing horizontal curves or eliminating irregularities in profile. Regrading or resurfacing to introduce or increase super-elevation on curves, or to improve sight distance where such work does not exceed 1,000 feet per mile.	None.	
Right of Way Additions None, some, or all.	None or some.		
Stage Construction Where above operations are performed by stages, widely separated in time, each may be classed as construction, betterment, or maintenance, depending on its characteristics.	Where above operations are performed by stages, widely separated in time, each may be classed as betterment or maintenance, depending on its characteristics.		

SUMMARY TABLE CONTINUED
Distinction Between Construction and Maintenance

Construction		Maintenance	
Construction and Reconstruction	**Betterment**	**Physical Maintenance**	**Traffic Services**
Incidental Items Where overall improvement is classed as construction or reconstruction, incidental operations of the maintenance or betterment class should be considered part of the construction or reconstruction.	Where overall improvement is classed as betterment, incidental operations of the maintenance class should be considered part of the betterment.		
Unusual or Disaster Operations Extensive repair or replacement for damage as a result of storm, flood, or military operations may be charged to extraordinary maintenance, betterment, or construction, depending upon scope of work.			
Drainage All work incidental to above. Building flood control, flood prevention, and earthwork protective structures. Installations or extensions of curb, gutter, or underdrain for a length of 500' or more in any 1 mile.	Extending old culverts and replacing inlets and headwalls. Replacing culverts with those of greater capacity. Installation of additional pipe culverts or additional structures with spans not greater than 10'. Installation or extension of curb, gutter, or underdrain of less than 500' in any 1 mile.	Replacement (using approximately same design) of curb, gutter, riprap, underdrain, and culverts. Cleaning and repairing culverts, inlets, etc.	Same as above.
Structures All work incidental to above. On structures having a span of 20' or less, complete reconstruction to a higher standard. Complete reconstruction or additions to bridges of more than 10' span. Widening of bridges over 100' long. Extensions or new installations of walls involving over 80 cubic yards of structural material.	Replacement of bridge rails and floors to higher standard. Widening of bridges 100' or less between abutments. Extensions or new installation of walls involving 80 cubic yards or less of structural material. Replacement of walls to higher standard.	Cleaning, painting, and repairing. Replacements (using approximately same design) of rails, floors, stringers, or beams. Replacement of walls in kind. Repair of drawbridges and ferries.	Operation of ferries, including cost of power, operators, and periodic replacements. That part of operation of drawbridges charged to highway traffic.
Traffic Control and Service Facilities All work incidental to above. First erection of traffic signs and direction and route markers.	Replacement of all major signs with superior set, or individual installations of specially erected signs. Installation of traffic signals, railroad protection devices, lighting systems or extension of same. Extension or new installation of guardrail for 500 continuous feet or more. Nominal channelization. Installation or complete replacement with superior design of facilities for roadside rest areas.	None.	Painting, repairing, and replacing in kind of signs, guardrails, guide markers, signals, lighting standards, etc. Addition of small numbers of traffic control devices. Maintenance and replacement in kind of rest areas. Servicing of and furnishing power and light bulbs for lighting and traffic control devices. Policing, roadside cleaning operations, operation of roadside areas, towing service, information booths, etc.

Chapter 1 Integrated Maintenance Management

HIGHWAY CONSTRUCTION DEFINITION
Adopted June 30, 1986

General Criteria
Work must be: substantial in scope, and substantially extend the service life of the facility or component thereof, and/or enhance safety, and/or replace or renovate a failed component of the highway facility that has served its useful life.

Pavement and Shoulders
All pavement or shoulder work on new or existing alignment that is significant in scope or that is incidental to roadsides, structures and drainage or traffic services work; that would provide improved traffic capacity; that would provide improved traffic safety; that would improve or restore adequate skid resistance; that would improve or restore the profile or cross slope of a pavement; that would increase the structural capacity; that would seal pavement surfaces to prevent water intrusion; that would substantially extend the service life of a pavement; that would replace all or part of a failed pavement or shoulders with an equal or higher standard material; that would stabilize, restore, or replace failed pavement base material; and/or, that would replace or restore a failed highway component that has served its useful service life.

Roadsides
All roadsides work on new or existing alignment that is significant in scope or that is incidental to pavement as shoulders, structures and drainage or traffic services work; that would provide for the stabilization or restoration of significant roadway section landslides; that would provide for the installation of landscape materials; and/or, that would provide for improved roadside safety.

Structures and Drainage
All structural and drainage work on new or existing alignment that is significant in scope or that is incidental to pavement and shoulders, roadsides or traffic control, and traffic services work; that would provide for improved traffic capacity; that would provide for improved traffic safety; that would restore or increase carrying capacity or hydraulic capacity; that would restore the structural integrity of bridge components; that would protect bridge reinforcing steel against corrosion; that would provide for the replacement of a failed protective coating system with an equal or higher standard material; that would provide for the installation of pavement and/or shoulder drainage systems; that would provide for the replacement of a failed structural or drainage component with a higher type of material; and/or, that would replace or restore a failed component that has served its useful service life.

Traffic Control and Service Facilities
All traffic control or service facilities work on new or existing alignment that is significant in scope or that is incidental to pavements and shoulders, roadsides or structures and drainage work; that would provide or replace traffic control devices, including pavement markings and highway lighting; that would provide for the installation of new guardrail or other traffic safety devices to protect traffic from hazardous areas; that would provide for the upgrading to current standards of existing guardrail or other devices, or the replacement of guardrail or other devices; that would provide for the installation of monitoring devices of facilities for speed monitoring or weight enforcement; that would provide for the installation, restoration or replacemnt of facilities for roadside rest areas; that would provide for emergency communcation equipment, and/or, that would replace, restore, or rehabilitate a failed highway component that has served its useful service life.

1.2.1.1 Organization for Maintenance

While there are similarities in how the various states are organized to manage and conduct maintenance, each state has unique functions and organizational characteristics. When the organizational chart is developed outlining the maintenance manager's responsibilities and authority, the scope and depth of activity covered will be defined partly by the overall organizational philosophy of the parent DOT (or department of highways). A decentralized DOT with the central office having only a coordination role in the planning and programming of a maintenance district may outline a much broader range of responsibilities and authority than a centralized one. The degree to which a DOT is organized with functional activities grouped for program planning and implementation or is organized with functions grouped by the mode of transportation with which each activity is associated profoundly impacts the range and scope of a maintenance manager's responsibilities and authority. Generally, most DOTs are organized by mode, with highway maintenance being the sole responsibility of the manager of roadway maintenance. However, the New York DOT has a history of emphasizing organization by function.

In most cases, anything that is not construction or traffic operations on existing highways is a maintenance responsibility. The maintenance organization may be responsible for other activities such as operating ferries, operating toll collection systems, issuing and monitoring permits, operating and maintaining a departmental motor pool, and maintaining public service facilities. In some DOTs, traffic operations is an integral part of the maintenance responsibility.

In some states, maintenance and construction are part of the same organization. Many states are divided into districts with some districts under the supervision of operations or maintenance engineers. In some districts within a state, the operations or maintenance engineer is a subordinate in a more general organization structure that sometimes includes design engineering and facility construction.

Iowa Department of Transportation

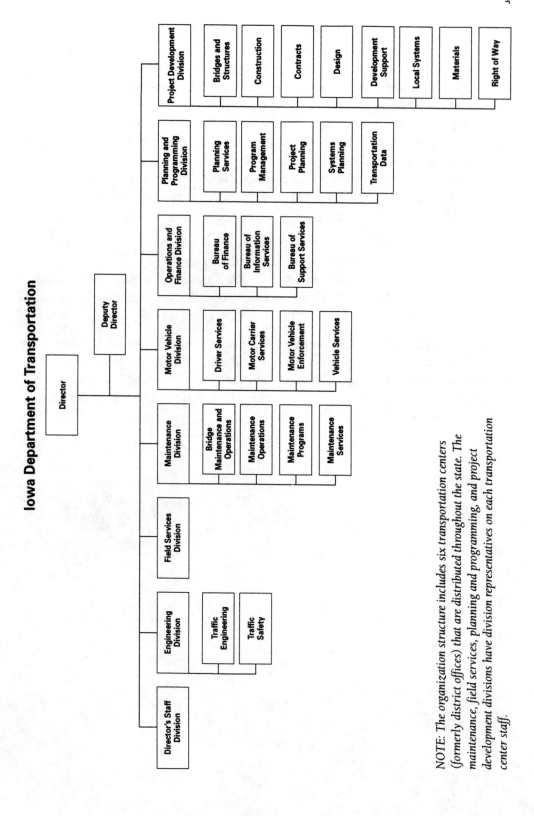

NOTE: *The organization structure includes six transportation centers (formerly district offices) that are distributed throughout the state. The maintenance, field services, planning and programming, and project development divisions have division representatives on each transportation center staff.*

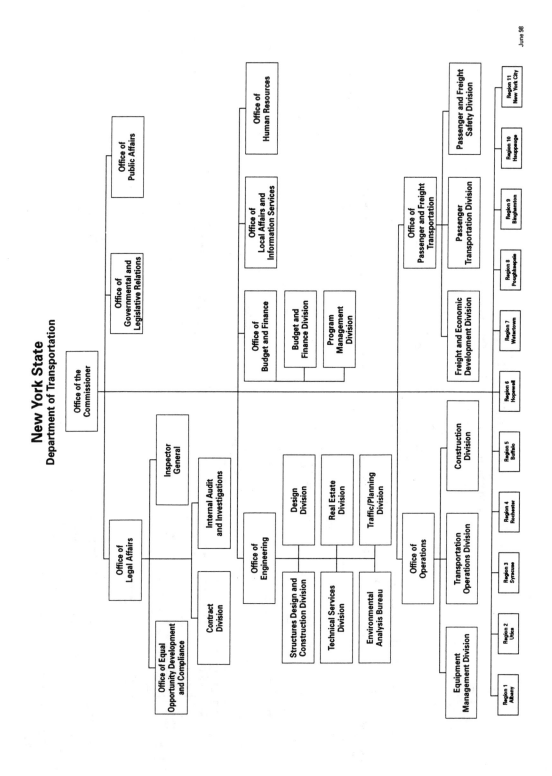

Chapter 1 Integrated Maintenance Management **1-15**

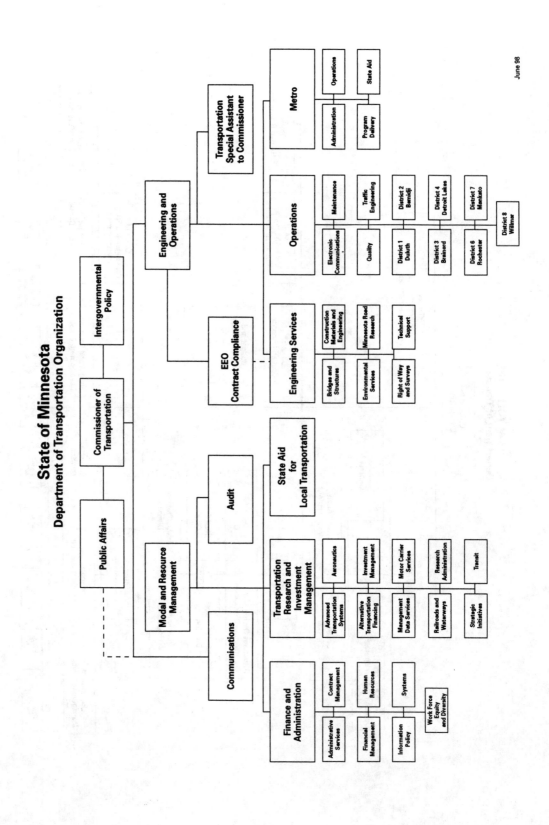

1-16 .... *Chapter 1 Integrated Maintenance Management*

1.2.1.2 Planning and Budgeting for Maintenance

Planning and budgeting for maintenance are interconnected activities. However, planning should focus more on assessing the functional activities necessary as a result of the highway network configuration and the distribution of the highway features with respect to the distribution of resources (personnel, equipment, and materials). Planning for maintenance should provide an estimate of the amount of maintenance needed to sustain the highway network at various levels of service (repair and replace for safety needs only, sustain the value of present investment in the network, upgrade the value and functional utility of the network, etc.). Then the budgetary requirements for the various levels of service can be set after maintenance performance and quality standards have been applied to yield estimates for the personnel, equipment, and materials needed to accomplish a certain level of maintenance service. A wide variety of tools and methods are available to do this type of planning to various degrees. Each state has somewhat different information management needs and capabilities. Thus, each state should carefully consider what seems to be useful and helpful in any method or technique to make their maintenance planning compatible with program planning across the DOT.

A variety of management systems to develop budgets are in use in various states. Selection of a budgeting (and planning) system by each state should largely be done on the basis of what works best and which system encourages the best lines of communication within the organization.

Highway maintenance budgets can be formulated from past performance data or needs based on established levels of service. Because budgets based solely on past performance can perpetuate both good and bad practices, they should be prepared with caution. There is growing interest in doing both planning and budgeting from a perspective of preserving the public investment in the highway network, similar to the capital investment analysis that a private corporation would perform. Adoption of such a budgeting philosophy will likely require some adjustment to the current budget development process in place in most DOTs to incorporate more of a "present net worth" analysis.

Maintenance management should work closely with those responsible for development of the total highway budget to accomplish the following:
- Recommend which highways and bridges need reconstruction in consideration of projected maintenance requirements. Use of pavement and structures management systems is a systematic way to define needs and establish priorities.
- Adjust maintenance requirements in view of approved construction projects and schedules.
- Determine future construction schedules in order to adjust maintenance requirements on roadways that may be reconstructed or constructed beyond the next budgetary period.

Budgeting should also include the monitoring of expenditures, ensuring that the intended level of service and allocation of funds is not exceeded. Agencies seeking to improve their budgeting process as part of an overall effort to improve the quality of completed maintenance activites should consult NCHRP Project 14-12, *Highway Maintenance Quality Assurance Final Report*. It contains recommendations for linking the budget process, resource monitoring allocation, and evaluating maintenance quality; it also has a model application to implement the process. This project report can be reviewed on the Internet at <www.nap.edu/readingroom> under the heading "transportation."

1.2.1.3 Scheduling Maintenance

Seasonal or annual events and routine planning produce most maintenance schedules. The need to clean, repair, or perform maintenance and service functions is dictated considerably by seasonal variations in weather. The most effective maintenance is based on anticipating a need and scheduling the proper resources and corrective actions to achieve the best results. The continued evaluation of SHRP test sections for pavement maintenance treatments should provide equations that can be used to estimate the effectiveness of a given pavement surface treatment based on the condition of a roadway surface, the traffic it is subjected to, and the general climatic zone in which it is located. Aids like these that are being developed have the potential to project preventive maintenance beyond an annual cycle for investment preservation.

New requirements are constantly increasing the scheduling responsibilities of maintenance managers. More sophisticated traffic control devices, mechanized equipment with greater capabilities and cost, the future introduction of robotics equipment, a wide range of new materials, increased training and safety requirements, public concern for ecological and aesthetic aspects of highways, energy availability and cost, and tort liability are some of the factors influencing the maintenance manager's role. As the TQM process moves more of the managerial decisions down to the field level, maintenance teams and crews will increasingly perform the scheduling of maintenance. These people need to be trained in how to make rational trade-offs in setting priorities with conflicting objectives. These conflicts result from the factors listed above and others, including political considerations, demands from various constituencies, and general public concerns. It is unfair to thrust decisionmaking in such an environment onto field personnel without also equipping them to process such conflicts.

A number of scheduling factors influence the manner and effectiveness of highway maintenance activities. Many of these factors are subject to a wide range of variables. Some of these factors are as follows:

- Distance of work sites from the base of operations and the time and expense to transport personnel, materials, and equipment to site. (At the upper level of the maintenance organization, planning needs to consider how the location of maintenance garages and stockpiles limit team and crew scheduling.)
- Weather conditions that can influence whether a repair can even be made or the nature of the materials and equipment that will be required. (The developing research started in SHRP on the use of weather information systems and more precise weather forecasting applied to maintenance has the potential to reduce the weather's disruption of maintenance schedules significantly.)
- Availability of skilled personnel, equipment suited to the task, and the proper materials (in sufficient amount and appropriate quality).
- Size and grouping of each work package, if it is within the scope of resources available, and if it will result in a high unit cost because of transportation distance.
- Problems caused by competition for resources between work projects, influence of unanticipated events, preparation requirements, etc.
- Influence of budgets on priorities and amount of work that can be accomplished within a budget period.
- Traffic conditions at different time periods and the degree to which maintenance operations can be expected to trigger congestion, delays in traffic, and accidents.
- Outside influences such as labor contracts and political requests.

Management systems to ensure effective highway maintenance should employ scheduling methods that, in consideration of the foregoing factors, will accomplish the most work with the resources available—as balanced against the most important needs of the highways in the network. To determine the most important scheduling needs or priorities, management should have a reporting and evaluation system to identify maintenance requirements.

Highway maintenance has become more mechanized and less dependent upon hand labor. Few maintenance functions remain that are completely performed by hand labor. Equipment-oriented personnel have become the basic employees in maintenance, replacing common hand laborers. Remote-control and robotic equipment is the coming technological transition in maintenance. The use of more expensive equipment and a larger work potential imposes greater obligation on maintenance management to schedule work efficiently.

A state maintenance organization manages personnel, equipment, facilities, and material valued at many millions of dollars. In every respect, it has a large responsibility that justifies the most effective use of resources. For instance,

- A poorly trained employee or an investment in equipment that is seldom used represents a loss to the organization. They may also affect indirect losses in the use of other resources. Obsolete or underused equipment should be disposed of to recover any remaining value. Equipment use should be regularly monitored.
- The considerable dispersal of personnel, equipment, and facilities over a large area makes efficient control difficult. Highway maintenance departments have difficulties in planning for efficient use of resources, difficulties that exceed similar problems in most other industries. This is one reason there is hope that empowerment of field-level supervisors through the TQM process may lead to significant increases in efficiency and effectiveness, since these persons are at the level where trade-offs and sharing of resources can have the most impact.
- The relationship of material costs to transportation costs presents a special problem. There are few situations in other industries where the cost of basic materials can double because of transportation costs.

Travel time is a large contributor to high unit costs in highway maintenance. If every pothole or crack were within 1 mile of the maintenance yard and the stockpile, the productive rate of a maintenance unit could be much higher. To schedule labor, materials, and equipment to travel 40 miles to repair one pothole is the extreme of poor scheduling in terms of cost-effectiveness. There may be times, however, when such scheduling will be done and cannot be avoided for traffic safety or other reasons. The most effective means for maintenance to improve efficiency and to achieve a high level of production is through scheduling and, particularly, reducing the ratio of travel time to production time. Some maintenance organizations find that four 10-hour days per week are more cost-effective than five 8-hour days. Again, as field-level units of the maintenance organization become responsible for determining how maintenance tasks are to be carried out to a larger degree than has been true in the past, the crews may improve production by making radical schedule adjustments.

Standby time is difficult to avoid in maintenance. However, scheduling based on experience and planning can foresee and avoid much lost time caused by delays in delivering materials, work bottlenecks, breakdowns, etc.

1.2.1.4 Data Systems to Support Maintenance

As the management of maintenance moves more to employing the techniques from operations research, management science, industrial engineering, and business management, it will require refining the current data systems available to support maintenance management. Bridge management systems, pavement management systems, and the existing maintenance management systems all share much common data, which is good. Maintenance management also interacts with payroll systems, material and equipment inventories, and so on, all of which are resident within the entire DOT information system databases.

What is evolving now is the way in which these various databases can be polled or questioned for specific items (if it is possible to do so in a particular department's information management system) to provide input to mathematical programming models, statistical quality control models, engineering economy models, and other optimization methods. Cost-benefit models have been used for decades. Now cost-effectiveness models are beginning to be used. Before embarking on a management strategy that depends on a particular optimization model or optimization strategy, determine what data needs apply to the model or strategy, then assess if it is economically feasible to retrieve the necessary data from the existing databases.

1.2.1.5 Incident Management and Emergency Response in Maintenance

Incidents are unplanned, unscheduled events for which highway maintenance is expected to respond in some way. Incidents may include hurricanes, tornadoes, floods, or fuel tanker accidents on a freeway. Proper management for maintenance response includes making certain that highway maintenance managers are included in the immediate information control and planning loop for response to any incident. Emergency response and disaster response teams must have some person from highway maintenance as a liaison to the maintenance organization if maintenance is expected to be an effective player. Persons who are not aware of the capabilities and limitations of highway maintenance to respond may make commitments on behalf of the maintenance organization that are nearly impossible to keep.

Some events that the general public sees as incidents are the result of phenomena for which it may be possible to obtain enough advance warning to be prepared to respond with the appropriate people, equipment, and materials to do the job expected of maintenance. Storm-tracking capabilities can give some lead time on when, where, and with what intensity a snowstorm is expected to hit. Likewise, a hurricane can be tracked sufficiently to give advance warning for response planning. Stream-gauging stations can provide estimates of when critical bridges may be threatened by rising flood waters.

Some events are random but have a statistically predictable pattern, such as the relationship between accidents and traffic flow-density patterns. If maintenance is supposed to respond to certain levels of accidents in the traffic stream, then traffic operations and surveillance centers need to inform maintenance managers when the traffic is reaching levels that increase the probability of a major accident.

If maintenance crews are to be involved in the response to hazardous material spills, their involvement must be clearly spelled out in response policies. Crews must be provided training in what they can and cannot do and must be properly equipped to carry out any role assigned to them.

1.2.1.6 Work Zone Safety

The fundamental guide for all maintenance managers, at all levels, in the control of maintenance work zones is Part VI of the *Manual on Uniform Traffic Control Devices* (MUTCD). Each state may have its own version of the MUTCD that contains some local special features, but all states have the essential elements of the MUTCD in their own manual.

Keep in mind that the principles and guidelines, including the examples shown in the MUTCD, were never intended to be all-encompassing or totally definitive. The MUTCD is intended to portray the synthesis of generally accepted research and engineering practice, and to be applied with good engineering judgment to the specific situation that the maintenance forces encounter at each site.

For commonly encountered work zone situations, many agencies adopt a traffic control plan that gives specific guidance to be followed in the absence of any highly unusual nature of the site or the traffic conditions.

Work zone safety has two aspects that are sometimes not kept in balance when there is too much pressure for production efficiency or there is too much fear of a lawsuit resulting from a traffic accident. Each maintenance worker is entitled to have a work environment that is reasonably safe, considering that working on the roadway has a certain level of risk. Each motorist (and any passenger) is entitled to have a traffic environment that is reasonably safe, considering the fact that driving on the roadway has a certain level of risk. Generally speaking, a work zone that is organized (including placement of equipment, storage of materials, placement of workers) for worker safety will be a work zone that is controlled for traffic safety (has sufficient warning devices, is properly delineated, etc.)

This manual contains two sections on traffic control for work zones. One has been provided for roadway maintenance and one for bridge maintenance. While there are significant similarities between the two topics, there are also some unique requirements for each.

Maintenance personnel may have an innovative idea that will create a better or an improved technique for traffic control or for worker protection in the work zone. If the innovation or improvement involves a change in signing, marking, or signalization from that approved in MUTCD (or the state manual), consult the state traffic engineer to obtain formal approval to test and evaluate the idea in the field. (Such approval will limit agency liability if motor vehicle collision occurs while the test and evaluation are being conducted.) If the innovation or improvement involves a barrier, attenuator, or other crash safety device, remember that NCHRP Report 350, *Recommended Procedures for the Safety Performance Evaluation of Highway Features*, discusses how all such devices should be crash tested before they are placed on the roadway.

As frequently as possible, maintenance engineers and maintenance managers should encourage law enforcement personnel to be present in work zones. Some agencies have adopted formal policies of reimbursement to law enforcement agencies for increased presence beyond routine patrol activity. The presence of law enforcement personnel near work zones increases safety and motorist compliance with traffic control regulations.

1.2.2 Human Resources
1.2.2.1 Motivation
A variety of management training courses emphasize learning how to motivate employees to produce superior work. This aspect of human relations was largely treated, in the past, as a matter of recognizing the human needs that can be provided by an employer and exchanging some measure of their fulfillment for a measure of successful work. The development of TQM principles has shifted motivation to providing employees with more responsibility, authority, and control over their work environment. Recent investigations by general news media reporters found that young workers are not motivated by the older "needs"-based traditional incentive programs. Young workers are more often responsive to "relational" motivation incentives. Thus, maintenance organizations must be aware of the different motivational styles that are appropriate to younger versus older workers if high-quality maintenance is to be achieved.

Maintenance engineers and maintenance managers will increasingly be asked to function as leaders rather than as managers who allocate resources to solve maintenance problems. One list of key leadership qualities that has been compiled includes the following [12]: self-confidence, fairness, objectivity, sensitivity, empathy, self-control, decisiveness, high energy, drive, courage, character, and integrity.

To provide maintenance managers with tools to become effective leaders and properly motivate the people for whom they are responsible, one organization has developed the following motivation guidelines [13]:

- Set clear standards of performance that are difficult but attainable. Remember that employees must have a high expectation of achieving the goal in order to be motivated.
- Rewards should always be contingent upon performance. If rewards are given when they are not deserved, they lose their value.
- Do not give too much reinforcement. Too much reinforcement is almost as bad as none at all.
- Reinforcement depends upon the individual. What reinforces one person may not reinforce another. Find out what reinforces employees and use it at appropriate times.
- Reinforce good performance as soon as possible.
- Do not unnecessarily threaten or punish. Threats and punishment are negative and encourage a person to avoid performing a certain task rather than changing the way that the task is performed.
- Make the individual employee feel important and recognized. Exhibit interest in and knowledge of each employee.
- Provide employees with the responsibility for task completion. Do not just assign duties, assign responsibility.
- Exhibit confidence in an individual's abilities. A great deal of research supports the contention that people who are expected to achieve will do so more frequently than people who are not aware of any expected performance.
- Communicate the job requirements and ensure that all employees understand them. We all know what we mean when we say something, but often others do not. Unclear expectations can lead to confusion.
- Listen to, and deal effectively with, employee complaints. Handle problems and complaints before they get blown out of proportion. In addition, people feel more important when their complaints are taken seriously. Conversely, nothing hurts as much as when others view a personally significant problem as unimportant.
- Demonstrate personal motivation through behavior and attitude. Nothing turns people off faster than supervisors who do not practice what they preach.

1.2.2.2 Training

The wide range of tasks, the need for special skills to maintain certain activities or equipment, and the requirements for safety precautions justify good training programs for all levels of maintenance employees. Training programs can be developed for special purposes, limited to a few selected personnel, or developed broadly for all maintenance employees. The resources of the National Highway Institute (Federal Highway Administration) and the network of resources that can be accessed through Local Transportation Assistance Program centers should be used to maximize the impact of the training budget. While training has long been recognized as important for highway maintenance agencies, it continues to be a deficiency in many agencies.

Technology changes rapidly in material quality control testing, surface deterioration measurement, environmental quality measurement, traffic control techniques, instrumentation for weighing trucks, and communication and signal equipment. These areas are some in which the training of personnel is key to effective use of new technologies. Implementation of TQM techniques will require an extensive commitment to training personnel in setting priorities, in group dynamics, and in-team-building skills [14].

Personnel who complete training courses satisfactorily should be recognized for their accomplishment. Where it is appropriate to ensure an interest in quality maintenance, completing appropriate training courses should be a condition for advancement.

Certain maintenance tasks require special skills or involve performing operations in a hazardous environment. Where maintenance personnel involved in these tasks have to be certified as qualified to conduct these operations or to operate certain equipment units, the certification process should be conducted in conjunction with other training programs, minimizing operational downtime for training.

In many states, an effective network of community colleges and vocational-technical colleges exists. Highway maintenance agencies should not overlook these organizations as a training resource.

1.2.2.3 Personnel Safety

Personnel safety involves the safety of maintenance personnel while using the equipment and materials required to conduct a maintenance task and their protection from through traffic entering the work area.

The safety of maintenance personnel while using their equipment and materials relates to two kinds of training: (1) proper training and proper qualification to operate any equipment assigned to a worker and (2) safety training in materials handling and in equipment safety.

Protecting workers from hazards associated with traffic through the work zone requires providing them with signing, barricades, attenuators, or shadow vehicles as needed to separate the workers from the traffic and to warn the traffic away from the workers. Getting a maintenance activity done quickly and efficiently is important in minimizing the time that workers are exposed to traffic danger.

Refer to the section on work zone traffic control in the chapter on roadway maintenance and the chapter on bridge maintenance containing discussions relating to personnel safety. Further discussion of personnel safety can also be found in the section on environmental aspects of roadway maintenance and in the section on environmental aspects of bridge maintenance. (See the index for page references.)

1.2.2.4 Recruitment and Development

Highway maintenance organizations need to create a profile of the kind of person they want to have in maintenance. Do you want more females, more minorities, high school graduates, or technical school graduates? Do you want people with welding skills, general mechanic skills, etc.? Once you have a sense of the type of persons that you desire, work with the human resources office to recruit them. This long-term personnel development strategy targets high schools, technical programs, and job placement organizations. Agencies that wait for people to come to them will always have less motivated persons in their ranks.

Develop a process of selecting individuals who seem to have promise and can be trained in skills needed in the maintenance organization. Then create a training opportunity that upgrades these individuals to the needed skill levels. If skill upgrades demand some significant investment on the agency's part, perhaps a minimum-term contract that requires the individual to stay in the agency after completing the training is needed. An in-house effort to upgrade people is as important as a recruitment effort.

1.2.3 Technical Resources

Management often focuses on people relationships. It is equally important to consider the technical, nonhuman resources in managing maintenance.

1.2.3.1 Equipment for Maintenance

The equipment appropriate for maintenance varies as much as it does for the construction industry. The most important aspect of maintenance equipment is to determine the right size and capability of equipment to do the task needed. For instance, in utility maintenance, trenches for communication cable used to be excavated with a tractor-mounted backhoe. However, the backhoe outriggers often damaged private property outside the narrow easement limits of the utility corridors, creating undesirable private property damage claims. In response to this problem, equipment manufacturers began offering a "miniature crane" bucket digger that is crawler track driven. In the same way, if a standard piece of construction equipment does not accomplish your maintenance task, talk to other maintenance engineers whose personnel perform the same activity. If you need specially modified equipment and several maintenance areas will use it, someone will eventually fill that market need.

1.2.3.2 Materials for Maintenance

The same materials used in construction are available for maintenance repair. However, it behooves maintenance managers to review the materials science literature continually for nontraditional materials that could be used to maintain roadways and bridges. Sometimes a material that is too expensive for initial construction is cost-effective for maintenance because it reduces traffic delays and worker exposure times by enabling the job to be completed in less time or increasing the life of the repair. While maintenance is cost and budget driven, the lowest initial cost for a maintenance activity may be false economy with respect to maintenance materials.

New materials are constantly being developed. Before any new material is accepted for local testing and evaluation, it should be reviewed by any "new products committee" the agency may have that coordinates the study of new products. Maintenance engineers and maintenance managers should periodically check the list of new materials that may be under evaluation (or that have been evaluated) by the National Transportation Product Evaluation

Program (NTPEP) at AASHTO (on the Internet at <www.aashto.org/prog_svce/ntpep>) and in the technology evaluation activites of the Civil Engineering Research Center (CERF) at the American Society of Civil Engineers (ASCE) (on the Internet at <www.cerf.org/evtec/eval>).

1.2.3.3 Models Available for Management of Resources

Maintenance engineers should begin to investigate the application of industrial engineering and management science models to maintenance activities. For instance, queuing theory models can estimate the traffic delays that can be expected at various traffic volumes when a lane is closed off. Queuing analysis can also estimate the efficiency of a trucking operation in moving material from a stockpile to the roadway location where it is being placed. Linear programming and other mathematical programming techniques can assist in the development of a strategic plan for the location of material stockpiles, the location and distribution of maintenance garages, and the selection of equipment phase in-out strategies. Network analysis methods that urban transportation planners use can be applied to selecting the best detour routes around road sections that are closed because of a bridge washout or similar localized disaster.

1.3 1991 ISTEA AND IMPACTS ON MAINTENANCE

The 1991 Integrated Surface Transportation Efficiency Act (ISTEA) initially required that a number of management information systems be developed and implemented within transportation agencies (see section 1.4). A maintenance management system was not included in the requirements. The various mandated management information systems were also not required to be integrated databases. However, these requirements have a strong relationship to maintenance and to the information systems needed to manage maintenance properly.

1.4 INTEGRATED SYSTEMS

The 1991 ISTEA originally mandated that all states begin developing and implementing management systems. While the degree to which each of these management systems will be implemented may vary in a given state in any given year, these systems (as a whole) indicate the trend in the management of transportation information systems. The specified management systems were as follows:
- Traffic congestion.
- Public transportation facilities and equipment.
- Intermodal facilities and systems.
- Bridge.
- Highway safety.
- Pavement.
- Traffic monitoring for highways.

Even though such activity is now optional, agencies that developed operational management systems have recognized their benefit and moved forward with such systems.

1.4.1 Advantages and Benefits

Pavement management and maintenance management have been interconnected for some time. Bridge management is an integral part of bridge maintenance. The other systems that manage the information flow in discrete areas of the transportation network are not as clearly related to managing the maintenance function. However, when these information systems are up and running, elements of information in them will benefit maintenance managers and enable them to refine maintenance management.

1.4.2 Limitations and Costs

When all of these various separate management systems are integrated, the information flow needs to be seamless if maintenance managers are to use the databases easily. Each of these management systems will foster a data collection and data maintenance process that will ultimately be related to the maintenance of increasingly sophisticated roadway and bridge networks. In some cases, highly sophisticated instrumentation and communication networks are being incorporated into the state highway network. For instance, one state is installing an extensive fiber optics network in the highway right of way to serve (among other things) a roadway weather information system. In such situations, the impact of any future maintenance requirements on the state highway agency must be carefully considered so department maintenance personnel can be prepared to assume responsibility for keeping the communication network functioning.

1.5 DEVELOPING ISSUES
1.5.1 Sustainable Development

There is, and will continue to be for years to come, a push for all transportation infrastructure to achieve "sustainable development." The ASCE has attempted to make the principle of sustainable development a part of its code of ethics. However, the term means different things to different people. In any case, most definitions of sustainable development encompass maximizing the capability to recycle components of the infrastructure and to minimize use of nonrenewable resources. This philosophy will be a driving social and political force that maintenance engineers and managers will have to consider in the foreseeable future while planning maintenance methods and processes.

1.5.2 Designing for Maintainability

Designing a weapons system that can be easily and effectively maintained has been a principle of defense weapon systems procurement for many years. Likewise, highway transportation agencies are beginning to provide formal interaction between maintenance engineers and design engineers to ensure that the designer's concept can be maintained safely and effectively. For hints on how to set up such a process or for ideas on how to effectively participate in such a process, see NCHRP Report 349, *Maintenance Considerations in Highway Design*.

1.5.3 Environmental Sensitivity

Maintenance is a highway agency's most exposed activity with respect to environmental pressures. Painting, sanding, salting, herbicide application, mowing and brush control, landscaping, and maintaining drainage are activities that can raise environmental objections. All material handling can have environmental safety implications for the workers and the general public. As environmental and health research continue to advance, both areas will further the need for an effective information dissemination program within maintenance forces. A forthcoming NCHRP report, *Synthesis of Highway Practice on Best Maintenance Practices for Environmental Maintenance* should be consulted for a concise overview of recommended practices.

1.5.4 Application of TQM Principles

Product manufacturing industries have demonstrated the benefits of applying TQM principles to their operations. As a result, educational institutions, service industries, and government agencies are now applying TQM to their operations. Maintenance engineers and managers need to consider learning TQM and seeking ways to apply it. However, they also need to be prepared to revise the concepts to fit their own organization's unique needs.

1.5.5 Litigation Involving Maintenance

The United States continues to be a lawsuit-oriented society. As the design and construction of new roadways and bridges continues to decline as a proportion of the total transportation budget, maintenance will share an increasing proportion of the litigation burden on transportation agencies, indicating that maintenance engineers and managers need at least minimal training in serving as an effective expert witness. Every maintenance crew should carry with it every day some means of making a photographic record of an accident and should have at least some minimal instruction in how to use it to document the conditions surrounding every incident (e.g., road surface conditions, traffic controls, etc.), even accidents involving maintenance workers only. Acceptable photographic equipment includes a 35-mm camera and film, video camera and tape, video minicam and cassette, or disposable camera. Perhaps the best of these alternatives might be the disposable camera, which contains film, eliminating the problem of inventory control.

1.5.6 Rational and Probabilistic Models to Predict Maintenance Needs and Effects

Pavement research continues to develop statistical models and damage equations that refine predictions of pavement deterioration for pavement management systems. Most bridge management systems are based on neural network theory, which enables users to predict the future date at which a specified level of damage will appear. These and other advanced mathematical techniques are foreign to most maintenance engineers and managers. However, maintenance engineers and managers will need a rudimentary understanding of these concepts to integrate the predictions from such models in other management systems into the maintenance management system.

1.6 REFERENCES

1. Highway Research Board. *Iowa State highway maintenance study*. Special Report 65. Washington, D.C.: 1961.

2. Highway Research Board. *Iowa State highway maintenance study*., Special Report 65. Supplement I: Washington, D.C.: 1961.

3. Highway Research Board. *Maintenance management*. Special Report 100. Washington, D.C.: 1968.

4. Highway Research Board. *Maintenance operations and applied systems engineering*. Highway Research Record 359. Washington, D.C.: 1971.

5. Woodward-Clyde and Associates. *Manual for the selection of optimal maintenance levels of service*. Report 273. Washington, D.C.: National Cooperative Highway Research Program, Transportation Research Board, 1984.

6. Smithson, Leland D. *Basic management activities within highway maintenance*. Paper presented at a workshop held by the TRB. Performing Highway Maintenance Using Total Quality Management, Transportation Research Board, Whitefish, MT, May 21–23, 1995.

7. Roy Jorgenson Associates. *Performance budgeting system for highway maintenance*. Report 131. Washington, D.C.: National Cooperative Highway Research Program, Transportation Research Board, 1972.

8. Markow, Michael J., F. D. Harrison, P. D. Thompson, E. A. Harper, W. A. Hyman, R. M. Alfelor, W. G. Mortenson, and T. M. Alexander, *Role of highway maintenance in integrated management systems*. Report 363. Washington, D.C.: National Cooperative Highway Research Program, 1994.

9. Strategic Highway Research Program. *Rating preventive maintenance treatments*. SHRP Manual 3033. Washington, D.C.: 1993.

10. Strategic Highway Research Program. *Specifications for preventive maintenance treatments*. SHRP Manual 3034. Washington, D.C.: 1993.

11. Strategic Highway Research Program. *Pavement maintenance effectiveness*. SHRP-H-358. Washington, D.C.: 1993.

12. Sheeran, F. Burke. *Management essentials for public works administrators*. Kansas City, MO. American Public Works Association, 1979.

13. Kilareski, Walter P., Robert E. Brydia, and Vincent P. Lewis. *Maintenance practices for local roads*. Vol. 2: *Personnel supervision*. Pennsylvania Department of Transportation, June 1994.

14. Hoffman, Gary L. *How maintenance activities might benefit from TQM*. Paper presented at a workshop held by the TRB. Performing Highway Maintenance Using Total Quality Management, Transportation Research Board, Whitefish, MT, May 21–23, 1995.

2.0 ROADWAY MAINTENANCE AND MANAGEMENT

2.1 ROADWAY MAINTENANCE

Maintenance of the roadway is the primary way in which a highway agency carries out its goal of providing a safe, efficient transportation system on a day-to-day basis. Roadway maintenance includes maintaining the traveled lanes, the shoulders, minor drainage structures and drainage surfaces, safety hardware, and other appurtenances. Bridges are covered in a separate chapter of this manual, although certain aspects of roadway maintenance relating to bridges are covered in this chapter.

2.1.1 Introduction

To some people, planning, design, construction, and materials research are activities that are more sophisticated and perhaps more important than roadway maintenance. However, maintenance and maintenance operations are an activity that is the most visible to the roadway user, the most urgently needed in permitting the public to use the roadway, and the activity that requires the most ingenuity to be done effectively.

2.1.1.1 Preventive Roadway Maintenance Philosophy

Preventive maintenance is a planned strategy of cost-effective treatments to an existing roadway system and its appurtenances that preserves the system, retards future deterioration, and maintains or improves the functional condition of the system (without substantially increasing structural capacity) [adopted by the AASHTO Subcommittee on Maintenance].

The key is moving from a "don't fix it if it isn't broken" attitude to an "a stitch in time saves nine" attitude. See NCHRP Synthesis of Practice 153, *Evolution and Benefits of Preventive Maintenance Strategies*, for more details on this concept [1]. As an example of an agency's effort to increase its focus on preventive maintenance approaches, the Colorado Department of Transportation hosted a pilot SHRP workshop on pavement preventive maintenance [2].

2.1.1.2 Interaction With Bridge Maintenance and the Management of Other Functions

Frequently, the engineers and managers responsible for bridge maintenance and for minor construction repairs to both roadways and bridges are in separate operating units with different maintenance engineers or managers. Regular coordination and liaison among these groups can be invaluable in coordinating the schedule of preventive maintenance activities, minimizing the traffic impact of work zones and the exposure of maintenance personnel to traffic hazards.

For example, if bridge maintenance is planning to rehabilitate a bridge deck in a section of roadway for which roadway maintenance has future plans to do rigid pavement joint repairs, perhaps the priority for the joint repair can be shifted so the work will be done simultaneously with the bridge deck repair. Then one work zone traffic control pattern can be used to protect both workers and the motoring public for two activities.

On a broader scale, interaction, coordination, and liaison with other agencies responsible for maintenance can often permit performance of maintenance through contract relationships in ways that can get the whole job done with fewer crews and fewer traffic control setups. This is particularly the case in urban freeway maintenance [3]. See NCHRP Synthesis of Highway Practice 170, *Managing Urban Freeway Maintenance*, for examples and details [3].

2.1.1.3 Interaction With the Design of Facilities

Most highway agency managers and administrators believe that designers consider all aspects of planning, construction, materials, operations, and maintenance in the process of creating a roadway or bridge design. However, as a practical matter, most highway agencies continue to construct roadways and bridges from designs that create difficulties in maintaining and operating the roadway and bridge network. It may be something as simple as a W-beam guardrail anchor placed in a structure location so that a maintenance worker has to be dangerously exposed to traffic when checking the anchor bolts. Many of these problems seem to stem from lack of cross-function experience between design engineers and maintenance engineers.

Many highway agencies have a regular process whereby various divisions of the agency participate in periodic reviews of a roadway or bridge design as it progresses. In highway agencies where such a process exists, if maintenance personnel are not represented among the design review participants, the top-level maintenance manager should volunteer the division's staff to participate in such reviews, communicating to all parties the long-term value to the public of such input. An in-depth study of processes by which maintenance issues could be integrated with the design process was developed and pilot tested in the Utah DOT [4]. Any highway agency beginning to work through how to incorporate maintenance input and feedback into roadway and bridge design processes should see NCHRP Report 349, *Maintenance Considerations in Highway Design*, for ideas and helpful guidance [4].

2.1.1.4 Role of Roadway Maintenance in the Safe, Efficient Operation of Facilities

As stated earlier, roadway and bridge maintenance is the primary means by which a highway agency keeps the transportation network a safe, efficient facility for the public road user. While the trend is to increase preventive maintenance practices, "there is a practical limit or point of diminishing returns for the preventive activities that can be afforded or justified" [5]. Maintenance assumes a primary role in monitoring the safety and efficiency of the roadway network by responding to short-term maintenance needs such as potholes, damaged guardrails, washouts, knocked-down signs and signals, traffic accident cleanup, explosion damage, material and cargo spills, flood damage, landslide damage, rock falls, blizzards, and debris from windstorms.

Maintenance engineers and managers must have sufficient flexibility in their planning, budgeting, and scheduling of preventive maintenance to suspend or shift preventive maintenance activities and perform maintenance and repairs to "fix" short-term maintenance needs as they arise. The roadway user and general public will judge the maintenance organization's effectiveness largely on the basis of how it responds to these needs, because they directly and immediately affect the safety and efficiency of roadway use.

See NCHRP Synthesis of Highway Practice 173, *Short-Term Responsive Maintenance Systems* for a detailed discussion of methods and processes that effectively and efficiently deal with short-term maintenance needs [5]. The basic rudiments of dealing with short-term maintenance needs exist in almost all highway maintenance agencies [5]:

- Have formal inspection processes for bridges, roadways, roadsides, and appurtenances.
- Incorporate informal inspections (i.e., from in-house resources).
- Verify or incorporate external observations.

Policies must define what the immediate, on-the-spot emergency response will be for critical, short-term needs. For example, if a maintenance patrol is placing a Type II barricade with a flasher unit on it and encounters a crushed, breakaway cable terminal (BCT) guardrail end section, then each patrol must always leave the maintenance garage with sufficient Type II barricades for the expected number of BCT hits in their patrol area for one shift.

All highway agency personnel should be informed of the procedure to report a roadway or bridge defect and should be encouraged to submit reports of all incidents observed. Thus, personnel doing field traffic counts, planners traveling to public hearings, and administrative officers traveling to meetings become additional "eyes and ears" of the maintenance forces, similar to the concept of military scouts.

External observations are reports from private citizens and from organizations outside the highway agency. Because of wide variation in precision and credibility of these observations, an agency may choose to verify them to determine the urgency of response or may choose to use them as significant input. For example, some cities and counties depend almost exclusively on citizen calls to identify the need to respond to down traffic signs.

2.1.1.5 Value of Roadway Maintenance to Agency

A method that can estimate the intrinsic value to the highway agency of conducting roadway maintenance is greatly needed. Lacking any means of estimating the value of doing roadway maintenance, the importance of not doing roadway maintenance is easily demonstrated. In Transportation Research Board State of the Art Report 6, *Relationship Between Safety and Key Highway Features,* three of the seven sections deal with roadway features for which maintenance is the primary way to retain safety of an existing roadway [6]. If any agency examines its litigation pressures as measured by number of actions filed, situations related to maintenance will figure prominently in the total. These two points are very general but suggest that maintenance is a key activity of a highway agency in carrying out proper public service.

Two things that can be done, as a minimum, to begin tracking the value of roadway maintenance to the agency are the following:

(1) Coordinate with the legal staff responsible for defending the agency in tort liability actions to develop a history of the number of maintenance-related claims and the amount of damages awarded for them.
(2) Coordinate with the staff responsible for pavement management to develop an index of the effect of maintenance on preservation of the paved roadway network.

2.1.1.6 Value of Roadway Maintenance to the User Community

In a TQM context, the "internal customer" of roadway maintenance is the agency itself. The "external customers" are the roadway users and general public. The Minnesota DOT is one of the farthest agencies along in attempting to define the value of roadway maintenance to their user community [7]. Their process included the following significant steps:

- Established a Maintenance Business Plan Steering Committee to formulate a definition of quality of maintenance in terms of what the public saw as important.
- Through the Maintenance Engineers' Group, identified seven products or services that maintenance had to deliver to satisfy their maintenance mission.
- Conducted a "market research" study with focus groups and an individual interview sample to (1) gain an understanding of current travel practices and road use by various user groups, (2) identify particular likes and dislikes relating to state highway maintenance that area residents and business people may have, (3) create a list of customer expectations of the Minnesota DOT within the service area and rate the department on the basis of those expectations, (4) evaluate the appropriateness of performance dimensions created by the Minnesota DOT, and (5) develop recommendations for future Minnesota DOT customer service research for other areas of the state.
- Analyzed the customer survey data for maintenance activities that could be improved, enhancing customer satisfaction. The goal was to define the outcome the department wanted to achieve in the customer's terms, then determine what outputs were needed to produce the desired outcome, and finally decide what activities and inputs the DOT must generate to produce the required outputs. In this way, the department is seeking to create a maintenance organization that is driven by desired outcomes and not programmed inputs (traditional model).

2.1.2 Maintenance of Roadway Surfaces

Roadway surface maintenance accounts for the major segment of the maintenance budget. The roadway surface is usually described as consisting of two structural zones: (1) a pavement structure that may consist of layers including a subbase course, one or more base courses, and one or more surface courses; and (2) a subgrade structure that is the graded portion of a highway (roadbed) prepared as a foundation for the traveled way pavement structure and the shoulders. Surface courses may be untreated, treated with admixtures, or paved. Pavement structure is usually categorized as flexible (bituminous-type pavement), rigid (portland cement concrete), or granular (aggregate surfaced or gravel).

The maintenance of all roadways should be consistent with the functional classification of the roadway, the traffic volumes, and the roadway condition.

Flexible pavement is a pavement structure that distributes loads to the subgrade through aggregate interlock, particle friction, and cohesion. Rigid pavement is a pavement structure that distributes loads to the subgrade through a portland cement concrete slab.

Maintenance policies and procedures for roadway surfaces should be designed to offset or minimize the effects of age, wear, loading, and various types of distress. The type of distress and the maintenance methods used are largely influenced by the type of pavement structure, materials used, weather, and traffic.

The range of variables that affect the life of a roadway surface can be extensive. Temperature changes, moisture content, structural materials, design, and the type and amount of traffic are some of the factors that can influence the length of time before a roadway will require resurfacing, replacement, or other maintenance. The quality of maintenance directly affects the life of a pavement and its serviceability. Quality of maintenance is based upon a combination of proper materials, their proper use within the context of the appropriate maintenance method, and timely application of the method and material.

Many surface failures may result from either improper construction or inappropriate design. A knowledge of the root cause of these failures is essential to proper maintenance and repair. Failures because of inappropriate design or increased traffic should be referred to design engineers for correction through a new design and construction project. Surface failures that can be attributed to poor construction usually require increasing amounts of corrective action by maintenance personnel over the life of the pavement.

2.1.2.1 Planning and Budgeting

Development of maintenance budgets for roadway surfaces should be based on actual needs as determined from the highway system inventory and adjusted for the history of maintenance in previous years. Budgeting is the process that relates work to be accomplished with resources (labor, equipment, and material) and funds required. Any adjustments to the recommended budget can then be reflected in a changed level of service for maintenance activities affected. Previous years' requirements should be qualified by considering

- additions or deletions, including new routes;
- improvements that reduce maintenance costs;
- new equipment and materials adopted to reduce costs;
- deteriorating conditions or other factors that will increase requirements over that of previous years;
- additional training required to implement new processes (such as adopting TQM);
- increased cost to implement new equipment systems, new methods, or new materials associated with increased quality standards; and
- changes in priorities associated with changing the balance between preventive maintenance and short-response maintenance.

A highway maintenance organization should have a system, preferably automated, that will project its maintenance requirements for a sufficient number of years, avoiding situations that will cause severe fluctuations in maintenance employment. If projections show large increases or decreases in labor from one budget period to the next, it may be necessary to adjust maintenance requirements to ensure a consistent level of labor requirements. Changes in projected labor requirements are also opportunities to assess the impact of shifting selected activities to contract maintenance.

Using maintainable roadway features and performance standards combined with service-level values permits development of the annual work program and budget. Service levels are used to quantify magnitudes of maintenance performed on a specific roadway feature. Because the amount of features is fixed and the cost of resources can only be slightly altered, service level is the only variable that can be used to change the annual cost of performing a given activity.

The first cut in developing the annual maintenance budget and the initial work program is developed using target service-level values. This initial budget identifies the cost of maintaining the roadway system if there were no budget or resource constraints. When actual legislative appropriations have been determined, service-level values must then be adjusted through an iterative process, bringing the work program projection in line with the appropriation. Developing a work program projection that optimizes the available resources and the funding is a very important component of the maintenance management process.

To avoid large expenditures on roadways scheduled for reconstruction, close coordination with the construction plans for the budgetary period is recommended. Routine maintenance funds should be designated only for improvement or for the restoration and repair of roadways to the safe, usable conditions to which they were originally constructed. Extensive corrective work that involves increases in thickness of pavement, overlays over 1.9 cm (3/4 inch) in thickness, and over 153 m (500 feet) in length should be classified as "betterment."

Planning and budgeting are activities that have a significant impact on a maintenance organization's ability to meet the quality goals and expectations for maintenance. Agencies seeking to improve their planning and budgeting process as part of an overall effort to improve the quality of maintenance activities should consult NCHRP Project 14-12, *Highway Maintenance Quality Assurance Final Report*. It has a model application to implement the process and contains recommendations for linking the budget process, monitoring resource allocation, and evaluating maintenance quality. This project report can be reviewed on the Internet under the heading of "transportation" at <www.nap.edu/readingroom>.

2.1.2.2 Scheduling

Preventive maintenance and repair scheduling begins with long-range planning. On the basis of structural design and performance, the date when a highway will require resurfacing or replacement should be predicted. Routine maintenance and periodic repairs should also be anticipated and planned. The effect of variables that can influence long-range plans and schedules should then be considered. If traffic loads increase or weather conditions are extreme, the need for repairs and resurfacing may occur sooner than predicted. The most practical reaction to a change in the factors that influence roadway surfaces is to increase the number of inspections. When inspections show that surface distresses are increasing, a schedule reevaluation is in order. If proper maintenance measures are not taken in a timely manner, neglected repairs can result in areas of major distress. The personnel responsible for inspections should be the most qualified available. In the long run, their ability to detect problem areas in their earliest stages can save a great deal of time and money.

The schedule plan flows from the budgeting process in allocating resources as the final budget permits. As soon as final budgets are approved, the work program and resources (labor, equipment, and materials) to accomplish the work must be allocated. Care must be taken in the allocation process to ensure that imbalances do not occur. If maintenance needs exceed available resources, imbalances can usually be corrected through personnel shifts, contract maintenance, or formation of special crews to accomplish specialized maintenance. The scheduling process is laying out work for the future and developing work schedules for various levels with the agency division that conducts maintenance. Schedules are generally very specific at the lower levels of the organization and less specific at the upper levels. Scheduling will include any specialty crews. A maintenance supervisor's schedules will specify location, time, date, and assigned crew members. Scheduling needs to conform to seasonal requirements and performance standards so the schedule can be consistent with reasonable and possible levels of effort. The scheduling process needs to contain flexibility so that plans made by higher administrative levels or responses to emergency conditions can be met without unduly burdening the maintenance organization.

2.1.2.3 Maintenance Needs by Type of Surface

Flexible Pavements: Flexible pavements are constructed using plant mixes of graded aggregate and bituminous materials. Bituminous materials include asphalts derived from petroleum and tars derived from destructive distillation of materials such as coal and wood. Tars have been little used in highway construction and maintenance in recent years because of their relative lack of availability. In addition, tars contain carcinogenic compounds that present a potential risk to workers and to the general population. Tars in the form of emulsions are used for areas where there is possible spillage of oil, gasoline, or kerosene substances, however, because tars are relatively insoluble. Because maintenance uses for tars are required only in special instances, the term "bituminous materials" when used will generally refer to asphalt products.

Deterioration repair by recycling asphalt surface

Slurry seal surface treatment

Wheel Ruts: Rutting is the formation of longitudinal depressions in the roadway surface from traffic wear in the wheel path. Ruts are usually the width of a wheel path and are generally caused by traffic wear and displacement of the surface course or base by heavy traffic loads. Inadequate compaction during construction or an improper mix design is also sometimes a cause. The rutted area may be repaired by leveling with asphalt mix overlay, cold milling, heater planing, or recycling the surface and replacing it. Maintenance patrols should be alert for ponding of water along the wheel path as a sign of rutting. Removal of ruts will reduce the possibility of hydroplaning or the development of ice patches in freezing temperatures. SHRP validation studies on the Superpave asphalt binder specification suggest that using Superpave in leveling courses may significantly reduce wheel rutting problems [8].

Potholes: Potholes are steep-sided holes of varying sizes in pavement resulting from localized disintegration often caused by poor drainage and aggravated by traffic loads. Potholes and localized failures should be patched as soon as practical after they are reported. Potholes that exist in pavements during dry weather should be repaired by patching before the beginning of any extended inclement weather. Either temporary or permanent patches can be applied, depending upon which is most appropriate. Permanent patching is generally preferred when the hole is squared up to sound material. If the roadway section is scheduled for general repair through reconstruction or overlay, then temporary patching is most appropriate. For urban streets, the temporary patch is cost-effective when the patch is thoroughly compacted into the hole.

Findings of an SHRP study of pothole patching suggest that it needs to be rethought with respect to the conventional logic. When high-quality materials were used, the throw-and-roll procedure (that is, dumping the patch material into the hole with no preparation, then driving the truck over the patch several times) was found to be just as effective as the semipermanent procedure (defined as "doing it right," by squaring up the sides of the hole and cleaning and drying the hole before filling it). What seemed to matter more was the quality of materials and workmanship, not the procedure used to fill the hole [9]. Maintenance engineers and managers new to the field and those persons seeking to update their background in the sealing and filling of cracks in asphalt pavements should see SHRP-H-348, *Asphalt Pavement Repair Manuals of Practice* [10]. The second half of this document is a manual, *Materials and Procedures for the Repair of Potholes in Asphalt-Surfaced Pavements*. The major grouping of the topics covered is as follows:

- Need for pothole repair.
- Planning and design, including materials, repair techniques, patching costs, and overall cost-effectiveness.
- Construction, including traffic control, safety, winter patching, and spring patching.
- Evaluating repair performance.

Pavement patched until roadway can be rehabilitated

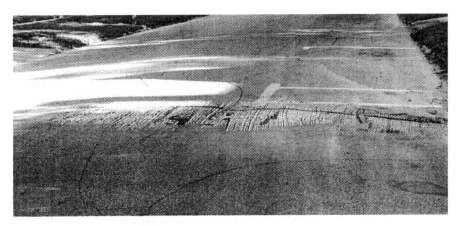
Milled asphalt surface at rough joint (bump)

Cracks: Cracks may result from structural weaknesses or from the normal aging process of the asphalt surface or the underlying subgrade layers.

Structural Weakness Cracks: Cracks in the roadway surface that result from problems within the underlying pavement structure include the following:
- Reflection cracks appear in an overlay caused by joints or cracks in the underlying pavement. They may occur longitudinally or transversely in relation to the roadway centerline.
- Edge cracks occur parallel to the pavement's outer edge and usually are a foot or two inside from the shoulder. These cracks and edge line cracks occurring between the pavement and the shoulder result from inadequate shoulder support. Inadequately compacted shoulder material or failure to provide lateral support allows the weight of heavy vehicles to displace the base or subgrade material. Sealing of these cracks should be regarded as a temporary measure unless continued inspections show that the shoulder has become stabilized and the cracks are no longer widening. Permanent repairs may require restructuring the subgrade and replacing shoulder material. If the subsidence is caused by excessive moisture under the outer edge of the pavement, installation of edge drains effectively lowers the water table in this area.
- Alligator cracks are interconnected cracks forming a series of small polygons in the pavement surface that resemble an alligator's skin. They are caused by fatigue due to traffic loading and inadequate support of the pavement structure.

Shrinkage Cracks: These are random interconnected cracks caused by a reduction in the volume of the pavement because of improper materials or proportions and aging of the asphalt mix. Asphalt normally shrinks with age and loses its ductility. In street pavements where an asphalt surface course has been placed over a stabilized soil base, cutting off all moisture with the pavement can produce shrinkage cracks in the base that penetrate the surface course.

Corrective Measures: A continuum of activity exists for the practice of filling cracks, from not filling any cracks in asphalt paving, to filling only large cracks, to filling all cracks. The permeability of the base material is used as a criterion to determine the extent to which cracks are filled by some agencies, while other agencies pursue a fixed policy. In general, crack filling and sealing is considered a good maintenance practice.

When cracking is caused by poor surface paving materials, replacement of the surface course may be the only adequate remedy. Structural repairs should be scheduled for completion by fall; however, repairs should be made as soon as possible when cracking has progressed to the extent that failure of the roadway surface is imminent. Removal of wet subgrade or base material and installation of proper drainage are usually required to prevent a recurrence.

Crack filling and sealing can be scheduled for the cold weather season when it is dry. Cold pavement temperatures open the cracks for better filling and sealing.

Some agencies have reported good results in significantly reducing cracking problems by using the SHRP-generated Superpave process for a leveling course [11].

Maintenance engineers, managers new to the field, and those persons seeking to refresh their background in sealing and filling cracks in asphalt pavements should see SHRP-H-348, *Asphalt Pavement Repair Manuals of Practice* [10]. The first half of this document is a manual, *Materials and Procedures for Sealing and Filling Cracks in Asphalt-Surfaced Pavements*. The major grouping of the topics covered is as follows:

- Need for crack treatment (includes evaluating the cracking condition, determining the type of maintenance, deciding whether to seal or fill, and deciding when to seal and when to fill).
- Planning and design (includes selecting a sealant or filler material, determining placement configuration, selecting procedures and equipment, estimating materials requirements, and performing cost-effectiveness analysis).
- Construction (includes traffic control, safety, crack cutting, cleaning and drying, material preparation and application, material finishing and shaping, and material blotting).
- Evaluating treatment performance.

Corrugations and Shoving (Waves, Sags, and Humps): Corrugations are transverse undulations at regular intervals in the asphalt surface consisting of alternating valleys and crests, generally less than 10 feet apart. When more than 10 feet apart, they can be called waves. Corrugations add to the discomfort of motorists. Small cars and motorcycles may experience difficulty in retaining control when crossing a severely corrugated surface. Corrugations usually are caused by traffic over unstable mixtures in which the content of fines or asphalt is too high.

Shoving is also caused by traffic but characteristically occurs locally at intersections and on hills where vehicles are accelerating or braking. The first indications of shoving are slippage cracks that are crescent-shaped cracks with the apex of the crack pointing in the direction of the shove. To correct such distress and prevent recurrence, the surfacing should be removed down to a firm base and replaced. SHRP studies suggest that the use of Superpave in replacing the pavement layers should significantly reduce the problem.

On high-traffic roads, planing or complete removal and resurfacing are recommended actions. A heater-planer or a cold milling machine can be used on thick asphalt surfaces, followed by a seal coat or overlay. On low-traffic roads, cutting off high spots and adding leveling material to low spots will take advantage of the roadway compaction already present. A seal coat or overlay may also be required.

Flushing (Bleeding): Flushing is a condition where asphalt concentrates at the pavement surface. It can result from a mix that is too rich, traffic loading that causes an increased combination of aggregate, too much asphalt in patches, an unstable mix, or other factors. Flushing occurs in hot weather, and if not treated, may cause slippery surfaces. When underlying aggregate is lacking, flushing often results in surface irregularities. Slight to moderate flushing without surface irregularities may be repaired by blotting with sand or porous aggregates. Generally, permanent repair of flushed areas requires removal and replacement, overlaying with a seal coat or hot plant mix having a suitable asphalt content, or planing the area.

Polished Aggregate: This condition is associated with heavy traffic wear and is caused largely by the type of aggregate used in the mix. Smooth, rounded aggregate or soft aggregates, such as some limestones, may wear or polish to the point that the surface becomes slippery when wet. Any such surface should be treated to restore skid resistance. Applying a thin overlay with an open-graded asphalt friction course, applying a chip seal, applying a slurry seal, or overlaying the area with hot plant mix are all appropriate treatments.

Raveling (Pitting): A pavement surface can gradually disintegrate from the surface downward by losing the surface aggregate particles; larger-sized aggregates are more subject to being dislodged. Raveling is usually caused by construction conditions, such as inadequate compaction, not enough asphalt in the mix, wet weather during lay down, overheated asphalt in the mix, etc. Repairs with a fog seal, sand seal, chip seal, or slurry seal should be adequate except for roadways under very heavy traffic volumes operating at high speeds. If a raveling section is not treated before the raveling becomes extensive, an overlay will be required.

Settlements and Upheavals: Excessive moisture rising through capillary action in the subgrade, swelling soils, surface cracking, and poor drainage can all cause pavement heaving or sagging. This condition adversely affects pavement ride and structural performance, especially when freezing weather produces frost heaves. Other conditions affecting the subgrade and distorting the pavement are depressions produced by improper backfill of utility cuts across the roadway and poor compaction of the roadbed. Improper materials used in construction may cause depressed areas to develop that collect water. Repairs to the subgrade are essential to avoid continued distortion of the surface. Unsuitable material should be removed and replaced with material that has good drainage qualities. Underdrains may be necessary. Repairs to the subgrade should be well compacted, and the base and surface courses replaced. Slab jacking or grouting may correct localized failures or distress areas.

Rigid Pavements: Rigid pavements have either a base or surface of portland cement concrete. Rigid pavement with asphalt surfaces may be subject to many conditions that affect flexible pavements such as rutting, potholing, cracking, shoving, flushing, polishing aggregate surfaces, and raveling. In fact, many motorist complaints about spring potholes in the snow and ice belt are in areas where the freeze-thaw cycle has broken pieces of an asphalt overlay loose from the portland cement concrete pavement underneath. Some conditions that affect rigid pavements with asphalt surfaces are influenced by the depth of the asphalt course. Ruts in asphalt overlays are limited in their depth, and the depression is usually confined to a problem in the surface course. The same limitations apply to potholes. Potholes, which usually form in areas with 2- to 3-inch layers of asphalt, are characterized by the breaking away of the asphalt in the pothole area down to the concrete base.

Asphalt surface cracks over a rigid pavement are generally a reflection of cracks or joints occurring in the portland cement concrete base. Reflection cracks may occur as longitudinal cracks formed by shoulder and edge joints or as transverse cracks following expansion joints. They also result from breaks in the underlying portland cement concrete base having other causes.

Shoving occurs in the same travel areas as it occurs in flexible pavements but can usually be attributed to a slip plane created by water, oil, or contamination that impairs the bond between the asphalt surface course and the portland cement concrete base. Lack of an adequate tack coat may also be a cause.

Extensive cracking in concrete pavement that has been sealed

Spalling along joint repaired by patching

Badly deteriorated concrete pavement with larger cracks and holes repaired

Concrete pavement with all holes and cracks repaired

Surface Texture Failures: Surface texture failures of portland cement concrete occur as map cracking and scaling. Excessive wear that results in a polished aggregate may also occur. Map cracking appears as fine interconnected lines resembling the lines on a road map. The cracks are usually shallow. If they are wide enough to allow water to enter, they can cause additional damage during freeze-thaw cycles or allow chlorides to reach reinforcing steel. Scaling can result from the freeze-thaw cycles when water enters small surface cracks in the concrete and leaves the top surface rough with exposed, embedded aggregate. Scaling usually is traceable to improper air entrainment, inadequate curing, improper finishing, or the wrong concrete mixture. When a considerable area of portland cement concrete has been affected by scaling or excessive wear, a surface treatment of hot plant mix or slurry seal is recommended. Small areas may be repaired with an epoxy surface patch.

Cracks and Defective Joints: Portland cement concrete slabs expand and contract with heat and cold. Temperature variations between summer and winter can cause severe cracking in larger slabs. Usually this is controlled by including contraction joints in the design. Contraction joints are added to a sufficient depth in the concrete slab to encourage the slab to crack at the joint. Most agencies choose to fill contraction joints with a material that expands and fills the joint during periods when the slab contacts, inhibiting the entrance of water and incompressible materials. Contraction joints must be sawed into the pavement very early after the concrete is placed. Sawing should begin as soon as the concrete surface is firm enough to support the weight of the sawing machine. Expansion joints are generally designed into the roadway structure. They extend the full depth of the concrete slab and should be filled with a material that will fill the joint and exclude water during both hot and cold weather.

When a concrete slab is placed, the original joint material inserted should be examined and considered for removal and replacement if it does not have the capability to expand and contract with the concrete and keep the joints sealed against the entry of water and incompressible materials. Initial shrinkage during the curing of concrete pavement is greater than any subsequent expansion or contraction of the cured concrete caused by moisture and temperature changes.

The joint material compresses during periods when the concrete expands. If the joint filler is inadequate for its purpose, the base of the subgrade is likely to be damaged as a result of water entering through the joints. Because it is easier to check for a broken bond between the edge of the slab at the joint and the filler material in cold weather, many agencies conduct joint sealing and repair operations during dry winter weather.

Some cracks may result from horizontal stresses caused by poor pavement design, such as not incorporating the proper joints or an adequate number of joints. Other causes may be improper mixing or curing and unusual vertical stresses on the slab caused by an inadequate base or subgrade. Roadway joints and cracks should be checked periodically to ensure that they are properly sealed. Suggested times include winter during a time of maximum concrete contraction, before the wet season, and summer to ensure that the filler is not being forced out of the joints or cracks by expansion.

Different temperature ranges between the top and bottom of a slab can cause a slab to distort and curl. This curling action can loosen some solid sealing materials and cause them to be ejected. Cracks caused by inadequate base support usually occur first in the corners of the slab. Recommended methods of repair require that the broken corner area be removed and the base properly stabilized before the corner is replaced either with portland cement concrete or asphaltic concrete. Defective or inadequate joint filler material should also be removed and replaced.

If the slab is steel reinforced, fine cracks in portland cement concrete pavements may significantly affect performance. The action of water or chlorides entering the concrete can corrode steel reinforcement, causing damage that can lead to a need for major repairs. In response, various agencies have used impervious materials, linseed oil, chip seals, slurry seals, and other sealing materials to treat fine cracks. Large cracks exceeding 1/8 inch in width should be filled with a material that will permit expansion and contraction of the concrete.

Maintenance engineers and managers new to the field and those persons seeking to refresh their background in the repair of joint seals in concrete pavements can refer to SHRP-H-349, *Concrete Pavement Repair Manuals of Practice* [13]. The first half of this document is a manual, *Materials and Procedures for the Repair of Joint Seals in Concrete Pavements*. The major topics covered include the following:

- Need for joint resealing (includes seal condition, pavement condition, climatic conditions, traffic level, and determining the need to reseal).
- Planning and design (includes selecting a sealant material, backer materials, primer materials, and joint reservoir dimensions; selecting preparation and installation procedures; selecting equipment; estimating material, labor and equipment requirements; and analyzing cost-effectiveness).
- Construction (includes traffic control, safety, joint preparation, material preparation, and installation).
- Evaluating joint seal performance.

Spalling: Spalling occurs as chipping or splintering of a localized area of portland cement concrete and is generally caused by expansion. Expansion can result from heat or from corrosion of reinforcing steel affected by the presence of moisture or chlorides. The admission of incompressible material into unsealed cracks or joints that have widened when concrete contracted in cold weather may, when the weather turns hot, cause excessive pressure between two slabs. Spalls should be temporarily repaired as soon as practical; permanent repairs should be made when weather permits. When joint spalls exceed the limits that can be adequately controlled by means of routine joint-sealing methods, high-density concrete or other high-quality repair materials should be used.

Maintenance engineers and managers new to the field and those persons seeking to refresh their background in the repair of spalls in concrete pavements should see SHRP-H-349, *Concrete Pavement Repair Manuals of Practice* [13]. The second half of this document is a manual, *Materials and Procedures for the Rapid Repair of Partial-Depth Spalls in Concrete Pavements*. The major grouping of the topics covered includes the following:

- Need for joint partial-depth spall repair (includes pavement condition, climatic condition).
- Planning and design (includes assessing existing conditions; selecting a repair material; selecting accessory materials; selecting dimensions of the repair area; selecting patch preparation procedures; estimating material, equipment, and labor; analyzing overall cost-effectiveness).
- Construction (includes traffic control, safety, initial joint preparation, removing deteriorated concrete, cleaning the repair area, final joint preparation, pre-placement inspection of the repair area, mixing the bonding agent, mixing the repair material, applying the bonding agent, placing the repair material, consolidating and compacting, screeding and finishing, curing, joint sealing, cleanup requirements, opening to traffic, inspection).
- Evaluating partial-depth patch performance (includes data required, calculations).

Blowups: Blowups are localized buckling or shattering, usually occurring at a transverse crack or joint. They are caused by expansion of slabs during hot weather, creating excessive longitudinal pressure, when there is inadequate room between slabs to expand. On interstate highways and other high-speed arterials, blowups that create raised transverse ridges over 25 mm (1 inch) high should be repaired as soon as practical. If a ridge is over 50 mm (2 inches), it should be posted for a lowered speed limit as soon as practical. Blowups on highways with lower speed limits should be repaired as soon as practical if they exceed 50 mm (2 inches) in height and should be posted and repaired without delay if in excess of 75 mm (3 inches).

Measurement of transverse deviations such as ridges caused by blowups or vertical displacements caused by faulting and settlements should be accomplished with either a 3-meter (10-foot) straightedge or string line. Measurements should be made longitudinally, and deviations at any point should not exceed the amounts noted above.

Agencies having a program to replace continuous reinforced concrete pavements with unreinforced concrete pavements and having a program to regularly clean and seal joints have generally found the incidence of pavement blowups greatly reduced.

Frost Heaves: Frost can cause differential upward displacements of concrete pavement slabs. When frost heaves cause a permanent displacement of the pavement slab resulting in a condition similar to a blowup, corrective measures should be taken. Pavement repairs require that frost-susceptible material below the pavement slab be removed and replaced with the appropriate depth of frost-free material. Subdrain systems have been found to be a less costly remedy in some areas.

Faulting and Settlements: Faulting and settlements are differential vertical displacement of slabs adjacent to a joint or crack. The cause is usually due to a problem with the subgrade. Poor drainage or inadequate compaction should be suspected first. On interstate highways and other high-speed arterials, slabs with vertical displacement beyond 6 mm (1/4 inch) at joints or cracks should be considered for repair. Faults on other highways with lower speed limits should be repaired as soon as practicable when the faulting is 13 mm (1/2 inch) or more. Faults over 50 mm (2 inches) should be repaired as soon as discovered. Slab replacement as a unit, slab jacking, and spot asphaltic concrete overlays should be considered as repair measures. If the slabs have stabilized after the initial faulting, milling the area may be an appropriate repair.

Pumping: Pumping is the ejection of mixtures of water and clay or silt, or both, along pavement edges or along transverse or longitudinal joints or cracks. Pumping is caused by downward slab movement activated by the passage of heavy axle loads over the pavement, after free water has accumulated in the subgrade. Slabs observed to pump on interstate highways, principal arterials, or other major highways should be repaired as soon as practicable. On these highways, displacement up to 13 mm (1/2 inch) may be tolerated except that repairs normally should be made when 50% of the slabs on any 305-meter (1,000-foot) section of roadway exhibit characteristics of pumping or displacement exceeding 13 mm (1/2 inch). Materials engineers should be consulted before corrective measures are undertaken.

Aggregate (Granular) Surfaces and Earth Roads: Unpaved or earth-aggregate-surfaced roadways with treated or untreated surfaces may serve as service roads; detours; access roads to quarries, stockpiles, gravel pits, roadside rest areas, and primitive park areas; and other temporary connections. There may also be situations in which aggregate-surfaced roadways may serve as low-volume principal routes in the farm-to-market network. They may be surfaced with crushed rock or gravel or native materials such as pit run gravel, sand, quarry rock, clay, crushed slag, or talus. The roadway surface should be maintained by blading a crown of 42 mm to 83 mm per meter (1/2 inch to 1 inch per foot) sloping outward from the centerline. Blading should be done as soon as practicable after a rain or when the surfacing material is moist. (Avoid blading, except to drag loose material into surface irregularities, when the surface is dry and compacted.) Good drainage is essential to maintaining unpaved roads. Dust palliatives should be added whenever dust becomes a traffic hazard or a detriment to abutting properties. Frequent inspections and regularly scheduled maintenance are more essential to this type of roadway than to paved surfaces because of their tendency to deteriorate rapidly under traffic or adverse weather. Stabilized earth-aggregate roadways offer a firm load-bearing surface that resists raveling and loss of aggregate, are resistant to water and frost action, and can be maintained at a reasonable cost. It is important when choosing a stabilizing material and method to make a laboratory determination of the soil to be stabilized and to consider the desired qualities of the stabilized roadway. Clays, sand, silt, and so on vary in their composition and will react in different ways depending upon the stabilizing materials used and the manner of their application.

Low-volume aggregate-surfaced road

Stabilization of unpaved surfaces can be accomplished by a number of methods and materials. The usual method is to pulverize the soil to be stabilized and mix it on-site with a stabilizing material such as asphalt, portland cement, calcium chloride, sodium chloride, lime, or a trade name product. The mix is then shaped and compacted, preferably with a rubber-tired roller. Earth-aggregate roadway surfaces are more likely to develop problems that require maintenance action than paved roadways.

Blading aggregate-surfaced roadways is a continuous, routine maintenance activity. Proper blading of aggregate-surfaced roadways will prevent or extend the time interval between many major maintenance activities required on aggregate-surfaced roadways. The essential elements and principles of proper blading of aggregate-surfaced roadways to be communicated to beginning motor grader operators and to include in refresher training for experienced operators is contained in a booklet titled *Blading Aggregate Surfaces* published by the National Association of County Engineers. A very effective slide-tape presentation titled *Maintaining Granular Surfaced Roads* is available through the LTAP Center network [14].

Corrugations: The progressive growth of corrugations in earth-aggregate or low-type surface roadways is caused by traffic. The only satisfactory method to repair corrugations is to scarify to an appropriate depth, pulverize, shape, and compact. When mixed in place with the pulverized materials, suitable stabilizing materials can substantially improve the ability of the roadway to resist the formation of corrugations. If corrugations develop on an earth-aggregate roadway having a surface treatment, it is best to remove or scarify the surface treatment, reshape the base, and after application of a priming coat, apply a new surface treatment. As a temporary measure, dragging and blading can be used to repair corrugations. Proper dragging and blading at the proper time can limit development of corrugations.

Ruts: Longitudinal depressions, occurring in pairs in earth-aggregate or low-type surfaces, are caused by traffic and usually follow the pattern of vehicle wheels. They can direct vehicles' front wheels and therefore be a potential problem when a surface has hardened after a rain storm. Filled with water, they can create splash or icing problems. Ruts are usually evidence of aggregate being dislodged from the roadway or to either side. Inadequate compaction or inadequate grading of surface materials can also cause ruts to develop. Dragging and blading while the roadway is damp, followed by compaction, is the usual method for repairing ruts. Material is bladed from the outside of the roadway toward the center, preserving the crown section.

Soft Spots: Settlements of earth-aggregate roadways can be attributed to either drainage or soil problems. A drainage problem requires correction of the cause before permanent repairs can be made. Drainage problems can be caused by a high water table, hydrostatic head, capillary action, underground stream, or groundwater seepage.

Saturated subgrade material should be removed and replaced, if practical and warranted, with a subgrade of coarse-graded materials. Subdrain systems may provide a more practical and cost-effective remedy.

Settlements may also be caused by subsequent settling of subgrade materials resulting from traffic loads. If this is the case, filling, compacting, and regrading should provide an adequate subgrade for a repaired surface course.

Dust: During dry weather, dust can become a serious problem on earth-aggregate roads. The condition can create visibility difficulties for drivers, become an inconvenience to motorists and residents along the road, create an airborne particulate pollution problem, and contribute to excessive vehicle wear. The use of dust palliatives in the presence of such conditions may be necessary on roads having a significant amount of traffic.

Potholes: The repair of potholes in earth-aggregate surfaces is a continuing process. Materials used to temporarily repair potholes by filling them should be similar to the existing road surface. More permanent repairs consisting of blading or scarifying and mixing should be performed when moisture is available. The condition of a road at the end of the dry season is the best indication of the need for upgrading the roadway to a more stable structure.

The success of pothole repairs can be directly related to good preparation of the holes, the quality of the repair material, and the amount of compaction applied. The best rolling equipment available should be used. The wheels of a loaded dump truck may be used for individual pothole repair on low-volume roads when better equipment is not available.

2.1.2.4 Materials for Roadway Surface Maintenance

Maintenance engineers and managers are not expected to be materials engineers. There is, however, a general overall background that maintenance engineers (who do not have specialized education, training, or experience in materials engineering and science) should acquire to relate the various typical materials to the maintenance process. Maintenance engineers and managers should continually seek more understanding of existing and new materials that may potentially reduce costs, increase the life of repairs, increase the safety of operations, or improve productivity. It will often be necessary to consult with materials engineers to ensure that the proper materials are being used in any maintenance method applied to correct or repair a problem.

The following brief sections are for general overview description and familiarity. Reference should be made to the current AASHTO *Standard Specifications for Transportation Materials and Methods of Sampling and Testing* for detailed characteristics and qualities of appropriate materials.

Asphalt: Asphaltic materials are currently classified as cements, liquids, and emulsions. The asphalt cements used for hot mixes are available under the following three classification systems:
- Penetration (40-50, 60-70, 85-100, 120-150, etc.).
- Viscosity (AC-5, AC-10, AC-20, AC-40).
- Residue viscosity (AR-1000, AR-2000, AR-4000, AR-8000, AR-16000).

Penetration testing at 25°C (77°F) is conducted as the material comes from the refinery. Viscosity is graded at 60°C (140°F) also as the material comes from the refinery. Residue viscosity is graded by the 60°C (140°F) viscosity test after the material has been artificially conditioned to simulate the aging process that takes place in pug mill mixing. In the viscosity systems, the larger the number, the harder the material. In the penetration system, the larger the number, the softer the material. Handbooks are available from AASHTO and the Asphalt Institute for selecting, specifying, and testing asphalt materials.

The PG, or performance grade, asphalt-grading system will come into widespread use in the future. This classification system, a SHRP product, uses test procedures newly applied to asphalt cement to determine the high and low temperature susceptibility of an asphalt cement. A PG 58-28, for example, shows that an asphalt pavement designed and constructed using this asphalt cement will be able to withstand rutting up to a pavement temperature of 58°C (136°F). The -28 shows the same pavement can resist thermal cracking down to an air temperature of minus 28°C (18°F).

Under the existing classification systems, the liquid asphalts each have five grades according to their fluidity and three types according to their curing rates. Curing rates are designated as "rapid cure" (RC), "medium cure" (MC), and "slow cure" (SC). Rapid-curing grades are RC-30, RC-70, RC-250, RC-800, and RC-3000, with the smaller number being the most fluid. Rapid-curing asphalts are "cut back," or diluted, with naphtha-type solvents and can be flammable. Medium-curing asphalts are cut back with a kerosene-type solvent. Slow-curing asphalts are diluted with a heavy fuel oil-type solvent that is usually left in the asphalt during the refining process.

Because of concern about hydrocarbon emissions from asphaltic cement and cut-back asphalts, emulsions are becoming the preferred asphalt binder for many agencies in maintenance operations. Consult your own agency maintenance manual and your agency's material policy for details on asphaltic materials selection.

Three types of graded emulsions are commonly available according to their setting rate. Setting rates are designated as "rapid set" (RS), "medium set" (MS), and "slow set" (SS). If a "C" precedes the setting rate, the material is a cationic emulsion; unless otherwise designated, the emulsion is anionic. CSS-1 is an emulsified asphalt that is a cationic slow-set emulsion. The number following the letters designates the viscosity characteristics, with a larger number indicating a more viscous emulsion and a smaller number indicating a more fluid one. When an "S" follows the number, this indicates the emulsion is applicable for mixing with sand, such as CMS-2S. If an "H" follows the number, it designates an emulsion in which a hard grade of asphalt was used in manufacturing the emulsion. Some precautions to be observed in ensuring that an emulsion binder will perform well in a maintenance application are as follows:

- Laboratory testing is recommended using the actual aggregate and the emulsion that is to be used on the roadway.
- Selection of grades in conformance with AASHTO materials guidelines is desired.
- Strict adherence to the specifications and guides for usage is recommended.
- Careful handling of the emulsion to prevent contamination, settlement of the asphalt droplets, or premature coalescence is desired.
- Consultation with the emulsion manufacturer's representative is suggested when special or unusual problems occur.

General uses of some common grades of asphalt emulsions are listed below. For anticipated application of other asphalt emulsions, the emulsion supplier should be consulted for application control requirements.

Asphalt Grades for Typical Applications

Asphalt-aggregate mixtures
• Pavement bases and surface
    Hot plant mix (MS-2h, HFMS-2h)
    Cold plant mix
        Open-graded aggregate (MS-2, HFMS-2, MS-2h, HFMS-2h, CMS-2, CMS-2h)
        Dense-graded aggregates (HFMS-2s, SS-1, SS-1h, CSS-1, CSS-1h)
        Sand (HFMS-2s, SS-1, SS-1h, CSS-1, CSS-1h)
• Mixed in place
    Open-graded aggregate (MS-2, HFMS-2, MS-2h, HFMS-2h, CMS-2, CMS-2h)
    Dense-graded aggregate (HFMS-2s, SS-1, SS-1h, CSS-1, CSS-1h)
    Sand (HFMS-2s, SS-1, SS-1h, CSS-1, CSS-1h)
    Sandy soil (HFMS-2s, SS-1, SS-1h, CSS-1, CSS-1h)
    Slurry seal (HFMS-2s, SS-1, SS-1h, CSS-1, CSS-1h)

Asphalt-aggregate applications • Single-surface treatment (chip seal) (RS-1, RS-2, CRS-1, CRS-2) • Multiple-surface treatment (RS-1, RS-2, CRS-1, CRS-2) • Sand seal (RS-1, RS-2, MS-1, HFMS-1, CRS-1, CRS-2)
Asphalt applications • Fog seal (MS-1, HFMS-1, SS-1, SS-1h, CSS-1, CSS-1h) • Prime coat, penetrable surface (MS-2, HFMS-2, SS-1, SS-1h, CSS-1, CSS-1h) • Tack coat (MS-1, HFMS-1, SS-1, SS-1h, CSS-1, CSS-1h) • Dust binder (SS-1, SS-1h, CSS-1, CSS-1h) • Mulch treatment (SS-1, SS-1h, CSS-1, CSS-1h) • Crack filler (SS-1, SS-1h, CSS-1, CSS-1h)
Maintenance mix • Immediate use (HFMS-2s, SS-1, SS-1h, CSS-1, CSS-1h)

Asphalts for surface maintenance are applied to roadway surfaces by the following three methods: (1) using premixed materials from a hot or cold mixing plant or stockpile mix, where the asphalt comes mixed with the aggregate and other ingredients; (2) applying liquid or emulsified asphalts directly onto the roadway surface and covering with selected aggregate (seal coat); and (3) mixing in place (cold mix base and surface) or mixing during application (slurry seal).

A number of additives can give strength, stability, and durability to asphalt pavements. Some additions include ground glass, hydrated lime, and other materials marketed under trade names that may include epoxies, silicones, latex, and ground rubber. Polymer-modified asphalt is now widely used. Future asphaltic material classifications and standard material specifications for applications can be expected to include special notation of appropriate polymer-modified asphalts for maintenance use.

The use of asphalt rejuvenation materials may be warranted if improved performance is cost-effective.

Portland Cement: The standard AASHTO specifications for portland cement provide eight types of portland cement classified according to their general uses:

Type I	For general construction when special properties are not required.
Type II	For general concrete construction exposed to moderate sulfate action where moderate heat of hydration is required.
Type III	When high early strength is required (such as in pavement patching).
Type IV	When a low heat of hydration is required.
Type V	When high sulfate resistance is required.
Type IA	For general concrete construction when special properties are not required (air entrained).
Type IIA	For general concrete construction exposed to moderate sulfate action or where moderate heat of hydration is required (air entrained).
Type IIIA	When high early strength is required (air entrained).

There are also special cements and additives with special qualities such as fast set times that are available under trade names. Other types are also specified if pozzolanic materials are added or if slag is added to the cement.

Epoxies, Resins, and Polymers: New materials are constantly being developed and proposed for use in roadway maintenance. Each new material is intended to provide some quality desired in maintenance that is not available in the current stream of material technology. Qualities that can be obtained by some of the newer materials, which are predominately trade name products, are as follows:
- Rapid setting and curing times.
- Ability to bond to adjoining materials.
- Convenience of application.
- Wearing and structural capabilities for special purposes.

In the past, evaluating claims of proprietary products has been a problem. Testing and evaluation of products by the Highway Innovation Technology Program and the National Transportation Product Evaluation Program now provide objective comparisons and assessments. Maintenance engineers and managers having questions about the validity of product claims should inquire about the availability of a test and evaluation report from these organizations. Most state DOTs and highway departments have a "product evaluation committee" that can provide assistance.

Chapter 2 Roadway Maintenance and Management

Aggregates: The quality of asphalt roadway structures strongly depends upon the aggregate used. Some of the controlling factors in determining the suitability of an aggregate are as follows:

- Aggregate types include talus, pit run sand and gravel, quarry rock, crushed rock, and slag.
- The grading (sizing) of an aggregate is measured by passing a representative sample of the material through a series of sieves having a specified range of square openings. Refer to the current AASHTO's *Standard Specifications for Transportation Materials and Methods of Sampling and Testing* for the sieve sizes required for various tests. Grading accomplishes the following:
    - Controls top size (i.e., the maximum size aggregate permissible for the application).
    - Controls quantity of fine aggregates (which may improve drainage where water or frost action is a potential problem).
    - Controls shear strength and stability.
    - Controls surface texture.
    - Controls voids and ensures proper proportion of asphalt.

Aggregate stockpiles

To a large extent, gradation controls the voids present in an aggregate assortment and thereby the quantity of asphalt required. Too much asphalt will result in flushing, too little will yield raveling. Density of aggregate in a mix is largely determined by the relationship of the sizes of the aggregate particles. Smaller sized aggregate is included to fill some of the voids between the larger particles.

- Shape of the aggregate particles is important to the stability of the mix. A high percentage of crushed aggregate is preferred in asphalt mixes. Aggregates generally should be produced from sources or by processes such that the overall maximum dimensions do not exceed the minimum dimensions by more than 3:1. High percentages of rounded aggregate particles have a tendency to cause mix instability; too many elongated particles may make initial compaction more difficult and result in delayed compaction by traffic loading.
- Wear quality is determined by testing for resistance to abrasion. Aggregate not sufficiently hard will be more likely to wear or polish under traffic. It will also be subject to developing ruts in the wheel track.
- Fracture can be estimated by visual examination of a representative graded sample. An appropriate percentage of aggregate having at least one fractured surface is essential to the stability of asphalt cement concrete. Proper aggregate interlock cannot be achieved with worn, polished surfaces.

- Portland cement concrete aggregates will typically make up 60% to 80% of the volume of portland cement concrete. Gradation of aggregates may vary considerably in producing portland cement concrete that is both workable and strong for maintenance purposes. The volume of cement required will be directly proportional to the void content of the aggregate mix. AASHTO's *Standard Specifications for Transportation Materials and Methods of Sampling and Testing* provides for the various types of aggregates allowed and gradations for mixes that are appropriate for maintenance applications.

Crack and Joint Fillers: Materials for sealing cracks and joints in concrete to prohibit entry of water and foreign material should have qualities related to their purpose and the climate in which they will be used. A good sealant material has the following qualities [11]:

- Withstands horizontal movement and vertical shear at all temperatures to which it is exposed.
- Withstands environmental effects such as weathering, extreme temperatures, and excess moisture.
- Resists stone and sand penetration at all temperatures.
- Maintains complete bond to concrete joint sidewalls at all temperatures.

Some commonly accepted sealant materials are as follows [11]:
- PVC coal tar (ASTM D 3406).
- Rubberized asphalt (ASTM D 1190, ASTM D 3405, AASHTO M 173, AASHTO M 301 0851, Fed SS-S-164).
- Low-modulus rubberized asphalt (modified ASTM D 3405).
- Polysulfide (Fed SS-S-200E).
- Polyurethane (Fed SS-S-200E).
- Non-self-leveling silicone (individual state specification).
- Self-leveling silicone (individual state specification).
- Preformed neoprene compression seal (ASTM D 2628, AASHTO M 220).

Small cracks less than 64 mm (1/4 inch) can be filled with lighter grades of asphalt heated to a pourable viscosity. Larger cracks can be filled with asphalt, rubberized asphalt compounds, or one of the proprietary compounds. When rubber gasket-type material is used, a good adhesive should bond it to the joint walls.

Impervious membranes for sealing fine cracks in portland cement concrete should be considered on bridge decks and steel-reinforced surfaces subject to corrosion spalling. Impervious membranes require wear-resistant surface courses to protect the membrane.

Premixed Materials: The two kinds of premixed materials are hot plant mix and stockpile.

Hot Plant Mixes: Temperature limits should be strictly observed when using asphalt cements. Too low a mixing temperature can result in poor coating; too high a temperature will damage the asphalt. Mixing temperatures should be adequate to achieve a kinematic viscosity of 150 to 300 centistokes. Normally this is equivalent to achieving a temperature between 135°C and 163°C (275°F and 325°F) using commonly available asphalts.

Hot plant mix transported to the site should be laid and compacted as soon as practicable after being taken from the mixer. The temperature of the mix should not have dropped below 99°C (210°F) at the time of compaction; above 105°C (220°F) is preferred. The level of density that can be achieved by compacting declines rapidly as temperatures drop below this level. Evidence of displacement, cracking, or shoving during compaction indicate that the temperature is too low.

Viscosity grades of asphalt cement should be selected on the basis of performance requirements and climatic conditions, e.g., AC-20 for warm climates and AC-10 for cold climates. Mixtures and materials should be evaluated and approved by a qualified materials laboratory if one is available.

Stockpile Mixes: Medium- or slow-curing asphalt can be used to prepare premixed paving material for stockpiling. Mixes using MC-250 or MC-800 may be stored for only short periods; SC-250 and SC-800 material may be stored for longer periods.

2.1.2.5 Methods for Roadway Surface Maintenance

Methods are discussed in two categories: (1) preventive maintenance and (2) repairs (or short responses). This section is not necessarily comprehensive for all of the respective preventive maintenance and repair functions that may be required for roadway surfaces. Methods that apply to preventive maintenance may also be applicable for repairs in some circumstances.

Preventive Maintenance: Preventive maintenance strategies flow out of pavement maintenance strategies. Thus, preventive maintenance can be thought of as selecting the methods and processes that will extend the life of the pavement sufficiently to more than justify the total cost of applying the treatment, when performed early in the life of a pavement (or early in the pavement deterioration cycle). NCHRP Synthesis of Highway Practice 223, *Cost-Effective Preventive Pavement Maintenance*, contains a concise compilation of practices that are available to conduct preventive maintenance [12]. In appendix A, the synthesis also includes a primer on preventive maintenance, which is a useful guide to developing a training program on preventive maintenance.

Slurry Seal: Slurry seal is a mixture of graded aggregates, mineral filler, and emulsified asphalt, with water added to achieve slurry consistency. The ingredients are mixed in the process of being applied and spread evenly onto bituminous or portland cement concrete pavement to fill cracks, repair raveling asphalt pavement, or to provide a skid-resistant surface. Equipment designed especially for mixing and applying is required. A light application of asphalt emulsion diluted with three parts water for a water spray may serve as a tack coat. Surface preparation includes removing excess grease, vegetation, and loose material.

Aggregates should be carefully graded. The range of sizes in a mix may vary from fine aggregates to 95 mm (3/8 inch). The formula will vary with the materials available or the qualities desired. Correct proportions are critical. Quick-setting slurry seal using accurately regulated quantities of mineral filler can set up for traffic operations within an hour. The average setting, however, requires 2 to 12 hours. Laboratory evaluation of materials relative to the type of surface will usually ensure a good result. Compaction with a pneumatic roller is recommended if slurry seals are over 64 mm (1/4 inch) thick or when excessive dehydration cracks appear in the spread material.

Microsurfacing (slurry seal using polymer-modified emulsion) is effective for preventive maintenance and for treating surface irregularities that are larger than those normally treated with regular slurry seals. Because of the short set time, microsurfacing has been successfully applied in multiple lifts. Multiple-lift applications have included correcting minor to intermediate surface irregularities and leveling wheel ruts.

Fog Seal: Fog seal is a light application of diluted asphalt emulsion, usually without a covering aggregate, used to restore bituminous pavement, fill cracks, and prevent raveling. It is applied at the rate of 0.02 to 0.07 gallon per square yard. Dilution rates for emulsified asphalts range from 1:1 to 7:1 (water to emulsified asphalt).

Sand and Aggregate Seals (Chip Seals): This is a light application of liquid cutback or asphalt emulsion applied to a bituminous surface using a sharp, clean sand or uniformly graded aggregate for a cover. Asphalt is applied at the rate of 0.1 to 0.5 gallon per square yard, depending upon surface conditions and the type of aggregate being used. Application rates will be proportionately higher when emulsified asphalt is used. Porous sections of the roadway may require priming to prevent the seal coat from being absorbed into the surface and not bonding to the aggregate. Soluble surface treatments may be used with the second aggregate course graded, filling the voids of the first aggregate course and forming a more dense surface having the same approximate thickness as the first course. Designs are available for applying a seal coat to a specific condition or range of conditions. SHRP Report PH358, *Pavement Maintenance Effectiveness* includes information and commentary on designing a seal coat application.

Dust Palliatives: Water is efficient as a temporary dust preventative but has the disadvantage of drying out quickly and allowing the particles that form the dust to lose their cohesion to one another.

As additives to dusty road surfaces, sodium chloride and calcium chloride have a longer lasting effect. Calcium chloride is more effective than sodium chloride in holding together clusters of small particles because of its greater ability to absorb and retain moisture. One and one-half pounds per square yard is recommended. Chlorides may be added either singly or in combination when the roadway surface is damp. They can also be sprayed on in solution. Possible long-term environmental concerns need to be addressed when considering the application of sodium chloride.

Other types of moisture-retention materials can be used such as magnesium chloride or lignin, a waste product from paper manufacturing. When using any waste product, a chemical analysis of the material should be obtained before making a decision to apply it. Heavy metals and other hazardous elements that could injure surface water quality may be present in a waste material.

Cut-back asphalt or asphalt emulsions (diluted from 3:1 up to 10:1, water to asphalt emulsion) can be used to prevent dust. When sprayed on dusty surfaces, they should penetrate the surface and coat the upper layers of fine dust particles. Avoid buildup of an asphalt surface to form a crust. For best results, the material and rate of asphalt application should be determined by a materials laboratory. Annual road oiling can be used to help build up a surface that will reduce dust. This is usually done as part of a multiyear plan to build up the surface.

Crack and Joint Sealing: Timing and scheduling the sealing of joints and cracks in portland cement concrete can be critical to success of the application. If possible, sealing should be done when inspection indicates the need for it has arisen. Sealing should not be postponed until the old sealing material becomes completely ineffective. Periodic inspection of joints and cracks should be scheduled, and when sealing material is found to be ineffective, it should be replaced. Sealing in the fall or spring when pavements are about halfway between maximum and minimum expansion is suggested. Special attention should be given to sealing operations before winter, usually beginning in the early fall, to ensure that the pavement joint is completely sealed to prevent water entry. NCHRP Synthesis of Highway Practice 211, *Design, Construction, and Maintenance of PCC*

Pavement Joints, contains a concise discussion of maintenance and rehabilitation of portland cement concrete pavement joints and further elaboration on how the problems encountered may relate to design and construction defects [105].

Concrete pavement joint sealing

When joint seals are to be replaced, they should be plowed or scraped out, taking care not to damage the edge of the concrete. Material not removed with a plow may be removed with a power cutter designed for removing residue from the joints and for refacing the joints to provide a rough concrete face to which the new sealer will adhere. The power cutter or a routing and refacing machine should be a self-powered machine operating a rotary cutter or revolving cutting tool that will not cause spalling or otherwise damage the edges of the pavement joint or crack. A concrete saw can be used to provide the proper joint shape. When necessary, a power-driven wire brush or a sandblaster may be used to clean the joints and cracks after the old joint sealer has been removed. Use of the wire brush is not recommended if a substantial amount of joint sealer remains in the joints. Joints should be blown out with oil-free compressed air immediately before sealer materials are applied. A hot-air lance provides the added benefit of heating the joint while cleaning it. The proper width-depth ratio should be established for each type of material used.

Hot-applied sealing compounds should be melted in a double-walled, indirect-fired heating kettle. The space between the inner and outer shells should be filled with a suitable heat-transfer oil having a flash point of not less than 277°C (530°F). The kettle should be equipped with a thermometer with a temperature range from 93°C (200°F) to 260°C (500°F). Means should be available for agitating the sealer to ensure uniform heating and melting. Materials should be heated to the temperature range recommended by the manufacturer and applied at the recommended temperature with a pressure hose attached directly to a pump unit on the melting kettle. The pressure hose should be equipped with a suitable nozzle for inserting the material into the joint. Hot-applied sealers that are allowed to cool generally should not be reheated for further use but should be discarded.

Joints may be filled flush or to within 64 mm (1/4 inch) of the surface, depending upon the joint design and the type of material being placed in the joint. For a detailed discussion of the various combinations, see SHRP-H-348, *Asphalt Pavement Repair Manuals of Practice* and SHRP-H-349, *Concrete Pavement Repair Manuals of Practice* [10, 13]. Conditions for which a backer rod is appropriate and for which it is not appropriate, conditions for which routing of the crack is recommended and for which it is not, and specific details of applying various materials are also discussed in these two manuals.

Sweeping and Cleaning: Cleaning of roadways has been generally limited to litter pickup in rural areas, collection of large debris from the roadway surface, retrieval of road kill, and the use of brooms for sweeping urban streets. Loose gravel and other debris on highways with high-speed traffic should be swept from traveled surfaces. Vacuum equipment is increasingly used to clean urban highways and streets because it reduces the amount of dust and better removes dirt from rough surfaces. Vacuum units are especially good for picking up sand and grit remaining after the snow and ice control season is over. Water flushing and spraying equipment offer considerable advantages for the removal of dusty material where there are good drainage systems. The use of a grader may be required to remove heavy accumulations of snow and ice control materials.

Roadway cleaning and sweeping should be scheduled with consideration of the following factors:
- Seasonal concerns such as reducing dust in dry weather and removing leaves and other materials that will clog drainage systems in wet weather.
- Special events that require improved appearance of the roadway (e.g., sweeping before and after a parade).
- Regular cleaning of highways that carry high volumes of high-speed traffic.

Hot Weather Sanding: Flushing or bleeding of asphalt pavements occurs mostly in hot weather when the asphalt becomes less viscous and flows to the surface more easily. Hot weather sanding is suggested as a method of blotting the excess asphalt. It is generally considered a temporary measure that will require repeated applications over time to minimize the problem effectively. Chip sealing with carefully controlled application of road oil or 13-mm to 19-mm (1/2-inch to 3/4-inch) overlays are more effective measures. If flushing persists, removal of the roadway material and resurfacing may be the only appropriate alternatives. If resurfacing is required, consideration should be given to applying Superpave.

Sealing Concrete Surfaces: Steel-reinforcing bars in concrete are susceptible to corrosion damage by water and chlorides that enter the concrete surface through cracks. Corrosion causes the reinforcing steel to swell; spalling of the concrete surface may result. Effective preventive maintenance methods to seal concrete surfaces are still being developed. Some research suggests that sealing of the concrete should begin with completion of initial construction.

Asphalt Rejuvenation: Preparations for rejuvenating aging asphalt materials should be evaluated in relation to various climatic and traffic conditions. Generally, laboratory testing is required to determine the proper materials and application rates. Under certain controlled conditions, cost savings may be realized. Milling and recycling should be evaluated with respect to cost and pavement life extension.

Repairs: In this section, the following types of repairs are discussed: tack coats and prime coats, patching, scarifying and blading, compaction, subsurface repairs, spall repairs, fills and overlays, and pavement burning.

Tack Coats and Prime Coats: A tack coat is an important preliminary step in bituminous repair work. It is defined as a bituminous material that is applied to an old roadway surface, creating a bond between the old and the new surface. In applying the tack coat, the surface is first swept clean of dust and foreign matter. In some instances, it may be necessary to flush away foreign materials and to break up hard dirt deposits with a power broom. Bituminous material is then applied with a distributor at a rate of from 0.05 to 0.15 gallon per square yard. The coating should be just enough to wet the surface. The tack coat should be allowed to become tacky or sticky before the surface course is laid.

Patching: These three methods apply to repairs of asphalt surfaces: (1) penetration-method patches for low-type surfaces, (2) patches for premixed patching material, and (3) patches for portland cement concrete.

Penetration-Method Patches for Low-Type Surfaces: The first course of aggregate should be spread uniformly over the area that is to be patched and then thoroughly compacted. It is desirable to tack the area before spreading the aggregate. Fill depressions with additional aggregate, remove excess material that might cause humps, and compact with a roller or vibrator. As noted earlier in the discussion of pothole patching, SHRP studies have shown that compaction with a loaded truck back and forth over the patch will suffice if the materials are high quality and careful attention has been given to preparing the hole and placing the materials [9]. Use a string line or straightedge to ensure that the patch is level with the road surface, especially on high-speed traffic roadways.

Apply bituminous material over the area that has been compacted. Generally, seal coat-grade materials should be used. Care should be taken to obtain a uniform distribution of asphalt so that an excess will not collect at the bottom of the patch. Excess asphalt could work to the surface under the combined action of traffic and hot weather, causing bleeding and distortion of the patched area.

After the application of asphalt, cover the area being repaired with crushed rock of smaller sizes than that used in the original asphalt base. Again, add an application of asphalt and proceed with enough layers to level off the surface. In penetration-type patching, good results can be obtained if tamping and compacting are done sufficiently to interlock the aggregate particles. Excessive rolling sometimes crushes the aggregate being used in the patching operations.

Patches with Premixed Patching Material: Failures such as potholes, upheavals, persistent flushing, and shoving in high-type asphalt pavements should be given prompt attention. Permanent repair of these conditions requires removal of all material down to a solid base, restructure of any damage to the subgrade caused by water, then installation of the patch. The following general procedures apply:
- Shape the hole or patch area to be square with the pavement. The edges of the hole should be neatly trimmed to sound material and cut vertically or undercut slightly. If the base material is found to be soft, loose, or unstable, it should be removed and replaced. If the bottom of the hole or patch area is wet, the wet material should be removed or dried by the addition of a small amount of dry portland cement, hydrated lime, or the use of a flame torch or an infrared heater.

- The bottom and sides of the prepared patch area should be given a light tack coat. If a hand sprayer from the asphalt distributor is used for tacking, do not apply an excess of asphalt because it will penetrate the patching material and contribute to flushing. If necessary, a brush may be used for even distribution. After tacking, allow some time to elapse, permitting evaporation of water from the emulsion or volatiles from the cutback to occur before the hole is filled. Small, portable hand spray units are excellent for applying tack coats in small areas.
- Cold premix should be placed in thin layers. Each layer should not exceed 3.8 cm (1.5 inches) after being compacted. Level the patch with the roadway surface, using a rake or lute. Use a straightedge or string line as a guide so the patch will conform to the road surface, especially on high-speed traffic roadways. If the hole is in an asphaltic concrete surface, it should be repaired with hot mix material; however, in an emergency, holes in this type of surface may be repaired with cold mix and replaced later when hot mix is available. Material should be applied in layers if the hole is more than 7.6 cm (3 inches) deep. The material should slightly overfill the hole and be compacted as thoroughly as possible down to the level of the roadway surface. If necessary, apply extra material to achieve an even surface.
- Patches should be thoroughly compacted before they are opened to traffic. Patches are usually compacted with rollers, vibratory compactors, or tampers. According to SHRP studies, small-area patches can be compacted satisfactorily by rolling with the rear dual wheels of a loaded truck, as noted previously [9]. When compacting is finished, the surface of the patch should be the same level as the surrounding surface. Avoid overfilling holes and building bumps into the pavement surface.
- Consult SHRP-H-348 for additional details on patching of asphalt-surfaced pavements [10].

Patches for Portland Cement Concrete: Patching of broken portland cement concrete slabs is subject to a variety of considerations. The main concern is to remove the cause of the original fracture, which may be due to a problem in the subgrade or subbase or due to a joint failure. When replacing the subgrade or subbase material, the process requires good compaction. In full-depth patches, replace dowels for load transfer across the patch to the existing slab. In drilling holes in the existing slab for the replacement dowels, avoid fracturing the existing slab. Contraction and expansion joints should be replaced where required. High-early-strength cement should be used to open the patch area to traffic as soon as possible. Keep the patch area covered and the moisture in the new concrete while it is curing, because there is always public pressure to open the patch area to traffic as early as possible.

Alternate measures to repair broken portland cement concrete slabs include replacement with asphaltic cement concrete, overlays with asphaltic cement concrete to level or strengthen depressed areas, and precast replacement slab sections. The principles contained in SHRP-H-349 for repair of partial-depth spalls should be considered in developing a preferred procedure [13].

Scarifying and Blading: Scarifying low-type stabilized and earth-aggregate surfaces is accomplished by breaking up the surface course of a roadway by mechanical means. For asphalt-stabilized surfaces, scarifying is usually done because asphalt has oxidized or there are extensive corrugations or disintegration. Breaking up the surface can be accomplished by either a tractor using from one to three ripper teeth or by a motor grade with scarifier teeth installed. The number of teeth usually depends upon how deep the surface needs to be broken up. Scarifying should be done during a season when the surface course contains adequate moisture to facilitate mixing, to aid in compacting, and to reduce the release of surface dust.

Minor grade corrections can be accomplished during scarifying operations. New materials for reconditioning are added to the scarified materials and mixed with disk harrows and motor graders. The surface is shaped to proper crown and thickness and compacted with pneumatic-tired or sheep-foot rollers. The surface course should then be restored.

Blading of earth-aggregate roadways should be done in the spring after the frost is out of the ground and while the structure of the roadway is still moist. Under traffic, gravel has a tendency to move toward the outer edges of the roadway and should be brought back across the traveled way. Large rocks that have worked their way up to the surface should be removed and their holes filled. Windrows should be removed after grading is completed.

Compaction: For earth-aggregate surfaces, an objective of any structural or maintenance effort should be to achieve maximum density of the materials being used. If maximum density is not achieved, subsequent settlement resulting from weathering and traffic loading or lateral material movement will produce more failures and more need for repairs.

For proper compaction, earth-aggregate materials should be moist. Too much moisture in the presence of a high percentage of fine aggregates or clays can cause the material to become unstable. After scarifying and mixing stabilizing materials into earth-aggregate structures, pneumatic-tired rollers or vibratory compaction equipment should be used for compacting.

Achieving maximum density of earth-aggregate materials requires proper equipment. Maintenance equipment for compaction may be as large as 8-ton pneumatic-tired or steel-wheeled rollers and can include power-operated tamping and vibrating equipment. Maintenance compaction may also include using the rear wheels of a loaded dump truck, with or without vibrating or tamping, for small patch jobs.

Although the objective is to achieve maximum density, it is not economically feasible to expend time and effort beyond a point where continued effort will produce only a small additional amount of compaction in a maintenance repair. To achieve the best results for a given amount of effort, such as six or eight passes of a heavy roller, the moisture content of the materials should be near optimum. Optimum levels of moisture will vary considerably for different materials and their gradations. The two best methods for maintenance personnel to use in determining optimal moisture are probably to rely on experience in working with native (or available) materials or to rely on laboratory determinations. If problems develop with particular earth-aggregate materials, they should be subjected to laboratory analysis to determine what stabilizing measures can be taken and what compaction effort is required.

Compaction of asphaltic cement concrete with steel rollers provides a smooth even-finished surface. The kneading action of pneumatic-tired rollers will produce a denser mat than steel rollers alone. Using a combination of pneumatic-tired and steel rollers will generally produce superior results. After the material is laid, rolling should immediately begin. The number of passes required by each roller, rolling patterns, and the temperature of the asphalt mix material need to be evaluated. A vibratory steel roller, with greater compaction than a static steel roller, may also be used. Tests for density include core samples and nondestructive testing for water permeability, air flow, and nuclear density. The nondestructive tests should be considered as relative tests, because a variation in reading can result from different aggregates and several other factors. If true densities must be determined by nondestructive testing, the readings must be calibrated against true density determinations made by core sampling methods. Calibrations can vary with the use of different aggregates or mixes.

Some voids should be included in compacted asphaltic cement concrete to allow for asphalt expansion in hot weather. Generally, asphalt expansion is 20 times that of aggregate. Too few voids can result in flushing. Excessive voids can make the pavement structure too permeable to surface water and subject to raveling and disintegration. Surface course voids of 4% to 6% using a mix with well-graded aggregates and the correct proportion of asphalt are usually recommended as optimum. A slightly higher percentage of voids may be acceptable in base courses. A dependable field method to measure density of compacted material should be adopted and used consistently.

To ensure that proper density is being obtained when standard testing is not feasible, a proper mix design must be used and the materials must be properly blended. The mix should be placed at the proper temperature with an adequate ambient temperature and using a good rolling pattern with the proper equipment. There should be no excessive cracking or displacement during rolling.

Vibrators are sometimes combined with rolling equipment or are available as separate vibratory plate units. One-person, gasoline-powered compacting equipment, such as impactors and rollers, is also available for maintenance purposes.

Compaction is equally important in preparing subgrades and other granular subsurface pavement structures. Many surface failures can be attributed to inadequate compaction during construction. Repair measures should ensure that failures do not recur for the same reason.

Subsurface Repairs: Grout pumped under pressure under portland cement concrete slabs can be an effective method of restoring settled slabs to their original position. Special equipment is required to drill holes through the slab and to pump the grout. Grout mixtures applicable to this process can vary. Native sandy loam soil mixed with portland cement is a low-cost material that has been used successfully in a number of areas. Proper procedures for applying grout under pressure include avoiding buildup under the slab and controlling the grout viscosity to prevent stooling. If underdrains have been installed in the pavement, grout pumping must be done carefully or the underdrain tiles will be filled with grout. Excessive grout pumped into a void under a slab can also saturate the granular subbase, solidifying it and destroying the capability of the subbase to drain water away from the underside of the slab. Undersealing of slabs with special asphalts or cement fly ash grouts has been used to eliminate water, to provide improved uniform slab support, and to reduce slab pumping.

Spall Repairs: Spalling adjacent to joints in portland cement concrete pavement is generally caused by poor construction practices or joint maintenance. If the cause is related to an inadequate joint, the joint cutting and sealing procedure to widen the area for expansion should be considered. If the spalled area is to be corrected temporarily, it may be filled with a bituminous paving mixture. Larger areas should be repaired with concrete or an epoxy compound to ensure a uniform, long-lasting riding surface.

Material considerations include selecting a repair material (portland cement concrete, polymer concrete, or bituminous material) and selecting the accessory materials (bonding agents, joint bond breakers, curing materials, and joint sealants).

To prepare the joint area, the old sealant must be removed, the joint must be sawed to delineate the sound areas clearly, and the joint inserts must be sawed out. The deteriorated concrete must be removed. These steps can be performed several ways including saw and patch, chip and patch, mill and patch, water blast and patch, and clean and patch. Once the deteriorated concrete has been removed, the repair area needs to be cleaned. Cleaning may include sandblasting, air blasting, and sweeping. Once the repair area is clean, final joint preparation needs to occur, including preparation of any transverse joints, any centerline joints, and any lane-to-shoulder joints. Then the bonding agent needs to be mixed; the repair material must be mixed (as required, depending upon the material selected); the bonding agent must be applied; and the repair material must be placed. The repair materials must be consolidated and compacted, then screeded and finished for pavement surface level and for skid-resistant finish. A curing compound or curing cover should be applied to any concrete material. When the repair has cured, the rebuilt joint area should be sealed and the area cleaned up, including removal of traffic controls, opening the area to traffic.

Consult SHRP-H-349 for additional details [13].

Fills and Overlays: To bring channels or depressions in asphalt surfaces to the level of the surrounding traveled way, a fill of premix (preferably hot plant mix), in-place mix, penetration mix, or cold milling should be used.

The area of depression should be marked out, using a straightedge, and the old surface should be thoroughly cleaned, loose gravel removed, and cracks sealed. If a grinder is available, a vertical edge to a depth of at least 64 mm (1/4 inch) should be cut outlining the area.

If a hot or cold premix is to be used, a light tack coat should be applied. The mix should not normally be applied to the fill area until after the tack coat has cured. The fill materials should be tamped and raked to the proper level. The best available compacting equipment should be used. After the area has been checked with a straightedge to ensure the proper level, a sand seal should be applied to exclude moisture. If possible, hot mix asphalt should be used on high-traffic roads.

If a mix-in-place or penetration method is used, the first course of aggregate should be spread uniformly over the area and compacted by rolling, vibrating, or tamping. Minor depressions should be filled with additional aggregate, and any excess material that might cause humps should be removed. A string line or straightedge can be used to ensure that the patch conforms to the cross section and grade of the road surface.

Apply liquid asphalt over the area that has been compacted. Be careful to use the proper ratios of materials and obtain a uniform distribution of asphalt so any excess will not collect at the bottom of the patch. Excess asphalt will work to the surface under the combined action of traffic and hot weather, causing bleeding and distortion of the patch area.

After applying the asphalt, cover the area being repaired with key rock of smaller sizes than that used in the original base of the patch. Again, add asphalt and key rock, applying enough layers to level off the surface. The greater the depth of the area to be patched, the larger the size of the aggregate that can be used in the repair work.

In penetration-type patching, good results can be obtained if compacting is carefully performed. Rolling and tamping should adequately interlock the aggregate particles. However, excessive rolling sometimes crushes the aggregate being used in the patching process.

Pavement Burning: Pavement burning, with planing where necessary, may be used to treat flushing of excess asphalt and to smooth corrugations. However, burning may not be permitted for safety or ecological reasons. In any case, pavement burning should never be undertaken without checking your agency's current environmental guidelines and worker safety policies. Air pollution regulations should also be checked before proceeding because they may be different for urban and rural areas.

Pavement burning may destroy the asphalt in the roadway structure because it reaches a high temperature. Asphalts cannot usually withstand temperatures in excess of 232°C (450°F) for even short periods of time. In fact, some asphalts can be damaged by even lower temperatures. If records of previous construction are available, check the characteristics of the asphalt used in the pavement construction. Where pavement burning is being considered, milling and in-place recycling is often an environmentally sound and cost-effective alternative.

Polymer Concrete Patches: Polymer concretes are a combination of polymer resin, aggregate, and a set initiator. The aggregate makes the polymer concrete more economical, provides thermal compatibility with the original pavement, and provides a wearing surface. Several polymer concretes are briefly described below [13].

Epoxy Concrete: Epoxy concretes are impermeable and make excellent adhesives. They have a wide range of setting times, application temperatures, strengths, and bonding conditions. The epoxy concrete mix design must be thermally compatible with the pavement; otherwise, the patch may fail. Deep epoxy repairs often must be placed in lifts to control heat development. Epoxy concrete should not be used to patch spalls caused by reinforcing steel corrosion because it may accelerate the rate of deterioration of adjacent sound pavement.

Methyl Methacrylate Concrete: Methyl methacrylate concretes and high-molecular-weight methacrylate concretes are polymer-modified concretes that could also be classified as cementitious materials. They have relatively long working times, high compressive strengths, and good adhesion. Many methyl methacrylates are volatile, and prolonged exposure to their fumes may pose a health hazard. As with all materials, material safety data sheets (MSDS) must be obtained from the manufacturer and followed to ensure the safe use of these materials. Supervisors need to ensure that all workers understand the MSDS instructions.

Polyurethane Concrete: Polyurethane concretes generally consist of a two-part polyurethane resin mixed with aggregate. Polyurethanes generally set very quickly (in about 90 seconds). Some manufacturers claim their materials are moisture tolerant; that is, they can be placed on a wet surface with no adverse effects. This type of material has been in use for several years with variable results.

Polymer Concrete: Polymer concretes have rapid hardening qualities that make their use attractive for maintenance of portland cement concrete, especially in high-speed, high-traffic-volume roadways. Higher material costs for repairing small areas can be offset by savings in labor time, shorter interruption of traffic, and the safety benefits derived from maintenance personnel being exposed to traffic hazards for shorter time periods. Epoxy surfaces may be considered for sealing portland cement concrete surfaces in selected areas exposed to poor drainage and high salt exposure. As a seal coating, the epoxy resins are a potentially effective means of resisting chloride damage.

Soil Stabilizing: Earth-aggregate roadway surfaces and substructures are most effective when they have achieved at least 95% of their compaction capability and can resist deforming and material losses. In addition, they should have good drainage qualities and not be subject to capillary action. Structures of large-sized granular material are usually not affected by water problems. However, fine materials such as clays and fine sand require a certain amount of moisture to preserve their cohesive qualities.

Many factors affect the stability and cohesiveness of different materials. A soil analysis by a materials laboratory is the best means of determining a method to stabilize the soil in an area causing a maintenance problem. A number of methods and materials can be used for soil stabilization. Additives of asphalt, calcium chloride, sodium chloride, portland cement, lime, and a number of proprietary compounds should be considered in relation to cost and effectiveness. In any case, the method to be used will determine the type of equipment required. Scarifiers, mixers, rollers, and spreaders should be selected to fit the job requirements.

Pavement Deflection Testing: Surface failures in asphalt pavements are often associated with problems arising in the underlying pavement structure. Drainage failures or other distressed conditions of the substructure that cause water to collect, and in freeze-thaw climatic zones, sometimes to freeze, can trigger a substantial amount of traffic damage. Deflection testing can determine if substructure problems can be expected. Several types of equipment can be purchased or rented for this purpose. The extensive testing of the falling weight deflectometer (FWD) in the Strategic Highway Research Program and its correlation with pavement structure conditions has encouraged many agencies to establish a program of inventorying pavement structure condition on the basis of FWD surveys. These data, where available, should be integrated into both the pavement management system and the maintenance management system to assist personnel in deciding where maintenance treatments may effectively forestall major reconstruction and where maintenance treatments are not cost-effective because the present condition is so poor that reconstruction is the only remedy. Such data can also be useful in determining seasonal traffic load limitations on roadways and overlay requirements that will optimize the maintenance effort.

Roadway Cross Section: Proper cross section is originally a design and construction concern, but sometimes maintenance operations will require that a crown or superelevation be restored. This is especially so when maintaining unpaved surfaces [14, 15]. Before extensive maintenance is scheduled for an unpaved roadway, the proper grading and cross section of the roadway should be established, usually by referring to the original plans and drawings or to standards. Sometimes the performance of a maintenance activity, such as ditch cleaning, requires restoring the proper cross section and profile as a latter stage of the maintenance process. Motor graders are often sufficient equipment to reshape the roadway to its original configuration.

In the absence of plans and drawings, drainage patterns should receive the first consideration. Water should run off the roadway into adequate ditches or channels. Superelevation may be required on curves to avoid a cross slope that will cause moving vehicles to slide toward the outside of the roadway. A general rule for the amount of slope required for crown and superelevation sections is not less than 4% or more than 8% across the roadway lanes. Crown sections should be sufficiently rounded that water cannot pool in the roadway. The roadway profile grade and the transition areas between normal crown sections and superelevated sections should be checked to ensure that they do not cause undesirable runoff patterns or cause water to pond on the roadway.

Care should be taken on the high edge of a superelevated section so that the changeover to the shoulder slope does not result in a sharp edge or that the area between the shoulder and the traveled way does not result in a flat spot where water will collect. If this seems to be a problem, an area about 0.3 meter (1 foot) with an intermediate slope to transition from the superelevation to the shoulder slope is recommended.

Inspections of aggregate-surfaced roads should be conducted with a vehicle having a "ball-bank indicator" to check the degree to which the superelevation is being retained properly with respect to the curve speed. Motor grader operators should be required to review the principles of proper blading of aggregate-surfaced roads annually, because improper procedures in blading can quickly build a poor crown and poor superelevation.

2.1.3 Shoulder Maintenance

The requirements for maintenance of shoulders and approaches are influenced by their design, usage, condition, and structural materials. Shoulders have two basic purposes in the roadway system: (1) they provide lateral support to the pavement structure and (2) they provide lanes for emergency or safety travel. To fulfill the former purpose, maintain the shoulder in such a way that the mass of material in the shoulder structure stays in contact with the pavement, helping it resist the lateral shoving pressure from traffic loads, and maintain the shoulder cross slope so water will flow from the pavement to the roadside ditch with little or no water penetrating under the pavement [2].

Shoulders are also an area parallel to the traveled lane that provide emergency or safety travel. To fulfill this purpose, maintain the area of the longitudinal joint between the pavement and the shoulder so no drop-off or elevation difference makes emergency movement off the pavement onto the shoulder hazardous. Also maintain the longitudinal surface profile of the shoulder so no potholes or ruts disrupt vehicles making an emergency movement onto the shoulder. Both of these requirements suggest that a shoulder needs to be maintained in reasonable conformance to the original design and construction configuration.

In all cases, shoulder surfaces should be maintained so that the surface is suitable for emergency vehicle parking, and where necessary, such as at sharp corners on low-volume highways, so that the shoulder can be used as a deceleration lane for short distances in preparation for turning. Slopes at a point 3.66 meters (12 feet) beyond the edge of the traveled way are not normally considered a part of the shoulder. Neither are the curbs and the areas beyond the curb in curbed roadway sections. For roadway cross sections without curbs, shoulder cross slopes usually range from 3% to 5% for paved surfaces, 4% to 7% for gravel- or crushed-stone-surfaced shoulders, and in the range of 8% or 9% for turf shoulders.

The width of the shoulder, the type, and cross slope are defined by the roadway cross-section design standards and policies in effect at the time of the roadway design and construction. Where the traffic volumes are low, the terrain is rugged, and high fill areas or deep drainage ditches are present, a shoulder wide enough to provide full parking off the traveled lane or to be used as a deceleration lane at intersections may not be possible. In all cases, however, shoulder maintenance should provide an orderly transition from the traveled lane to the embankment foreslope.

When performing shoulder maintenance, consider the multiple uses of the shoulder area. When conducting shoulder maintenance, if crossovers exist in a median or if turnouts exist for scenic views and historic interest points, any surface defects in these areas should be corrected at the same time, if practical. When maintaining rural road shoulders, note the shoulder's use for mail delivery and for mailbox locations. If such shoulder use seems to contribute to pavement edge drop-off in aggregate-surfaced or turf-surfaced shoulders, consider widening the shoulder near the mailbox or paving the shoulder in that vicinity. When maintaining rural road shoulders near an entrance, consider flattening the embankment side slope at the entrance to reduce the hazard associated with a vehicle leaving the main road and striking the entrance embankment. A 6:1 slope is desirable but may not be possible to provide without widening the right of way. In urban and suburban areas, sidewalks or recreational paths may exist within the right of way. Surface deterioration on such pathways is more appropriately corrected with the maintenance treatment applied to shoulders than with the maintenance treatment applied to roadways. In urban areas, make the curb cuts or special grade changes needed across a shoulder to meet the American Disabilities Act requirement for pedestrian movement by persons with mobility limitations.

2.1.3.1 Impact of Shoulder Type on Maintenance

Earth or Turf Shoulders: Earth or turf shoulders need enough granular material in the shoulder so that the shoulder will retain its graded, compacted shape and be stable in dry soil conditions. Turf should be maintained as native grasses, preventing the need for irrigation and other special maintenance activities. As much as possible, maintain turf shoulders in low-

Rut in turf shoulder at edge of concrete pavement

growing grasses to minimize mowing requirements. If blading is required to correct rutting, it is best to do the blading when the soil is moist and to do it frequently enough that the ruts can be filled with light blading so turf damage is limited. On aggregate-surfaced roads and on hard-surfaced roads where excessive sand is used in winter maintenance, fine material tends to build up in the shoulder vegetation. Periodically, the shoulder will have to be bladed off or disked down and reseeded to maintain the proper cross-slope drainage. Buildup of soil materials in the turf shoulder can create longitudinal rutting in an aggregate-surfaced road and can hold surface water at the pavement edge of hard-surfaced roadways.

Gravel or Crushed Stone Shoulders: When subjected to wheel loads, aggregate shoulders are more stable than turf shoulders. However, the aggregate surface tends to rut and displace under repeated wheel loads. Thus, it is necessary to regularly reshape the shoulder and to add aggregate to it. A deep rut in an aggregate shoulder at the edge of a hard-surfaced pavement can usually be corrected by reshaping the shoulder and foreslope with a motor grader. Motor grader reshaping needs to be followed by sweeping to remove any loose aggregate material off the pavement. When a drop-off in an aggregate shoulder begins to develop along the edge of a hard-surfaced pavement, the shoulder can be easily brought flush with the edge of the pavement by applying aggregate from a dump truck spreader box, followed by shaping the shoulder cross slope with a tractor-mounted drag, and finishing up by removing any loose material from the pavement surface with a power broom. Severe ruts and potholes in aggregate-surfaced shoulders may have to be repaired with stabilized material. If dust palliatives are used on an aggregate-surfaced shoulder to control dust in dry seasons, this material will serve as a bonding and stabilizing agent and helps to reduce the possibility that heavier stabilization may be required to correct severe ruts and potholes. Aggregate-surfaced shoulders may require some handwork around the base of guardrail posts and similar installations within the shoulder area to prevent a ridge of material from either altering the proper guardrail height or contributing to rutting in front of the guardrail. To control vegetation growth in an aggregate-surfaced shoulder, it may be necessary to apply herbicides, but only in accordance with environmental safety and strictly in accordance with proper safety precautions for worker safety.

Bituminous Shoulders: Bituminous-paved shoulders should be sealed and patched in the same manner as asphaltic-surfaced pavements. Because shoulder contrast with the pavement is often an important clue for the driver in maintaining proper lateral position on the roadway, as much as possible, use the same materials in patching as the surrounding shoulder pavement. Sealing open cracks in a paved shoulder and sealing an open longitudinal joint between the pavement of the traveled lane and that of the shoulder is very important to retaining the shoulder function for providing lateral support to the pavement. Intrusion of water along open cracks and joints contributes to early breakup of the shoulder pavement and to soft spots under the traveled lanes. A good program of crack and joint sealing will also significantly reduce the problems of weeds and grass growing in paved shoulder cracks and joints. If a bituminous-paved shoulder becomes badly oxidized, begins raveling, loses color contrast with traveled lane pavement, or develops high water permeability, it may be appropriate to treat the shoulder over an extensive area as you would a bituminous pavement (fog seal, chip seal, slurry seal, scarify and relay, etc.).

Portland Cement Concrete Shoulders: Portland cement concrete shoulders should receive the same maintenance that is applied to portland cement concrete pavements. However, repair of surface irregularities (faulted areas, punchouts) is not urgent as long as the cracks and joints associated with any shoulder deterioration are kept sealed.

2.1.3.2 Scheduling Shoulder Maintenance

Shoulders are generally maintained according to two methods. One is to schedule maintenance according to functional need so that the shoulder will continue to provide lateral support to the traveled lane pavement and provide a safety area adjacent to the traveled lane as originally designed and constructed. A second method is to schedule maintenance in response to risk management efforts to maximize the ability to defend maintenance policies and practices in the event of litigation arising from a collision where the shoulder area may have been a factor.

Scheduling for Functional Need: Earth or turf shoulders are most likely to be in need of maintenance immediately after a rain or when ground frost has thawed. Rutting and soft spots are most likely to appear then and need attention to retain a proper shoulder with minimum effort. Shoulder erosion will be evident after a rain; if the shoulder is inspected at some later time, vegetation may hide the eroded area. Eroded shoulder areas can reduce lateral structural support to the pavement and can contribute to difficulties in maintaining control of a vehicle driven onto the shoulder. Blading or disking and reseeding of a turf shoulder is preferable in the early part of the growing season to provide for good new turf growth before any winter dormant season.

Aggregate-surfaced shoulders should be inspected for winter damage after the spring thaw in the Snowbelt and after the spring rains in wet climates that do not experience freezing. Herbicides to control vegetation growth in aggregate-surfaced shoulders should be scheduled for application along the same applications guidelines that apply to agricultural applications, depending upon whether the herbicide is preemergent or postemergent in its botanical effect.

Open cracks and open joints in paved shoulders can be sealed quite effectively in dry winter conditions because the crack is open due to shrinking. Bituminous surface treatments to shoulders and patching work should be scheduled according to the same constraints as would be observed in scheduling pavement repairs and treatments. Inspecting for and sealing cracks and joints in shoulders paved with portland cement concrete is important because the shoulder may appear to be in good condition but fail to carry water away from the traveled lane pavement.

Scheduling to Prevent Litigation and to Respond to Nonfunctional Policies: In some states, litigation has established a legal definition of a safety hazard as any pavement drop-off exceeding 5.1 cm (2 inches). Numerous experiments have shown that this height of drop-off is not a significant hazard, but citing such studies in court proceedings in states where this legal precedent exists has largely proven to be unsuccessful. In jurisdictions where such litigation standards exist, it is important to have a regular pattern of inspection (it might be as infrequent as yearly) and a program of correcting shoulders with drop-offs exceeding the prescribed limit.

For other types of defects that are alleged or actually related to safety, it is important to document a program with inspection intervals and treatment follow-up. Shoulder conditions that may be included in this program are shoulder surface color changes, minor holes in the shoulder, limited areas of soft spots, and limited areas of uneven surface transition at the outside of the shoulder into the embankment foreslope.

The key to limiting risk exposure is having good records of both inspection and repair. Considerable latitude in engineering judgment can be allowed if maintenance decisions can be shown to be based on a rational engineering management plan.

2.1.4 Roadway Drainage Maintenance

A consistent program of inspection and maintenance is necessary for highway drainage systems to operate efficiently. In earlier times, most emphasis was placed on surface drainage. Now it is necessary to consider subsurface drainage in inspection and maintenance planning [16, 17].

Drainage systems for highways are designed to limit water damage to the roadway by controlling or directing the free flow of water over, under, or adjacent to the highway and to control the movement of water through the pavement's structural support where necessary. Factors considered in the design and development of highway drainage systems include rainfall patterns in the region, soil and vegetation characteristics in the area, land use and development patterns in the area, the water table in the area and its fluctuations, contour and topographic relief in the area, hydraulic energy of the water as it moves through the highway right of way, and frictional resistance to water flow along the flow path.

Roadway drainage maintenance focuses on retaining the intended design efficiency of the drainage system and on adapting the existing drainage system to accommodate environmental changes to the degree possible with minor changes. (Major changes in the drainage system imply a new drainage design and an associated construction or reconstruction project.) Maintenance activity that contributes to an effective, efficient roadway drainage system includes the following:

- Cleaning roadside ditches to provide a uniform flow line and consistent channel shape.
- Removing trees and other debris from natural water courses or ditches if they may restrict flood flows or become dislodged and accumulate against bridges and culverts, blocking water flow and creating the potential for washing out the structure.
- Correcting minor defects in the drainage system, such as by sealing cracks in pavements and culvert walls; repairing scour around bridge piers and abutments; stabilizing channel banks in the vicinity of bridges; regrading eroded foreslopes or backslopes to their original condition; and replacing small culverts at side road entrances and field entrances that have been damaged, restricting water flow.
- Anticipating drainage system changes by observing changes in roadway performance or in the drainage environment surrounding the roadway that suggest some future need for modified drainage. For instance, continued problems with roadway settling or pavement breakup following a wet season may suggest that subsurface drainage problems are the root cause; erosion around the downstream end of a culvert may be a sign of a crack or leak in the culvert barrel that is allowing water to pipe around the outside of the culvert during flood flow; or standing water in roadside ditches or property adjacent to the roadway resulting from wetland development may be an indication that overland drainage flow across the highway is becoming slower and the water table is rising.

Following major storms, flooding damage to the roadway system will frequently spur a reassessment of roadway drainage system maintenance. While it is appropriate to determine the extent to which diversion ditch maintenance at the top of a highway cut may have contributed to an earth slide or a rock slide or the extent to which lack of riprap material around a bridge may have contributed to excessive scouring around bridge piers and abutments, etc., it is important that engineers with expertise in hydraulics and hydrology participate in the assessment of storm damage associated with drainage. The storm may have been of such a duration, intensity, or area coverage to have grossly exceeded the original drainage design parameters. Perhaps the appropriate maintenance response is only an emergency repair until a new drainage design can produce a reconstruction that can handle a greater intensity storm.

As the control of surface and subsurface drainage has become more complex through surface water detention regulations, wetlands preservation regulations, agricultural irrigation regulations, and agricultural drainage regulations, it is important that the highway agency not allow abutting property owners to discharge water from the abutting property onto the highway right of way without processing a permit to do so. During the permitting process, drainage design engineers and maintenance engineers will review the impact of drainage from abutting property onto the highway right of way. Maintenance personnel who routinely patrol a roadway section should be trained in basic knowledge of which types of drainage from abutting property to the highway are permissible and which require a permit so that they are prepared to report any drainage activities that may need to be reviewed.

Good roadway drainage maintenance has a significant positive effect on roadway safety. Well-maintained foreslopes allow vehicles leaving the traveled lanes to be brought under control with minimum damage. Roadside ditches that are cleaned to prevent standing water minimize the possibility that a vehicle falling into the ditch after a storm will result in a drowning hazard for the driver or passengers. Embankment slopes and culvert ends that are well maintained minimize the potential to trip a vehicle that runs off the road and passes over the culvert end. Good maintenance patrol inspection of roadway drainage will identify culvert end sections or inlets that are potential hazards to vehicles and report these features to the engineering sections responsible for a redesign, mitigating the hazard.

Roadway drainage maintenance has to be scheduled considering the local climate demands, the availability of equipment in the context of other equipment demands, and the availability of personnel in the context of other maintenance activities. Generally, the most efficient time to perform roadway drainage maintenance is in advance of the rainy season. However, the change from the winter season to the rainy season may be a short, abrupt time interval. In such climates, it may be more effective to conduct roadway drainage maintenance in the dry season as a preventive maintenance strategy for future rainy seasons. As much as possible, roadway drainage maintenance should be performed with power equipment. If the roadway design and construction has produced a difficult drainage maintenance environment (limited space from which to operate power equipment, slopes too steep for safe equipment operation, etc.), this situation should be documented by maintenance engineers and forwarded to the design engineers, enabling future projects to produce roadways that can be efficiently and effectively maintained.

2.1.4.1 Maintenance of Roadway Surface Drainage

Curbs and Gutters: Curbs and gutters that direct or control the overland flow of water to ditches or inlets should be monitored for debris accumulation (sediment or trash) after storms. In areas where sand or other fine material is used in winter snow and ice operations, these areas should be inspected for debris accumulation shortly after the end of the snow and ice season. Curbs that have been damaged or broken due to snowplowing need to be repaired as soon as it is practical to do so. In moderate climates, curbs and gutters need to be observed for failure as a result of settling or subsiding material in back of or underneath the curb and gutter section. Any breaks in a curb and gutter line attributable to settling need to be repaired as soon as it is practical to do so [16].

Well-maintained curb and gutter with storm sewer inlet

Shoulder Inlets and Side Drains: Open ditches or pipes that direct surface water flow from the roadway into a roadside ditch or natural water course need to be inspected regularly during the rainy season to ensure that they are free of debris and silt deposits. They should also be observed to determine that they operate efficiently so no water backup occurs. Inspections for water backup can usually be performed by driving along the roadway, but they need to be done during a rainstorm. Shoulder inlets and side drains should be checked for broken pipe sections, leaks into the surrounding soil that can contribute to erosion, and excessive waterway scour at the outlet. In the case of a paved ditch, all cracks need to be sealed to prevent undermining that can lead to structural collapse.

Interceptor Ditches, Diversion Ditches, and Bench-Cut Slope Channels: These drainage features intercept the natural overland flow of water and carry it to adjacent fields, detention basins, or a less disruptive ditch or channel entrance. Check for silt deposits or erosion of the ditch cross section and profile. Correct conditions that could cause ponding in the ditch, because this may contribute to slope instability. Avoid unnecessarily breaking the sod surface of a grassed interceptor in maintaining the cross section or provide treatments to minimize any erosion of the ditch until the grass can regenerate the sod (e.g., silt fences, biodegradable mats, etc.).

Check Dams and Berms: Check dams and earth berms are most often used in roadside ditches and medians. They impound water runoff and act as settling basins for sediment, reducing erosion and silt deposits off the highway right of way while construction or major grading maintenance is underway and until seeding and vegetation can be established [18]. If the removal of check dams, berms, and silt fences is the responsibility of maintenance, the area needs to be inspected periodically to determine when the drainage surface is ready to withstand normal storm water flow without the temporary structures.

Roadside Ditches: These ditches will be open unless they cross under a side road, a driveway, or a walkway. Roadside ditches should be maintained as near as practical to the alignment, grade profile, depth, and cross section to which they were originally designed and constructed or as subsequently reconstructed. At periodic intervals, the roadside ditches should be inspected for and cleaned of fallen rock, heavy vegetation, sediment creating ponding, and other debris that may restrict design drainage flows. In rural areas near urban areas that have restrictive recycling policies, the roadside ditches need to be patrolled for dumpings of yard waste and other trash items. Ditches with debris that may impede proper water flow or represent a hazard to errant vehicles should be cleaned as soon as it is practical to do so. If any debris is encountered that is suspected of containing chemical or toxic materials, do not attempt to clear the debris until a properly qualified person has made an environmental safety determination.

Chutes, Flumes, Spillways, and Slope Drains: All paved or metal troughs (and pipes) carrying a rapid flow of water from a collector drain, a ditch into a roadside channel, or a natural water course should have good contact with the supporting material underneath, especially at the entrance, to prevent uplift and erosion. Open cracks should be sealed and settlement should be corrected. Patch and repair breaks and eroded areas with materials resistant to repeat failures. Open chutes may be an obstacle for roadside mowing operations; however, vegetation needs to be controlled around ditch paving and other drainage structures for hydraulic efficiency. Drainage structure aprons and other hydraulic energy-dissipating devices need to be inspected at regular intervals for damage and for erosion from failure to perform as intended. Substantial erosion around an energy-dissipation structure should be reviewed by an engineer with hydraulic engineering expertise to determine if the existing structure needs to be replaced with an improved design.

Natural Water Courses and Bank Protection: These are rivers, streams, gullies, arroyos, etc. that have continuous or intermittent stream flow and may include floodway channels within the highway right of way. The approach and exit to bridges and other major drainage structures in the channel should be kept clear of rubbish, brush, and other debris [19]. The actual channel must not be allowed to silt in, reducing the required design waterway opening and perhaps requiring dredging. Bank scouring on either side of a bridge can usually be minimized by adjusting the channel alignment to be as nearly straight through the bridge as possible. However, minimizing bank scouring may also require channel bank riprap, jetties, or other stabilization methods. Maintenance of the natural water course through private property is ordinarily the property owner's responsibility; however, highway agencies are more frequently being expected to contribute to such maintenance because the highway structure contributes to the hydraulic energy creating the maintenance problem. Bank stabilization and bank erosion control methods may include rock or stone slope protection (riprap), grouted riprap, sacked concrete riprap, concrete slope paving, gunite slope paving, pile revetments, jetties, retards, jackstraws, tetrahedrons, retaining walls, and cribs. Serious bank or water course erosion should be evaluated by an engineer with expertise in hydraulic engineering and geotechnical engineering to maximize the potential for successful bank stabilization. In emergency flooding operations, sandbags, temporary earthen dams, dikes, or levees may be necessary to protect the roadway or bridge from failure. (See chapter 3 for greater detail on these maintenance methods.)

2.1.4.2 Maintenance of Subsurface Drainage

Subsurface drainage consists of two types. One is traditional culverts and similar small structures carrying free water surface flows. A second is drains and other subsurface networks intended to collect porous media water flow into a concentrated flow that can be drained out of the roadway embankment structure to lower the water table and reduce the degree of soil saturation.

Culverts and Small Structures: This type of subsurface drainage consists of culverts and small structures. The two small structures include (1) catch basins and drain openings with grates and (2) storm sewers.

Typical reinforced concrete culvert headwall and end section

Culverts: While culverts may be reinforced-concrete box culverts, concrete pipes, corrugated metal pipes, timber box culverts, and small-diameter plastic pipe culverts, they all have some common maintenance requirements to ensure an efficient, effective roadway drainage system. Culverts need to be kept clean and unobstructed by removing sediment, debris buildup, and any brush or woody vegetation growing at either the inlet or outlet. The inlet or outlet flow channel alignment may have to be adjusted to prevent sedimentation from returning. Footing, headwalls, and cutoff walls need to be inspected for scour at culvert ends. Brush and debris may need to be removed upstream of the culvert to prevent or limit future storms from clogging the culvert. If state law does not give the highway agency right of entry to private property to maintain a stream channel, obtain the property owner's permission. To maintain good public relations, notify the owner in advance of the work even if statutory authority to perform channel maintenance exists. Changes in the upstream watershed caused by land development activities, farming practices, forest or prairie fires, etc. may increase sedimentation or debris deposition. In urban areas or near recreation areas where children might wander into a culvert, maintenance inspections should note the condition of any safety fencing, gates, or grating that discourages children from entering the culvert. If a culvert has settled, heaved, been pushed out of alignment, been broken, or suffered major damage, it should be scheduled for repair or replacement as soon as it is practical to do so [20]. Worn or corroded inverts of pipe culvert can be repaired with concrete grout, gunite, or asphaltic material as an invert line to extend the life of the culvert [21]. For smaller pipe culverts in which the original design provided sufficient excess drainige capacity, a smaller diameter pipe can be inserted into the existing culvert and grout can be pumped in-between the two pipes to seal the repair. Severe scour problems may require installing piles or large stone to dissipate the water energy upstream of the culvert. Minor cracks and joint separations in a culvert can often be sealed with high-pressure grout.

Catch Basins and Drain Openings With Grates: Drop inlets collect and pond water from gutters and drainage ways that flow into storm sewers and other drain pipes and act as connection units when the hydraulic engineering design requires a change in pipe size, the combining of several pipe flows into one, or a change in flow line grade. As a result of the energy changes in the water flow at these structures, sediment load may be dropped in the basin, requiring periodic (perhaps annual) inspection of the basin to determine if the sediment needs to be cleaned or flushed out. It is also important that the basin, inlet opening, or grate not be blocked by any debris to retain the hydraulic efficiency of the opening. Maintenance patrols after storms should be alert for and remove trash or debris lodged in inlet openings or basin openings. In urban areas where continuous curb and gutter street and highway sections are common, a regular street sweeping and cleaning program is necessary to minimize the clogging of inlets and storm sewers. The street sweeping program is especially important following any winter snow and ice control program where abrasive material has been applied to the street. Likewise, bridge catch basins should be inspected at the end of the winter snow and ice control season if abrasive material was used on the bridges.

Storm Sewers: Pipes (main lines are designated as trunks, and dead-end sections connecting to trunks are called laterals) carry the water from the inlets or catch basins to an outfall into a ditch or natural water course. Urban storm drainage for streets and highways is mostly handled through storm sewer drainage. Manholes are constructed at points where there are changes in storm sewer pipe alignment, in flow line grade, or in pipe sizes. If the manhole coincides with an inlet or catch basin, it may be a combined structure. Observing the maintenance practices noted above for catch basins and drain openings will minimize storm sewer pipe maintenance. If inspection during storm water flow indicates that the storm sewer may be clogged, this problem should be immediately addressed when the storm flow ends. Clogged pipes can often be cleaned with high-pressure water jets, but if tree roots or broken pipes are causing the clogging, more extensive service will be required. Flexible rotary cutters will remove roots intruding into a pipe. A broken pipe may be repaired by jacking an insert liner into the failed location; otherwise, the storm sewer pipe may have to be excavated and relaid to repair it. In the case of small-diameter storm sewers, it may be more cost-effective (depending upon site conditions) to excavate and relay the pipe for any break. Manholes and cleanouts should be inspected regularly (perhaps annually) to ensure that they are serviceable. Manhole covers that do not seat properly should be repaired; manhole covers in the pavement must be adjusted for finished elevation if a pavement is overlaid (as obvious as this seems, it is sometimes neglected by the construction process and left as a problem for maintenance). Anytime a manhole cover is removed, safety precautions that warn pedestrians and vehicles of the danger of the open hole need to taken. Where the design of storm sewers required pumping from sump areas, such as street and highway underpasses, the pumping systems should be tested in dry weather to ensure that they are working properly. Since draining underpasses is critical for vehicle safety, it is desirable to have a warning system for failure of such pumping systems.

Underdrain Systems and Associated Features: Pipe or conduit networks, under the pavement or at some critical area in an embankment, are often installed to draw moisture moving through the soil and granular material of the roadway structure into the pipe network and to release the water into the ditches or natural channels [22]. The pipe networks are open at the discharge end only. However, pavement edge drains may have frequent discharge points into the roadside ditch. The drain discharge points should be checked annually during the wet season to ensure that the drain system of perforated pipes or the discharge opening is not clogged [23]. If root and vegetation growth are clogging a drain, herbicides may be applied to kill the vegetation. Before applying any herbicide in response to a clogged underdrain system, assess what the peripheral impact may be (e.g., introducing possible toxins in the groundwater or killing trees that may be nearby and have feeder roots in the area). If the conduits are plugged with silt or granular material, it is probable that the drain network will have to be excavated and replaced.

Suction pumps, high-pressure water jets, and power rotary sewer cleaners can be used to clean out long sections of pipe that have been clogged with roots and sand if the original design and construction provided clean-out access points. If the pipe has to be excavated and replaced, placing a synthetic geotextile fabric between the pipe and soil or granular material will help prevent a return of the clogging condition.

Installation of an underdrain in the shoulder should have included backfilling the trench with a compacted layer of impervious material and capping the trench with paving material to seal its surface from infiltrating water to minimize future settlement [24]. When underdrains have been installed after the original pavement was constructed and in service, the additional joints created by the underdrain trench cut need to be inspected and sealed along with the longitudinal joint along the pavement and shoulder. In all joint and crack sealing, as well as any pavement undersealing operations, work near underdrains should be done in a manner that does not fill the underdrain pipe with sealant.

2.1.4.3 Other Considerations

The contour grading of roadsides is usually a part of the original roadway design and intended to impart a visual impact. However, if slope stability or roadside ditch stability is a continuing maintenance problem, it may be possible to change the geotechnical or hydraulic characteristics of a section by using contour grading in the repair process to alter the flow paths of rainfall and groundwater. Maintenance engineers should consult with persons having familiarity with the original hydraulic and geotechnical design if contour grading is an alternative, ensuring that the changes will be consistent with downstream drainage facilities.

In severe freeze-thaw climatic zones, frost heaves may result when a localized area of the subgrade freezes. In granular surface roads, a "frost boil" can develop. Improving the roadway drainage to lower the water table within the roadway support structure is the best long-term maintenance solution.

Earth slides are considered a drainage-related maintenance problem. Mud slides or other wet earth failures in a cut or fill area result from the moisture content in the soil structure lubricating the soil particles so the reduced friction between soil particles does not provide enough force to resist the gravitational pull (and combined pavement loading if the slide is under the pavement) on the earth mass. In pavement areas, seal the cracks and joints, limiting water intrusion into fills supporting the pavement. In cut areas, it may be necessary to flatten the back slope, reducing the dependence on internal soil particle friction to keep the slope stable, or to create diversion ditches at the top of the cut, bringing surface water away from the slope.

Dry earth slides can occur when the previously stable support to a fill or cut area is disturbed at its base or some geotechnical condition of the fill or cut slope has changed from the original design and construction condition. Maintenance engineers should seek the advice of engineers with expertise in geotechnical engineering to determine the cause of any dry slide and the appropriate way to remedy the situation. Correcting a slide condition may be a process beyond maintenance and may require a redesign and reconstruction. In all dry slide repairs, be careful not to start a new slide, which can create a worker hazard.

All slides may potentially create serious traffic hazards when any part of the roadway is blocked. In areas where rock cuts are expected to develop, the original design and construction will provide rockfall ditches from which collected rocks should be removed at least annually. When slide material closes all or a portion of the roadway, temporary traffic control should be immediately set up to warn and direct traffic away from or around the slide area. The slide material should be removed as soon as practical, but not until a safety assessment has been made to determine the degree to which the removal operation might trigger additional sliding. Slide material disposal needs to be done in such a manner that it does not add to any slope instability and cause further sliding.

In urban and suburban areas, special surface water handling facilities may be a part of the original design and construction. When surface water moves onto the highway drainage surface after coming from large, hard-surfaced parking areas (such as large truck stops, etc.), an oil separator basin may be required to pond the surface runoff and skim or collect the oil through a weir system before the surface water flows into a stream. If the highway includes such oil separators, they should be inspected annually to ensure that the structure is intact and will operate correctly. A better maintenance approach is to convince the design sections that agency policy should require oil separation of surface water flows before the water may be released onto the highway right of way. Settling basins to control sediment in highway surface flows and detention basins to slow the surface water peak flow are becoming more common because of environmental interest in improving surface water quality. If these drainage structures are on the highway right of way, the outlet to these structures should be inspected annually to ensure that the proper regulated flow is maintained. Again, the preferred maintenance approach is to have the agency's design and development policies state that sediment and detention basins associated with surface water moving onto the highway right of way are the abutting property owner's responsibility.

When the original design and construction of the roadway in combination with the surrounding terrain produces high-velocity ditch flows, ditches will typically be lined with rock or other material, or detention basins with drop culverts will be provided to lower the hydraulic energy of the flow. All energy-dissipation drainage structures and ditch lining intended to dissipate the energy of high-velocity flows should be inspected regularly (perhaps annually, depending upon the expected frequency of major storms). A failure of the energy-dissipation structures can produce significant roadway damage in a single major storm.

2.1.5 Roadside Maintenance

The "roadside" generally includes everything in the zone stretching from the outside edge of the shoulder to the right-of-way line (or fence if the right of way is fenced); therefore, it includes many of the features discussed under roadway drainage maintenance. Unpaved medians are included as a roadside feature since their maintenance is nearly identical with the outer roadside. None of the maintenance activity discussed in the drainage section will be repeated here, except as needed for emphasis or additional comment. While a majority of maintenance funds are required to maintain roadway pavements and bridges, the smoothness of the traveled lanes and the appearance of the roadside are the characteristics by which the traveling public most often judges the effectiveness of the highway maintenance effort.

The roadside should be maintained in a condition that contributes to safe travel and operation, presents an orderly appearance, provides convenient travel, enhances a pleasing travel experience, preserves the roadway investment, and protects the roadway investment, insofar as available resources permit rational allocation to roadside maintenance. This suggests that some objectives of performing roadside maintenance are related to the functional needs for highway maintenance and other objectives are most strongly related to aesthetic goals or cultural values.

Natural condition roadside vegetation requiring limited mowing

Riprap-stabilized back slope in a roadside

Functional Objectives: These may include, but are not necessarily limited to, the following:
- Ensure good drainage and minimize erosion. This may be a fundamental tenet of justifying a significant portion of the maintenance effort [25, 26].
- Minimize maintenance expenditures. If the public can be shown that reducing roadside maintenance efforts will allow more preventive maintenance on the roadway, more long-term maintenance results can be achieved with less resources [29].
- Minimize fire hazards. This may be particularly important in dry climates and in forested areas.
- Delineate the roadway. The roadside surface can be helpful to drivers in detecting changes in the roadway geometry far in advance of the need to adjust driving behavior.
- Screen headlight glare. Proper placement of and maintenance of tall vegetation in a median can significantly reduce oncoming headlight glare, increasing the potential for safe nighttime driving.

- Provide noise abatement. It is difficult to achieve significant traffic noise reduction through roadside maintenance unless the roadside includes noise walls or noise barriers. In any case, these noise attenuation structures must be properly maintained to provide the designed and intended effectiveness.
- Form snow barriers to restrict drifting. Living snow fencing can be created with properly planted and maintained shrub lines in the outer edges of the right of way. In some extreme snowfall areas, permanent man-made snow fencing may be installed that will require some maintenance as part of roadside maintenance. Design procedures have been developed for snow fencing (see the snow and ice control section).
- Eliminate shadows that will cause patch ice. Some roadways experience ice patches in the day-night, thaw-freeze cycle that drivers do not expect when trees or tall shrubs in the right of way cast daytime shadows on the pavement. Proper roadside maintenance can minimize these effects.
- Control roadside hazards and obstructions. Activities that remove large-diameter trees from the roadside, provide smooth surface transitions around culvert ends, produce gradual transitions among foreslopes to ditch bottoms to backslopes, etc. all reduce the potential for an errant vehicle collision to result in injury or fatality.

Urban paved ditch invert and roadside vegetation to screen view of the road (inlet cover may be a hazard)

Reinforced earth retaining wall and fenced urban right-of-way line

- Relieve driver tedium and monotony. Variations in landscaping, mowing pattern, the materials in the roadside surface that are visible to the driver, and the display or interruption of views through the roadside all can potentially contribute to the driver's visual stimulation. Contrarily, a roadside that consists of tall vegetation growing up to the shoulder's edge kilometer after kilometer (mile after mile) creates tension and tedium by appearing to hide possible wildlife and other potential hazards that may suddenly enter the roadway. In a similar vein, long, continuous lengths of the roadside mowed in a "lawn-like" effect out to the right-of-way fence contribute to monotony and boredom.
- Enhance pedestrian and motorist safety. All activities done to make it easier for a driver to recover control of a vehicle after leaving a traveled lane contribute to motorist safety. Maintaining a roadside in a condition that enables a pedestrian to walk comfortably and safely along the far outer edge of the shoulder contributes to pedestrian safety by allowing the pedestrian to stay away from moving vehicles.
- Reduce animal-vehicle conflicts. Measures to reduce roadkill need to start with the design phase of highway development. Maintenance activity cannot be expected to solve all of the problems associated with the public aversion to roadkill. However, maintaining the roadside such that ponding and standing water areas are minimized will reduce the attraction of wildlife to the roadside. Regular mowing of the roadside adjacent to the shoulder will increase the opportunity for drivers to see wildlife emerging from the roadside to attempt to cross the highway. Materials that simulate fallen trees can be placed from the right of way to the right-of-way line running under bridges and overpasses, encouraging wildlife to pass under roadways instead of across them. [27].
- Enhance sight distance. Typically, maintenance of the roadside can do little to enhance the vertical sight distance available to the driver. However, maintaining the roadside so headlights and taillights can be seen around the inside of horizontal curves can increase the horizontal sight distance available.
- Reduce pavement damage by plants and water. As discussed in the previous section, a well-maintained roadside is a well-drained roadside. A well-drained roadside increases the structural stability of the foundation supporting the pavement.
- Reduce deterioration of roadside hardware (posts, rails, etc.). Vegetation and soil around bolted connections on guardrails and sign supports (especially breakaway slip bases) contribute to corrosion. They can also contribute to decay in wood posts and corrosion of metal posts. Maintaining the roadside surface as nearly as practical to the designed and intended surface slope and condition will minimize the potential deterioration of safety hardware.
- Satisfy general public demands. If the public perception of roadway maintenance quality is used in setting maintenance policy, then roadside maintenance quality will bear heavily on the degree to which maintenance activity meets the quality of maintenance goals.
- Give proper perspective to intersections. In rural two-lane, two-way highway locations, assist drivers by maintaining the roadside so they have as clear a view as is practical when approaching the intersection and can see the foreslopes of the intersecting roadways when preparing to turn.
- Assist in controlling the spread of noxious weeds. While roadside maintenance expenditures can often be reduced by adopting a "naturalized" roadside policy, in the interest of being a "good neighbor" to abutting productive agricultural land, use of mowing and herbicides should be pursued with the goal of controlling weeds that have been legally declared noxious in the local region.
- Provide windbreaks where crosswinds present a hazard. Planting and proper maintenance of the appropriate size and type of trees can reduce the effect of troublesome crosswinds for high-ground-profile vehicles.

- Provide a backdrop for traffic signs. Traffic signs (and sometimes traffic signals) can be difficult to read (and see) on streets and highways that run on a due east-west direction when the sun is rising or setting at seasons when the sun is nearly aligned on the roadway azimuth. Vegetation can aid in sign (and signal) visibility by providing a background against which the traffic control device is more clearly visible. In urban areas, a vegetation background can help screen out the urban visual clutter beyond the right of way, which can reduce sign and signal visibility.
- Preserve natural features. A well-maintained roadside reduces the impact of erosion and other degradation processes on natural features in the area.
- Provide sanitary facilities (safety rest areas).

Aesthetic Objectives: These may include, but are not necessarily limited to, the following:
- Improve recreational potential. Where a route is intended to be scenic or provide scenic vistas, these features and qualities create public expectations of high-quality roadside maintenance.
- Introduce new plantings to an area. Normally, care should be taken to introduce only native plants into the highway right of way, limiting long-term maintenance costs. However, the highway construction process usually removes almost all traces of native vegetation. Roadside maintenance presents an opportunity to increase the diversity of the vegetative culture back to something closer to the original ecoculture [30, 31].
- Produce a pleasing effect through patterns, colors, and textures. The impression many visitors have of an area is initially formed by the roadside's aesthetic quality. Tourism and travel-based economic activity is supported by high-quality roadside maintenance.
- Screen distracting or unpleasant views. The same principles noted above apply here. If tall growing vegetation will not adversely contribute to snowdrift control, it can be maintained at a height to screen ugly views. Perimeter vegetation can also contribute to a good background for traffic control devices.
- Create buffer zones for adjacent residents. In urban areas, tall perimeter vegetation can screen highways from abutting properties. While tall vegetation will hide the road traffic from adjacent properties, it is almost impossible to significantly reduce the physiological effect of noise with vegetation in the highway right of way. However, getting the traffic out of sight will typically reduce the level of complaints regarding traffic noise from abutting property residents.
- Enhance a roadway's scenic qualities. Much of a roadway's scenic quality is determined in the design process when the geometric pattern and the three-dimensional surface of the roadway between the right-of-way lines are developed. However, the long-term scenic quality of the roadway environment can be enhanced by the way vegetation, drainage surfaces, roadside safety features, and litter control are maintained.
- Complement existing features and structures. While most engineers are not skilled in visual arts, they may find that plantings, water surfaces, and patterns of mowing can all either screen unsightly aspects of roadside features or draw vehicle passenger attention to pleasing structures and features.
- Preserve or conserve native plants. Conducting roadside maintenance in a way that preserves or stimulates the native plant ecoculture is one of the most profitable and positive public relations activities that maintenance can do. Wherever and whenever roadside maintenance contributes to the preservation or conservation of native plant culture, the highway agency should seek the interest and coverage of wildlife, conservation, and gardening writers for local media outlets.

Scheduling, Planning, and Budgeting: Local climate, native floriculture, and local public preferences will bear greatly on roadside maintenance scheduling, planning, and budgeting. However, some roadside maintenance activities significantly affect the potential to provide a safe driving environment and should, therefore, take a high priority in the overall maintenance program [28]. Activities significantly affecting safety include but are not limited to cutting brush back from the driver line of sight to traffic signs and traffic signals; trimming trees to expand the illumination pattern of street and highway lighting; cutting vegetation low in intersection sight triangles; and fall mowing of vegetation adjacent to the shoulder in snow regions, discouraging drifting snow.

Vegetation Management: Controlling the growth of vegetation to the intended or desired condition for optimum maintenance cost and efficiency by a variety of means is included in vegetation management. Activities appropriate to a locality may vary to achieve the desired result. Scheduling is extremely important to good vegetation management. Conducting any treatment at the right time is necessary for it to be successful.

Soil and Plant Conditioning: The soil environment may have to be improved if the roadside vegetation is to perform as needed with minimum long-term maintenance effort. Shrub areas may need to be cultivated. Newly planted trees and shrubs may need to be mulched. Plantings and turf areas may need fertilizer to become established. Small trees and shrubs may need to be protected from mowing damage or grazing by deer and other wildlife. Some turf and planting areas may need to be irrigated to become established, but vegetation requiring long-term irrigation should not be a part of any roadside maintenance program. Newly planted areas may need some short-term special protection against high-velocity water flows until the plantings are firmly established.

Mowing: Mowing of planted and natural grass areas controls excessive growth of grass, enhances driver safety, enhances scenic qualities in selected areas, improves water flow in drainage paths, assists in weed and brush control, and reduces roadside fire hazards in dry grass areas. Personnel safety is always an issue in mowing operations [28, 32, 33]. For instance, slopes steeper than 2.5:1 should not be mowed unless the mowing unit is side mounted and remotely controlled so the tractor (or power unit) remains stable on a fairly flat area. If mowing is normally done only once per year, fall is preferable. Do not mow newly planted roadsides until the second growing season. A plan for the frequency and extent of roadside mowing should be developed with input from highway design persons to consider the intended visual impact of the roadway and from landscape persons to consider the desired plant control effect.

Chemical Management: Chemical application to regulate and retard excessive vegetation growth can reduce the need for mowing. Herbicides can eliminate weeds and noxious plants from the roadside. Careful, selective use of herbicides is a cost-effective roadside maintenance practice [34]. However, there is always the risk of environmental sensitivity [35]. Not using herbicides in accordance with manufacturer's recommendations can lead to adverse impacts on water quality or to safety hazards to personnel [36]. Only personnel who have been properly trained and are properly supervised should handle herbicides and be involved in herbicide application [37]. Efforts to maintain the roadside landscaping will sometimes involve the need to apply pesticides to control insects and fungi attacking roadside plantings. Since these chemicals are also harmful to desired insect populations (such as honey bees), pesticides should be applied according to schedule and within the procedures specified by the state agency responsible for controlling the use of such chemicals.

Tree and Brush Trimming: Small woody brush (diameter less than 5 cm or 2 inches) can usually be mowed with a brush mower ("brush hog"). Trees and shrubs should be trimmed so they do not block the motorist's view of signs and traffic control devices. Dead trees, shrubs, and limbs in trees should be cut out, and the debris burned or composted, because the dead material may harbor harmful insect or fungi populations. Always cut any trunk in excess of 5 cm (2 inches) as near flush to the natural ground slope as possible; never leave any stump high enough to snag a vehicle's undercarriage [28].

Ground Cover Establishment: Good vegetation ground cover can contribute to cost-effective roadside maintenance by minimizing ditch cleaning and mowing. Because local soil and climate conditions produce a variety of factors that control the success of establishing roadside ground cover, horticulturists, landscape architects, and agronomists should be consulted if difficulties are encountered.

Burning: Burning is not viewed favorably in many areas because of concerns about air quality, the danger of drivers losing control in smoke-reduced visibility, possible adverse impact on ecoculture diversity, and possible adverse effect on small wildlife nesting in the roadside. When it is necessary to conduct controlled burning to reduce the potential for large fires or to kill off invading brush, take proper precautions to ensure safe traffic movement through the area, to protect workers conducting the burn, and to ensure the safety of abutting property.

Integrated Pest Management Principles: The principles of an integrated pest management program are related to having an effective vegetation management program. Threshold levels of vegetative condition could be established that dictate when treatments or actions should be taken, and a relationship to an overall maintenance level of service could be defined. Using these principles, routine monitoring of vegetative condition, evaluation of vegetative status, planting and establishing of competing beneficial forms of vegetation for long-term preventive maintenance, and conducting treatment at the proper time (critical for successful vegetative management and closely related to planning and scheduling for maintenance) all become part of the vegetation management program.

Other: Additional specialized activities in maintaining the roadside include maintaining safety rest areas as well as vector, litter, and erosion control.

Safety Rest Areas: The general rule of thumb should be, "Do not create them if you cannot expend the resources and the effort to provide high-quality maintenance." The buildings, picnic shelters, parking areas, stairs and ramps, plumbing fixtures, water supply, sewage treatment, lawn areas, plantings, lighting, signing and information displays, and refreshment and vending machines all need to be kept in good condition. When something breaks or deteriorates, it should be repaired as soon as possible to limit liability and to enhance user appreciation of maintenance services. This requires constant inspection and attention to maintenance details. Restrooms need to be kept clean, sanitary, and well lighted. Graffiti, especially gang-related insignia, should not be allowed to remain. Litter, insects and vermin, and the conditions that contribute to their presence, need to be eliminated. Highway lighting, traffic signing, pavement markings, and pavement surfaces associated with rest areas need to be well maintained for vehicle, pedestrian, and pet safety.

Historical Sites, Viewpoints, and Scenic Overlooks: Historical markers along the roadway may be a maintenance responsibility. While historical markers may not seem very significant to the maintenance engineer, lack of attention to them can be an emotional issue with interest groups. Routine maintenance of markers, such as cleaning, touch-up painting, and close trimming of vegetation, may be best handled by soliciting "adoption" of the markers by a local historical society or civic group. Scenic overlooks and viewpoints can be concentration points of litter. If an overlook involves any safety rail or fence, such safety features need to be inspected regularly for any necessary repair.

Vector Control: The roadside should not become a breeding ground for insect populations, rodent populations, or weeds that abutting property owners are trying to control. Vector control should be coordinated with other state or local authorities having responsibility for pest control. When pesticides are used, common precautions include not saving or reusing empty containers, not touching any spray equipment part with the mouth, never smoking while handling pesticides, not spraying when there is significant wind (7 miles per hour or greater), not applying pesticides near water wells, following manufacturer's instructions, avoiding prolonged contact with materials, wearing rubber gloves, washing gloves and all exposed skin areas thoroughly after handling materials, disposing of used containers as soon as possible per manufacturer's instructions and as per any applicable local regulations, providing personnel with the label-recommended personal protective equipment, and insisting that such equipment be used.

Litter Control: Roadsides should be kept free of litter and debris to maintain the highway in a neat, clean, and attractive manner. In several states, bottle and can deposit laws have quite effectively reduced the extent of roadside litter. Some states have found the "adopt-a-highway" program has reduced roadside litter by increasing public awareness of the need for a clean highway environment. Regardless of public information programs and other efforts short of collecting materials, it is generally necessary to clean the roadside of trash and rubbish once in the spring and periodically until winter, as needed to meet agency and public expectations of roadside cleanliness. (Debris that falls on the traveled way and roadkill should be removed as soon as it is known to maintenance forces, maximizing travel safety.)

Erosion Control: Erosion control should be designed into the roadside through the design and construction process. However, if erosion problems develop within the roadside after it has become the responsibility of the maintenance division, special efforts may be needed to halt erosion problems. Cleaning and repairing ditches and culverts, repairing and reshaping foreslopes and backslopes, and reseeding slopes may halt erosion caused by unanticipated changes in water flow. Mulching and applying fiber mats when reseeding is often helpful in establishing vegetation repair on slopes. Geotextiles have proven to be helpful in erosion control problem areas [38].

2.1.6 Maintenance of Safety Features

Safety features include guardrails, attenuators, fencing, pedestrian and bicycle overpasses and underpasses, and traffic islands and curbs. Maintaining these features and similar ones has become increasingly complex with the increasing sophistication of impact attenuators.

Guardrail break needs temporary warning device until repair can be performed

Guardrails and Barriers: Guardrails and barriers have somewhat different maintenance requirements depending upon the materials from which they are constructed. For instance, reinforced-concrete ("New Jersey-type") barriers have relatively limited maintenance needs. These types of barriers need to be inspected for an impact pushing individual sections out of alignment; for failures in connections between sections; for clogged drainage points along the barrier; and if appropriate, for needed minor surface repairs from impact damage or spalling. Major damage, such as that produced by a large crash fire, will likely create a repair project beyond maintenance's scope. If glare screens or delineators are attached to a run of concrete barriers, these items should be periodically inspected because oversized commercial loads frequently damage them. Earth berm barriers should be inspected for possible drainage problems from settlement or impact damage. Replace any hazard markers associated with earth berms as soon as practical after noting the damage. Steel beam and wire cable guardrails should be observed every time a maintenance patrol passes in case damage hits warrant planning immediate repair (for example, a deep pocket in the rail line from a hit, a crushed end section, sections torn loose from posts, an anchor at either end of a run broken loose, a W-beam rail section flattened) [39]. W-beam guardrails should be regularly inspected for loose bolts, missing bolts, posts losing soil support, soil ruts in front of the rail, and material buildup on aggregate or turf shoulders in front of the rail. Wire cable guardrails should be regularly inspected for proper tension in the cables; for posts losing soil support; for posts pushed off line, which reduces the restraining capacity of the cable line; and for missing reflectors on the posts. Maintenance patrols finding a badly damaged run of guardrail should immediately place vertical panels or Type I barricades at the damage spots to warn drivers to stay clear of that location until a repair can be made. Steel components that are losing galvanizing or paint should be painted with metal-based paint to retard rusting and corrosion.

Impact Attenuators: A variety of impact attenuator (or crash cushion) designs are in place, including mechanical (or hydraulic) energy-dissipation devices, plastic barrels filled with sand or pea gravel, clusters of vertical cylinders filled with water, clusters of empty steel drums cut and weakened to collapse upon impact, lightweight concrete panels or cells formed to provide a structure that collapses upon impact, and cable restraint systems. Because each impact attenuator, both fixed and mobile, is designed to dissipate the energy of a selected design vehicle at a selected range of design velocities over a selected distance, it is important that all maintenance and repair be conducted in accordance with the designer's and manufacturer's guidelines. An improperly maintained or repaired attenuator may function worse than none at all. Maintenance patrols should note any damaged attenuator and report it immediately for scheduled repair as well as mark the damaged attenuator with a Type I or II barricade to warn drivers away from it. In climates expecting winter freezing temperatures, attenuators that incorporate sand- or water-filled cells should be checked for antifreeze preparation (an environmentally safe liquid or calcium chloride mixture, for example) during the fall. Routine maintenance patrols should check for lost or damaged hazard markers at attenuators and replace or repair any damage or loss as soon as practical.

Traffic Island and Curbs: Traffic islands and curbs for safety and delineation of intersection areas should not require much maintenance. Annually check for any need to repaint curbs or island areas that have lost their visibility and target value because of snow and ice control or environmental deterioration. Check also for broken spots (especially in asphalt curbs) that will reduce drainage or traffic channelization effectiveness. Weeds and other vegetation growing in curb sections and island areas that are not turfed should be treated with herbicide to reduce the long-term cracking and spalling damage vegetation causes and to prevent interference with surface water drainage flow along the curb or island.

Fences: Since fences are provided along streets and highways to restrict pedestrian and animal access to the right of way, they need to be periodically inspected (perhaps once every 2 years) for breaks in the fence line. If breaks in a fence line appear to be associated with vandalism or criminal activity, maintenance personnel should alert law enforcement agencies. When repairing a fence line, consider the fencing need so that the type of fence used is consistent with the intended control of access to the right of way. When the fence is only supposed to keep horses and cattle off the right of way, a tight fence of three barbed wires will suffice, but in an urban area where pedestrians are to be channeled to cross-street intersections and over- or underpasses, a 4- or 5-foot high chain-link fence is needed. As land use abutting a street or highway changes, it may be necessary to change the type of fence to continue to provide proper access control.

Truck Escape Lanes or Ramps: Truck escape lanes or ramps and their associated arrestor beds are a part of the roadside and also a special safety feature. When design of a highway has included runaway truck escape ramps, it is important that these facilities be regularly inspected for use or damage. Each arrestor bed should have small-diameter poles installed that indicate the design depth of the loose material in the bed, making it easy to check the bed's depth and grading. Maintenance offices should have a cooperative arrangement with law enforcement agencies and towing firms so any call regarding a truck in an escape ramp will also be reported to maintenance. The approach to the arrestor needs to be checked for ruts or other significant surface irregularities that might make it difficult for a truck driver to retain control of a runaway vehicle. Inspect advance warning signing and signing marking the escape ramp to ensure it is all in place. In climates experiencing freezing weather in the winter, check the drainage of the arrestor area for any potential to retain water in the bed material, because a frozen bed is not effective. Whenever trucks using an escape ramp experience difficulties, maintenance personnel should report their observations to highway design engineers to see if a redesign and modification is needed. No change from the design configuration of a truck escape ramp should ever be made without an energy-dissipation analysis that assesses the impact of the change on truck operation into the escape ramp.

2.1.7 Building Facilities Related to Roadway Maintenance

Highway and street maintenance organizations are usually responsible for maintenance of all capital investment necessary to conduct highway and street maintenance, i.e., of buildings and equipment. Building facilities often include offices, employee lockers and restrooms, barracks, employee on-site residences, equipment sheds, paint and sign shops, bulk storage facilities, store rooms, tool rooms, and equipment repair buildings. Several factors influence the degree to which a significant amount of resources needs to be expended in building maintenance. If the maintenance agency operates most of its equipment fleet on a rental basis, the rental can include most major equipment maintenance, reducing the need for buildings to provide it. If significant levels of maintenance activity are conducted through contracting, then a reduced level of maintenance resources needs to be associated with buildings to support agency maintenance personnel.

Truck weigh station

If maintenance building facilities are located in rural areas or in urban industrial areas, building maintenance can be directed to keeping the facilities functional rather than having to be overly concerned about appearance. For example, if a roof needs to be patched, any shingles available can be used without worrying about if the color matches the original shingles. However, when maintenance buildings and facilities are in urban areas not zoned industrial or in areas with recreational and scenic value, building and ground maintenance must consider public expectations of neatness and appearance for a high-profile agency. In these cases, grounds landscaping may also be a significant part of the building and facility maintenance. In all cases, however, the first priority of building and facility maintenance is to provide the buildings and equipment needed for maintenance personnel to perform their tasks effectively and safely.

The location of maintenance garages, equipment support facilities, and material stockpile facilities is often a matter of history more than rational engineering analysis. In long-range planning for maintenance services, facility location should be reviewed with respect to optimizing facilities to support maintenance for the most efficient operations. Applying transportation network analysis, along with routing and scheduling algorithms, to analysis of garage and storage locations as source nodes may provide efficiency guidance in phasing changes of maintenance support facilities that can reduce overall maintenance costs and increase overall productivity.

District Maintenance and Field Offices: All supervisory, engineering, and management personnel should be provided with the necessary office space and facilities needed to conduct their duties effectively and efficiently. Even though such personnel may spend a significant part of their duty day in the field, each person needs a desk with adequate file space, computer access, and telecommunication access. Support personnel (clerical, staff, senior technicians, etc.) also need proper office space. While the electronic office does not require that all personnel be in the same office, the location of maintenance supervisory personnel should enhance their ability to operate as a team.

Locker Rooms: Each maintenance employee engaged in fieldwork should be provided adequate locker space in conjunction with lavatory, toilet, and shower facilities. Regular maintenance services should be based on local health, sanitation, and environmental regulations.

Storerooms: Storerooms should be designed and maintained with consideration given to the materials to be stored. Hazardous materials need special attention. Storage facilities built decades ago may not be adequate to meet current environmental and safety regulations without significant updating. Inspection of storage facilities should be conducted in light of the latest regulations in addition to checking for obvious facility maintenance needs.

Tool Rooms: Tool rooms should be organized to facilitate accounting of tools used infrequently or tools periodically checked out. Periodic inspection of a tool room should include an inventory check as well as checking the room's cleanliness and suitability for storing and servicing tools.

Equipment Sheds: Buildings in which vehicles and mechanical units can be stored out of the weather can save long-term maintenance expense on the equipment itself. In snow and ice regions, it is helpful to have heated facilities for mounting and servicing snow and ice attachments to trucks, which aids in the washing down of the equipment to minimize chemical corrosion and makes equipment inspection easier in cold weather. A building working temperature of 50°F (or even a little lower) is usually suitable.

Equipment Repair Buildings: The need for equipment service buildings is determined largely by equipment purchase and resale, rental, and contract maintenance policies and practices. For agencies that purchase new equipment and retain the items until a life-cycle cost analysis suggests that the cost of maintaining the equipment (including downtime and lost productivity for equipment maintenance) begins to exceed the cost of selling and replacing it, a well-maintained and well-equipped shop is a necessity. For agencies that rent most equipment, equipment service may consist of primarily routine lubricant service and small damage repair, so a modestly equipped shop with limited facility maintenance demands may be sufficient.

Paint and Sign Shops: Some agencies contract out work zone traffic control signing; some agencies contract purchase permanent signs; and other agencies fabricate all signs in-house. Agencies with a paint and sign shop for producing their signage need a well-maintained facility

Chapter 2 Roadway Maintenance and Management

with good lighting to facilitate proper sign layout and sign design; clean, well-organized sign and sign material storage areas to facilitate inventory control; well-maintained ventilation systems to remove dangerous fumes; and well-maintained loading/unloading areas to limit operation hazards.

Wood and Metalworking Shops: These types of shops need to be kept clean after each day's operation, limiting the potential for fire and explosions. Well-maintained ventilation systems are needed for worker safety. Good lighting needs to be maintained to aid in safe equipment operations.

Stockpiles: While a stockpile is not a building facility, the storage of materials, even in large quantities, may be associated with a covered structure. A properly distributed network of material stockpiles can be a time and cost-saving way of using a material inventory. If a material is needed in a short response time for a maintenance activity, the material can be stockpiled at strategic locations in off-peak activity times for later use when the need arises. In effect, a stockpile is a way of load smoothing the flow of material and the time and effort required to deliver and apply the material in times of high use. This is the engineering management principle behind the stockpiling of materials for snow and ice control. If the materials to be stockpiled are environmentally stable, no particular protection is needed for the stockpile. Generally, the materials that are suitable for stockpiling are those that are used in large quantities, are stable so they are not expected to deteriorate rapidly, are of relatively low unit value so that theft from the stockpile is not normally a serious problem, and are expected to be needed over a relatively wide area. Thus, stockpiling is usually limited to aggregates for asphalt and portland cement concretes, for base courses and shoulders and aggregate drains, and for road surfaces and surface treatments; granular material for snow and ice control; chlorides for snow and ice control; and paving materials for use in recycling projects. In most cases, chloride materials require protection from the weather, or moisture can cause a cementing action within the stockpile. Stockpiles should be located to minimize the total combined cost of material acquisition, transportation to the stockpile location, retrieval from the stockpile, transportation from the stockpile to the use site, and application of the material. Stockpile sites selected on the basis of maintenance economic criteria have to be reviewed for environmental sensitivity, especially for chlorides and recycled materials. Water draining from a stockpile containing a leachate or volatile hydrocarbons released into the air must not contribute to unregulated environmental pollution.

2.1.8 Special Pathways

Trails for walking, bicycling, horseback riding, snowmobiling, etc. have always been associated with highway development in recreational and wilderness areas. However, the Intermodal Surface Transportation Efficiency Act of 1991 created special provisions that placed more emphasis on joint development of highway rights of way to incorporate trails and nonmotorized transport paths in localities not previously expected to have such trails on any significant length of a highway right of way. Gradually, maintenance personnel will be expected to respond to more trails within the highway right of way as agency design offices receive requests for more such facilities to be an integral part of new highway design projects. Most of the trails will primarily be for bicycle traffic. However, almost all bicycle trails will also be used by hikers and pedestrians. Maintenance inspection of bicycle paths requires attention to much smaller pavement surface defects than is required in motor traffic pavements. Longitudinal pavement cracks in excess of about 2 cm (about 3/4 inch) can present control problems to touring-type bicycles. Loose debris (pebbles, dirt, small sticks, etc.) needs to be swept off hard-surfaced bicycle paths. If the bicycle path is designed for two-way traffic and is sufficiently wide to accommodate it (about 1.8 meters or 6 feet), then a centerline stripe should be maintained on any hard-surfaced pavement. Aggregate-surfaced bicycle paths should be inspected annually (preferably in the spring) to check for washed-out areas, significant surface irregularities, large-sized aggregates working up to the surface, and vegetation encroaching on the surface width. In urban areas, hard-surfaced bicycle paths are often used in snow and ice conditions by die-hard cyclists, suggesting that a maintenance policy should either remove snow and ice or close the paths for the winter season. Bicycle trail signing and traffic controls may require special observation to ensure that they are in place and in good condition because they are often not visible from the automotive roadway. To ensure that all vehicles at intersections have clear views, trees and shrubs at intersections where bicycle trails and automotive traffic intersect may require pruning beyond that needed for automotive traffic. If trash and litter along a bicycle trail becomes a problem, inventory the location of trash receptacles for convenience both in use and in servicing them. Trash receptacles at cyclist rest stops will be frequently used.

2.1.9 Snow and Ice Control

Historically, effective snow and ice control has been achieved through good planning, thorough preparation, and implementation of good tactical procedures. This continues to be so, but a new dimension has been added: an awareness of new developments in materials, equipment, and methods that may minimize the impact of snow and ice conditions upon transportation. It is important, both in communicating to the public and in defending the agency's response to a winter storm crisis, to have a policy statement that defines the manner in which the agency intends to respond to snow and ice conditions. In climatic zones where snow and ice conditions are infrequent or are of short duration, the policy might involve only passive measures such as warnings to the public regarding travel. These passive measures could perhaps be supplemented with salt and sand at the approaches to stop signs or signals. At the opposite extreme, a policy in climatic zones experiencing frequent winter storms and significant accumulations of snow or ice could stress attempting to provide and maintain bare pavement on all paved roads. In between these two extremes resides a compromise in which the agency establishes priorities for snow and ice control measures on the various segments of the highway and street network based on the storm's severity and duration; the relative traffic volume expected to use the road segments; and the importance of certain road segments to critical commerce activity as well as to fire, police, or medical services. In most cases, regardless of the general policy, features that are likely to create special difficulties for drivers during snow and ice conditions, such as bridges; intersections; long, steep hills; entrance and exit ramps with significant grades; and sharp curves will be given immediate attention or even anti-icing treatment in advance of a storm [40–42]. A more detailed documentation of snow and ice control is included in the AASHTO *Guide for Snow and Ice Control*.

Truck-mounted snowplow operation

Motor grader plowing snow

Planning and Budgeting: Once a policy has been established that defines the intent and scope of the agency's response to snow and ice conditions, estimates of the equipment, personnel, and materials needed to carry out the policy are required. An old caution is that "too much preparation is extravagant, and too little may be dangerous" in the planning and budgeting process. Historical records of snow and ice conditions and long-range weather forecasts can help formulate a planned program and a budget to implement the plan. However, just like the stock market, past weather conditions are no guarantee of future storm patterns, and most meteorologists have guarded confidence in any weather forecast beyond 24 hours. Thus, a storm response plan must incorporate enough flexibility to avoid being completely paralyzed by a storm larger than expected and have a budget that allows the transfer of resources from less critical maintenance areas if the storm season has been underestimated. Improved advanced planning was quite beneficial to agencies combating the 1996 blizzard [43, 44]. Kuemmel's *Managing Roadway Snow and Ice Control Operations* [45] contains a concise overview of the entire planning process and its impact on operations and budgeting with topical emphasis on personnel, materials, equipment, facilities, and technologies. Advance preparation by checking equipment status, having materials stockpiled at the appropriate locations, and having planned operating routes is important to successfully deal with snow and ice storms [46].

Training: Personnel with many years of experience in snow and ice control operations are highly desirable in an agency, but having such persons does not remove the need for continued training in operations. Each agency should evaluate the capabilities of its personnel and define training priorities to address the most critical needs [47]. Computer-enhanced and computer-aided training, such as simulations, have been helpful to some agencies [48]. Even experienced operators can often gain more knowledge and skill through the sharing of personnel tips and ideas that work in specific circumstances [49–51]. Some agencies have found that holding "snow rodeos" before the winter maintenance season is an effective way to promote safe operations.

Cooperating With Other Agencies: Agency cooperation is becoming increasingly important. Cooperative agreements can

- ensure that snow and ice control practices are consistent across jurisdictional boundaries, which limits surprises to drivers and promotes an integrated transportation network;
- maximize effect while minimizing effort in activities that cross jurisdictional boundaries, such as analyzing storm forecasts and providing public information reports;
- Provide advance authority for exchange of resources when storms temporarily exceed one area's capabilities, but a neighboring jurisdiction has excess capabilities.

Several states have documented the value of being able to apply cooperative relationships [52, 53].

Liabilities and Precautions: In snow and ice climates, the winter maintenance season can be a source of litigation and damages that generate claims against the agency. Personnel training should include topics designed to help all operating personnel be prepared to conduct themselves properly in a safety, damage, or injury incident. Some points to consider are as follows:

- Bridges and overpasses require special consideration. Falling ice or snow might injure persons below. When cleaning bridges, avoid dislodging icicles or allowing snow to fall over the side if traffic is below.
- Ice will form on the road surfaces of bridges before other parts of the roadway. Check that warning signs are clearly readable and visible to drivers when the use of such signs has been adopted.
- Use of snow removal equipment on certain bridges can present unusual problems. Decide beforehand what methods or materials are most effective for each bridge and overpass.
- Warning signs obscured by windblown snow should be cleaned as soon as practicable. Damage to traffic signs and signals should be immediately repaired.
- Damage to state property or to parked cars, mailboxes, fences, etc. should be reported immediately and the circumstances well documented (with witnesses, when available).
- If it freezes on the roadway surface, water from melted snow can create a greater hazard than the original snowfall. If at all possible, keep drainage open.
- Vehicle accidents on ice and snow can become cumulative. When a single vehicle obstructs a roadway, warn approaching motorists as far ahead of the accident site as possible.

- When a snow removal vehicle is operating ahead of motorists on the highway during a snowfall, give adequate warning to them. Operating proper warning lights on the snow removal equipment is imperative.
- Removal of stalled or parked cars that are obstructing snow removal operations should be handled through a regular procedure. Cities that experience regular winter snowfall should enact snow route ordinances.
- Photographs of accident scenes can help to settle liability claims.
- Contamination of the environment by using or storing chemicals associated with snow and ice control should be avoided or minimized.

Related Services and Programs: Drivers make decisions about travel during winter storms based on a variety of information sources. Travelers regard radio and television weather information bulletins as prime sources of information. Cable television subscribers use the Weather Channel to update trip planning. Households with access to the Internet or the World Wide Web can obtain real-time displays of radar weather data with forecast interpretations. Areas surrounding airports with certificated air carrier service typically have radio access to National Oceanic and Atmospheric Association weather data, which provides forecasts in great detail. Most agencies and state highway patrol offices maintain call-in numbers for prerecorded road condition reports. It is in the highway and street agency's best interest to have a public information effort that encourages travelers to use all possible sources of weather and road information for travel planning during winter storms. Informed drivers are more likely to make travel decisions consistent with maintenance objectives.

Roadway weather information systems, automated weather observing systems, and thermal mapping are all technological developments that may enhance the effectiveness of snow and ice control by enabling the maintenance treatment to be more precisely matched to roadway surface conditions [54–59]. Study of various technologies has estimated high benefit-cost ratio advantages to using the low-cost alternatives among these technologies [60]. Continued study of areawide deployments of selected high-technology alternatives is expected to lead to long-term maintenance and traffic operation benefits that may justify extensive deployment of such technologies.

Time-Phase Scheduling: Three major time phases affect snow and ice control. Starting with a cleanup phase after the winter maintenance season is over and beginning with preparatory planning for the next winter season, some of the first phase's principal activities are as follows:
- Evaluate the effects of the previous winter.
- Assess predictions for the next winter season.
- Begin planning for next winter: prepare local maps or databases showing resources, obstructions, drains, routes to emergency facilities, and critical or special problem areas; evaluate new developments in techniques, methods, materials, and equipment; and document procedures.
- Begin stockpiling materials.
- Store winter maintenance equipment, repair it if needed, and procure additional equipment as needed.
- Conduct training programs in winter maintenance for regular employees, as needed, and for all new and temporary employees.

A second phase typically begins about 30 days before the first expected snowfall. Some of the principal activities included are as follows:
- Review plans and preparations.
- Assign equipment and personnel.
- Install guides, temporary snow fences, special warning signs, and markers at obstacles.

- Condition equipment and facilities before use. Test all mechanical and hydraulic systems. If necessary, calibrate all material dispersal equipment.
- Remove obstructions and clean drainage facilities.

The third phase, the tactical or operational phase, begins with the first snowfall or the first warning of icy conditions if anti-icing action is to be taken. The principal activities are as follows:
- Road condition reporting.
- Sweeping.
- Plowing.
- Applying chemicals and abrasives.
- Reassigning resources to critical areas.
- Maintaining and repairing equipment.

(Note: Some agencies have found that early application of chemicals and abrasives is superior to treating existing snow and ice conditions. Evaluation of this anti-icing policy is recommended.)

Personnel Scheduling: Scheduling of operating personnel requires consideration of a number of factors, such as proximity of the personnel to the equipment to be operated and the area to be patrolled; employee status of the personnel, i.e., if they are contractual or regular employees; safety of the personnel and the motoring public; expected levels of sickness and injuries; union work rules; and the level of training for the task required. One city has found that a good advance plan enables them to make better use of personnel and equipment [61].

Materials Scheduling: Having enough material at the right place at the right time requires a well-thought-out distribution of stockpiles and proper protection of them so the materials do not deteriorate during storage. This also implies that the proper equipment in good working order be available at each stockpile to handle the materials. Some research is being conducted to increase both the storage and handling capabilities of salt sheds [62].

Equipment Scheduling: Advance planning and budgeting of resources will not provide an effective snow and ice control effort if an equipment fleet is not properly outfitted and in good working order. Thus, while it is desirable to schedule all equipment for snow and ice control activities, it is wise to keep some vehicles and equipment in reserve to replace equipment that will inevitably break down and if it becomes necessary, to execute a rescue during the storm. Advance warning from the weather information system can be especially helpful; it may provide time to fit multipurpose equipment with plows and spreaders. In-between storms, all equipment damage and failures should be addressed to ensure readiness for the next storm. Use all available equipment as efficiently as possible [63]; there is some evidence that global positioning systems may provide a high-technology aid to doing just that [64].

Urban Area Snow and Ice Control: Urban area snow and ice control is complicated by repetitive peak traffic flow patterns, intense concentration of economic and social activity that depends upon continuous traffic flow, and concentration of emergency service facilities.
- Snow or ice during night hours will require action at least several hours before the morning rush hour. Highest priority should be given to commuting routes and commercial traffic routes. Anti-icing treatments ahead of the storm may be an effective approach. Public information channels should be used to discourage unnecessary travel.
- Snow or ice during midday hours can lead to virtual gridlock of the urban network when the evening rush hour traffic begins unless efforts have been undertaken well before the rush hour to keep main arterials, commuter routes, and commercial routes passable. Again, anti-icing treatments ahead of the storm may be an effective treatment, but immediate plowing of snow is also necessary to allay the intense psychological pressure people feel to get home to maintain family care.

- Weekend snow or ice storms can be controlled more easily if a full, complete public information strategy is available to keep the population informed of road conditions. Priority should be placed on through area travel routes to minimize the impact on travel that is not local in nature.
- Snow storage and disposition is frequently a major problem and a major activity in urban areas. In commercial business districts, the snow must often be trucked away just to clear streets. In some areas, concern for surface water contamination from stored snow requires that snow storage areas be identified well in advance of the beginning of storms [65]. Industrial areas, commercial areas, and educational institutions with large parking lots where snow can be stored until the contaminants can be diluted by spring rains should be investigated as possible storage areas.

Interurban Area Snow and Ice Control: Routes where high speeds are expected require patrolling for patches of packed ice or snow that will surprise drivers after the initial clearing effort. Particular note should be given to curves, hills, stop sign approaches, signal approaches, areas subjected to daytime shading from adjacent trees, and areas that tend to drift in after being plowed open etc. Patrol units should be alert for motorist emergencies that make an integrated radio communication network important.

Mountain Passes: Each mountain pass needs to be evaluated and prioritized with respect to its importance for interregional travel on the highway network. If it is possible to close the pass because the route is not a critical link in the network, the best strategy may be to monitor road conditions through the pass and implement road closure procedures when using the route becomes hazardous. Reopening the pass in the spring will require excavation equipment and efforts similar to excavating earth.

Mountain passes that must be kept open as much as possible require the installation of snowsheds, patrols to monitor avalanche conditions, special snowplowing equipment, requirements for studded tires and tire chains on vehicles accessing the pass, and continuous patrol of pass conditions. Such passes will often have to be temporarily closed for safety reasons. Closing such passes needs to be coordinated with state highway patrol units that can enforce the closure and prevent drivers from ignoring it.

Bridges and Overpasses: Since bridges and overpasses often experience frost, icing, or slippery conditions before such road surface conditions exist on earth-supported pavement, drivers often do not exercise proper caution and are surprised by a slick bridge or overpass. Some agencies use the static signs "Watch for Ice on Bridge" or "Bridge May Freeze Before Road." Since these and similar static signs are always displayed, when road conditions become bad, not all drivers will heed the warning. Road weather information systems [66] and localized weather forecasts [67] are particularly helpful with anti-icing and de-icing efforts on bridges and overpasses. If ice control treatments can be applied just before the onset of frost or icing conditions, safe travel over bridges and overpasses is greatly enhanced. Even so, public information channels should be used to urge drivers to exercise caution when approaching bridges and overpasses.

Chemicals and Abrasives: The primary chemicals used to reduce the melting temperature of snow and ice are sodium chloride and calcium chloride. Environmental concerns, including corrosion effects on vehicles and roadway structures, have stimulated experimentation with reducing chloride application rates, recognizing that an increase in accidents may occur [68, 69]. Liquid application of chlorides, prewetting of a chloride and sand mix, and application of calcium magnesium acetate are treatments that promise to reduce the amount of chloride applied to the roadway while maintaining effective de-icing and anti-icing capability [70–72].

A variety of spreaders are used to apply abrasive materials, but the most positive development in spreaders is the development of zero velocity (or nearly so) drop of the material to the pavement surface, which reduces the amount of material that has to be applied to the roadway for effective snow and ice control [45]. Whatever technology is used to apply chemicals and abrasives to the roadway for snow and ice control, it is important that each spreader be calibrated for spread rate with respect to the truck's speed of operation so the equipment operator can apply the proper amount specified for the roadway condition being treated. The calibration rate needs to be available to the operator in the vehicle cab. This rate is especially important when the application method is intended to minimize the amount of salt applied but still retain effective snow and ice control [73].

Organic processes are yielding some byproducts that promise to be less corrosive and less toxic than traditional inorganic compounds. Biotechnology, agricultural and food-processing, and brewing industries all produce byproducts that, if they could be used in highway maintenance, would economically and environmentally benefit the industries' operations. A technical evaluation of one such product can be accessed on the Internet at <www.cerf.org/evtec/eval/iceban.htm>.

Anti-icing by chemical and abrasive application also requires some form of roadway weather information system, as explained below. To develop the "best" fit of the latest forecasting technologies to any given agency's snow and ice control needs, agencies must share their experiences in applying them; an example is reference [74].

Anti-icing should be considered a routine strategy for agencies with frequent snow and ice control demands. Weather information systems that employ advanced technology such as "super-Doppler" radar can help an agency estimate the onset of road surface ice so they can apply salt brine (or other anti-icing treatments) just before the storm or ice condition hits. Agencies that do not have such advanced technology systems may be able to contract for detailed weather forecasting services from meteorologists who do. Super-Doppler radar can provide data that enable properly trained meteorologists to produce precise computer estimates of storm movements. These forecasts can be used to initiate an anti-icing chemical application that significantly reduces the storm's adverse impact.

Agencies that have been applying salt brine rather than crystalline salt with sand to the roadway often develop an in-house capability to manufacture rather than purchase and store their brine solution. Because of the volume of material required, producing salt brine in-house is normally more cost-effective than purchasing it. The necessary mixing and holding tanks to produce salt brine should be appropriately distributed throughout the maintenance service area.

Equipment: The effectiveness of any snow and ice control policy averaged out over the variations in storm intensity and duration will be limited by the capability of the equipment available for personnel [63, 75]. Some snow and ice control equipment is specialized in function and requires storage during the warm seasons, while other general-purpose units can be converted by attachments to snow and ice control units. Equipment is intended to function in one of two ways. Either the equipment moves and displaces the snow and ice off the traveled way, or the equipment is used to increase the natural rate of melting so the snow and ice drain away as surface water or evaporate. Wing plows have become popular for moving snow clear of the shoulder in addition to the traveled lanes [76]. In urban areas, a snowplow or motor grader can roll a large windrow, which complicates clearing sidewalks and intersections. However, blade end gates have proven to be an effective modification to plows [77]. Urban snow control often requires equipment to load snow on trucks for hauling to a disposal site because storage areas within the right of way are often limited. Ordinary front-end loaders may be used, but when the hauling of large quantities of snow is routinely expected as part of the snow control strategy, conveyor loaders able to load trucks directly from a windrow are more efficient. A few minutes gained on a truck's loading cycle time can vastly increase the haul fleet's daily productivity. Several states in the Snowbelt have initiated a cooperative approach to developing a new, improved snowplow truck [78]. Significant interest in advancing the snow and ice control equipment state of the art has been spurred by dissemination of the results of a tour of European and Japanese technology and practices [79]. The experiences of the upper Midwest and the Northeast in dealing with the blizzard of 1996 also stimulated interest in adapting international methods and technologies to prepare for potential blizzards in 1997 [80].

Equipment systems are becoming more sophisticated for snow and ice control as maintenance engineers and managers begin to use technologies such as global positioning and geographic information systems as well as real-time computer control of the deployment of units [64, 81]. Cost-effectiveness analysis of the impact of these technologies shows the benefits of advancing the state of the art in snow and ice control. In some cases, the economic benefits can be significant [60, 82].

The maintenance of snow and ice control equipment has always been important. Equipment should be washed after each use, removing any corrosive chemicals and materials. Surfaces subject to corrosion should be treated with a petroleum-based material or other corrosion inhibitor. As more electronic components are installed, it is important that all connectors and sensors not become corroded, that seals to keep out moisture are frequently checked for breakage, that all radio and communication equipment is frequently checked and serviced, and that any on-board computers are not exposed to magnets or electrical fields, which can destroy programs or calibration data.

Snow Fences and Guide Markings: If the edge of the roadway is delineated for drivers, once a track is broken through the snowdrifts, the drifts may not have to be plowed back from the entire roadway to reestablish traffic flow in areas where people tend to have four-wheel-drive vehicles. Drivers also find delineators marking the edge of the roadway useful in maintaining their placement on the roadway during blowing snow conditions. Because any aid that keeps drivers from becoming stuck in a snow-filled ditch or median is not only a safety enhancement but a reduction in long-term maintenance costs, preparations for the snow season should include inspecting for broken and missing roadway edge delineators in areas where they have been installed.

Overpasses and interchange bridges can be especially difficult areas in which to control drifting. Wind tunnel studies by Ring [83] found that appropriate maintenance of landscaping in the interchange area could provide a "living snow fence" that substantially reduced the tendency for snow at overpasses to drift across the roadway. By some estimates, snow fences designed to minimize drifting across a road cost about one-tenth of repeated plowing in areas prone to repeated drifting. A comprehensive guide to designing and installing snow fences is available from the Strategic Highway Research Program [84]. This guide provides suggestions for permanent, temporary, and living snow fences. If some history of the location of problem drifting areas and a record of the storm characteristics that produced the drifting (amount of snowfall, wind direction, wind speed) are available, then a reasonably precise estimate of the drift reduction effect of a snow fence of a specified height, constructed of a specified material, and at a specified distance from the traveled lane can be produced. In this manner, an engineering economic analysis may be developed to decide if it is prudent budgeting and planning to place snow fencing in a particular area.

2.1.10 Traffic Control Devices and Roadway System Instrumentation

For many agencies, their main responsibility is maintenance of rural highway networks, signs, and markings. In urban areas, however, traffic signal maintenance is critical because of the safety implications of traffic signal failures. In areas where specialized traffic control systems have moved from the research mode to the routine agency operations mode, such as freeway traffic control systems and area wide traffic management systems, more sophisticated maintenance of traffic control is required. The level of training and expertise needed to conduct maintenance of traffic control and roadway system instrumentation will continue to increase with operational implementation of intelligent transportation systems and with roadway weather information systems. Traffic engineering personnel provide much of the expertise for traffic control device installation, operation, and maintenance. In agencies where roadway maintenance and traffic operations are not in the same integrated administrative unit, good communication between the two groups must exist with respect to maintaining all traffic control devices. Because the traveling public relies heavily on traffic control devices to provide safe, efficient travel, proper maintenance of such devices is very important.

Replacing a stop sign

Highway Signs: For maintenance inventory purposes, highway signs can be classified as either post mounted (roadside signs) or overhead mounted (signs on bridges or on sign bridges). Sign maintenance generally involves the following categories of activity for which personnel and resources need to be scheduled:

Clearing: This activity involves removing vegetation, or roadside surface obstructions, or sight obstructions that restrict sign visibility. It is particularly important in urban areas where city beautification efforts often conflict with the need to maintain a clear line of sight to the sign for the driver [85]. Reference [85] provides guidelines for instructing and training maintenance personnel in the proper trimming of brush and shrubs to provide sight distance to signs. For post-mounted roadside signs, tree trimming and brush cutting should be scheduled in early spring to midsummer. For overhead-mounted signs, tree trimming should be scheduled as required to ensure visibility or prevent damage to support structures.

Chapter 2 Roadway Maintenance and Management **2-67**

Cleaning: Sign legibility and reflectivity can be reduced by a dirty surface or surface contamination from vandalism. Annually cleaning signs to remove dirt maintains their design reflectivity. For post-mounted roadside signs, wash the sign face as required with a soap solution and then rinse; remove vandalism marks with approved solvents. For overhead-mounted signs, schedule sign washing as required; early summer is usually best. Use a soap solution then rinse.

Field Repairing and Repainting: Minor damage to signs and supports can be corrected practically and economically at the field site with materials and tools that can reasonably be provided on a sign maintenance truck. Reference [85] contains suggestions for materials, equipment, and methods appropriate to common field repairs. For post-mounted roadside signs, inspections should be scheduled every 6 months for damage needing repairs, and incidental damage should be repaired as soon as it is practical to do so. Repainting of sign posts and sign backs should be scheduled every 3 years or as required based on inspection reports; however, treated wood or galvanized materials may eliminate this maintenance. For overhead-mounted signs, inspection of the signs and supporting structures should be scheduled on a yearly cycle; damaged spots should be repaired as soon as possible. Recent results of a program that carefully inspected sign support bridges revealed that structural deterioration is easy to miss unless a detailed inspection is conducted [86]. Repainting of sign support structures should be scheduled on a 5- to 10-year cycle or more frequently if the climate and environment are especially hard on the structural coating.

Refurbishing Sign Faces: Some deterioration of sign face and legend can be refurbished through field efforts. For both post-mounted roadway signs and overhead-mounted signs, any effort to refurbish the sign face should be scheduled when the sign's visibility and readability is reduced to an unacceptable level as determined by nighttime inspection. When the inspection indicates that the sign cannot be effectively refurbished by a field repair, the sign face should be replaced as soon as possible.

Replacing Signs: If damage to a sign is uneconomical to repair or reflectivity is badly degraded, or the sign's legend no longer applies as needed, then the sign should be replaced as soon as possible.

Replacing Lamps and Cleaning and Aligning Illumination Units: Illumination units need to be periodically inspected to keep illuminated signs operating properly with good nighttime visibility and to reduce lamp failures. For both post-mounted roadway signs and overhead-mounted signs, a regular schedule should be established for lamp replacement, ensuring a minimum of lamp failures.

Checking Torque: Inspection of breakaway sign supports must include a check for proper torque resistance of all fasteners in the breakaway mechanism. Loose bolts or corrosion in the breakaway mechanism can change the breakaway force and thus reduce the intended safety level.

Preventive maintenance for signs includes some of the preceding activities. Sign preventive maintenance may include straightening a signpost or adjusting the position of a sign so it is better seen by drivers; spot-painting small damaged areas; cleaning the sign face and any reflectorized elements; clear coating a sign face; cleaning and repainting signposts; tightening bolts, lag screws, and other fasteners; checking torque on breakaway sign support connectors; and removing visibility obstructions. Replacing damaged elements or the entire sign assembly may be considered preventive maintenance by some agencies and response maintenance by other agencies. A regularly scheduled program of replacing lamps in illuminated signs to minimize the probability of burned-out bulbs is an appropriate part of a preventive maintenance program for signs.

Maintenance personnel conducting sign inventory and sign inspections should be periodically trained in identification of sign legends and messages defined in the *Manual on Uniform Traffic Control Devices* [87] for standard traffic control communication. A 3-year training cycle is probably sufficient to keep maintenance personnel aware of changes in approved signs and to learn to recognize sign messages that may be inappropriate or out-of-date.

Logo signs, tourist-oriented signs, and other guide signs are not usually a high priority in sign maintenance. However, because highway-oriented businesses depend upon these signs for a significant level of business and because travelers unfamiliar with an area need the information on these signs to find their way, such signs should not be allowed to go unattended until they reach a condition that renders them ineffective. Burnham [88] provides a compendium of policies and practices appropriate to the maintenance of logo signs, tourist-oriented signs, and other guide signs.

Changeable message signs are signs in function but have the maintenance characteristics of traffic signals. The practice of using this hybrid traffic control and driver communication device is well documented in Burnham [88], which has sections dealing with changeable message failures, maintainability, communication protocols, warranties, procurement, and testing, all of which are relevant to a maintenance engineer or maintenance manager.

Cunard has provided a compendium of maintenance management practices for street and highway signs [89]. It contains detailed discussion of organizing the sign maintenance effort; field inventories; appropriate facilities, equipment, and materials; personnel issues; costs and funding; and management control.

Markings: Markings are lines, symbols, words, and patterns painted or applied to pavements, curbs, and obstructions (or hazards) to function as traffic control devices that regulate traffic flow, warn drivers of potential hazards, or provide guidance information to drivers within the roadway field of view. Markings can be classified into these groups: pavement markings, curb markings, object markings, and delineators. A comprehensive description of markings and their recommended applications is contained in part III of the *Manual on Uniform Traffic Control Devices* [87]. Materials used for markings include the following, but research efforts are constantly seeking materials with greater durability, lower cost, and simpler application requirements:

Pavement Markings: Paint (including epoxy based) with glass beads is the most common material. However, thermoplastic, cold plastic, tape, ceramic markers, raised or recessed markers, temporary tape, instant dry thermo powder, and polyester materials are also used.

Curb Markings: Paint (including epoxy based) with or without glass beads is the most common material. Reflective markers applied to curbs with adhesives are also used.

Object Markings: Metal plates with surfaces of reflective sheeting and a painted background with plastic or glass reflectors are common. Also, metal, wood, or plastic posts with reflective markings or attached reflectors are used.

Delineators: Reflective elements of reflective sheeting or paint with plastic or glass reflectors are common. A metal back plate may be used to mount the reflectors, and typically, the delineators are attached to a small post.

One marking that might be considered a curb marking, an object marking, or a delineator is the small reflective markers that many agencies attach to individual sections of temporary concrete barriers placed in maintenance and construction work zones. To maintain a consistent approach to any possible litigation arising from a crash in a maintenance or construction work zone where such reflectors have been attached to portable concrete barriers, an agency should define which category of marking these reflectors are considered to be.

Marking Maintenance: Marking maintenance generally includes the following activities:

Cleaning: Cleaning removes dirt that impairs visibility. Traffic on haul roads, agricultural activity, floods, or dust storms may deposit soil on a pavement. Brooming the pavement, however, may suffice to make the markings visible. To maintain their reflectivity, object markers, such as those marking bridge pier hazards, should be washed on the same cycle as roadside signs.

Repainting or Reapplying: Repainting improves visibility impaired by wear or damage. For any roadway that has been in place for several years, the environmental deterioration and traffic wear on markings should be obvious enough that a regular schedule of repainting and reapplication can be determined that will maintain suitable marking visibility. In snow and ice climates, agencies that practice aggressive snow and ice removal must typically repaint and reapply pavement and curb markings each spring. Repainting and reapplying of pavement markings needs to be coordinated with patching, seal coating, overlay work, and other pavement surface maintenance to maximize work efficiency. Routine repainting and reapplying should be completed before the peak summer travel season whenever practicable, maximizing the safety benefit of new markings and minimizing the exposure of maintenance personnel to traffic conflicts.

Repairing or Replacing: Repairing brings damaged markings back to their designed configuration. For damaged object markings or delineators, have a guide that maintenance personnel can use to decide when markings have been so badly damaged that they need to be repaired or replaced. The Wyoming State Highway Department has developed a small manual that capably assists operations and maintenance personnel with the installation and maintenance of markings [91].

Marking Removal: Removing markings reduces the confusion that obsolete markings can cause. Removing markings without leaving a scar pattern of the old marking in the pavement surface requires careful attention to the removal process. Areas where markings have been removed need to be checked in both daylight and nighttime illumination conditions, ensuring that no confusing pattern remains. Where removal methods do leave confusing surface patterns, it may be advisable to apply a seal coat or other surface treatment that completely obliterates the area and then reapply the correct markings if a safety assessment indicates a long-term potential problem with the surface appearance.

Electrical and Electronic Equipment Systems: Electrical systems include any electromechanical traffic control devices and the electrical equipment necessary to supply and control all electrical and electronic traffic control devices. As technology has progressed into the electronic and digital age, most traffic control systems have evolved into digital and solid-state electronic devices. For the foreseeable future, however, many devices will remain whose components are essentially electrical and electromechanical in nature, such as flashing beacons and some sign illumination control units. Electronic systems were originally gaseous tube based but have now been almost exclusively replaced with digital solid-state electronic circuits. In many cases, this is even true of electrical power supply because the cost of solar power panels has dropped and the reliability has increased owing to continued space research and technology transfer of that research. Functional systems for which an agency's maintenance forces may be responsible include but are not limited to the following:

Traffic Signals: At street and highway intersections under the agency's jurisdiction, traffic signals need to be inspected according to the maintenance cycle recommended by the manufacturer of the equipment installed at the intersection. Inspection and maintenance needs to be performed by a person who is qualified by virtue of training in the unique characteristics of the particular traffic control hardware and software installed. Thus, if the agency does not have a large number of traffic signals that it maintains, perhaps a cooperative maintenance agreement could be negotiated with a qualified agency or organization. Because traffic signals are vital to safe, efficient traffic flow, the agency or the contracted maintenance provider must respond quickly to any traffic signal failure.

Flashing beacons at hazardous suburban highway intersection

Ramp Metering, Reversible Lanes, and Lane Designation Signals: Congestion management research has developed to the point that specialized traffic signal control systems are becoming common operational units. Preventive maintenance programs for these traffic control units must be conducted when the units are not operating and must provide a high state of reliability because the increased congestion resulting from hardware or software failure has widespread negative safety effects.

Roadway Weather Information and Ice Detection Instrumentation: Research from the SHRP and other separate studies have greatly expanded interest in and adoption of weather-sensing networks that aid in the planning, management, and execution of winter maintenance programs. It is not yet clear how such instrumentation networks are to be optimally maintained. It is important that they be inspected before the beginning of each winter maintenance season for proper operation, including calibration for proper signal and sensing response. If an agency with weather-sensing networks is not properly equipped to test the instrumentation or lacks qualified technical personnel, the agency should establish a cooperative agreement with an organization qualified to conduct the necessary maintenance according to the agency's schedule.

Television Cameras, Monitors, and Visual Security Installations: Any video-based equipment used for traffic control or surveillance purposes should be periodically inspected for proper tuning and balance. These types of equipment systems are sometimes targets of vandalism and need to be checked frequently for such damage. All electronic adjustment and maintenance should be performed by a person who is properly trained because electrical shock hazards may be present in the circuits.

Signal and Data Communication Equipment: Traffic detectors providing input to centralized signal control networks, fiber optic cable networks, remote telemetry traffic-counting stations, and weigh-in-motion stations are a few of the complex communication units that agencies are becoming responsible for maintaining. Equipment that transmits the data (e.g., multiplexing sending and receiving units), cable pull boxes, and other hardware units should be checked on a regular schedule, according to manufacturer and designer recommendations. Software that handles the data transmission should be designed to conduct parity and reliability checks on all data packets. Thus, if data transmission quality begins to degrade, the software will provide an early warning of possible maintenance needs.

Preventive Maintenance Program for Electrical Systems: A valid preventive maintenance program for electrical systems requires qualified personnel and a variety of hardware designs. Whether this type of maintenance is performed with agency forces or contracted out, the following activities would typically be included:

Contacts: Inspect for burning. If contacts are starting to pit, disconnect the relay and reface or replace.

Electromagnet: Inspect for quiet operation. Magnet noise may be caused by low voltage on the coil terminals, foreign matter on the core face, or corrosion of the hinge pin.

Service Cabinet: Clean, repaint as required, and check weatherproofing, grounding, and venting.

Wiring Circuits: Test circuits and tighten terminal screws. Check splices for weatherproofing and repair as required for proper test readings.

Photoelectric Controls: Clean each unit and burnish contacts. Check photo cell assembly and replace or tighten any components that appear to be loose. Photo cell switches should be calibrated to turn on at approximately 1.5 foot-candles of illumination and turn off at approximately 2.5-foot candles. Original conductivity of each circuit should be restored, which may require replacing faulty or damaged insulation.

Circuit Breakers: Check circuit breakers, surge protectors, ground fault interrupters (except fuses), and similar devices by turning them off and on to ensure that they are working properly.

Landscape Irrigation and Pump Controllers: Electrical controls for this type of equipment need the same preventive maintenance service as any other electrical hardware.

Controllers for Reversible Lane Gates and Barriers: Controllers and detection units used to regulate access to and from special-use lanes must be visually checked each time they are activated (once or twice per day depending upon traffic peaking conditions) for damage (vandalism or vehicle damage) and for proper control response operation.

Electrical service failures produce traffic control system failures; thus, maintenance service personnel must respond as quickly as possible and restore traffic control operation in the interest of safe, efficient traffic flow. Personnel conducting electrical service and electrical hardware repairs should be familiar with the National Electrical Code and applicable state or local codes, standards, or specifications, ensuring proper quality of repair and service work.

Preventive Maintenance Programs for Electronic Systems: If an electronic system still using gaseous tube equipment exists in an agency's inventory of traffic control equipment, it should be programmed for replacement with a solid-state digital unit as soon as possible to minimize maintenance difficulties. Gaseous tube equipment is subject to drift from calibrated settings, and obtaining repair parts requires a special order with limited availability of parts.

Solid-state digital control equipment has much longer preventive maintenance check cycles and is constructed with high-durability components. In either case, all maintenance should be performed by qualified technicians in accordance with manufacturer's recommendations.

Traffic signal installation on arterial highway

Traffic Signals: "A highway traffic signal is any power-operated traffic control device, other than a barricade warning light or steady burning electric lamp, by which traffic is warned or directed to take some specific action" [87]. Hardware installations included within traffic signals are as follows:
- Traffic signals controlling traffic flow at intersections.
- Railroad-highway grade crossing signals.
- Freeway ramp metering signals.
- Emergency traffic signals, including temporary signals in work zones.
- School and pedestrian crossing signals.
- Flashing beacons.
- Reversible lane control signals.
- Signals at toll collection facilities.
- Traffic signals for control of movable bridges.
- Hazard identification beacons.

Traffic signal failures disrupt safe traffic movement; therefore, maintenance personnel must respond to signal failures as soon as it is practical to do so. Damaged signal heads or unlighted lenses need to be given the highest priority for repair. Traffic signal heads that are not operating should not be left darkened but should be set in flashing mode or covered. If the flashing mode of operation is not possible, the intersection should be controlled by an emergency sign. Properly qualified technicians should complete the repairs to the standard of the National Electrical Code or better.

A concise description of the range of traffic signal control hardware and equipment is available in Yauch [92]. The description includes traffic signal controller assemblies, detectors, interconnected signal systems, and signal display equipment.

The Institute of Transportation Engineers has published a thorough manual that guides an agency in developing a traffic signal maintenance program [93]. It includes a discussion of common maintenance problems, low-maintenance signal designs (something all maintenance divisions should encourage traffic engineering design groups to achieve), inspection programs, preventive maintenance programs, response maintenance activities, administration of traffic signal maintenance, and risk management in the maintenance of traffic signals.

Chapter 2 Roadway Maintenance and Management

2.1.11 Roadway, Tunnel, and Bridge Illumination

Street and highway lights improve driver nighttime visibility and promote safer, more efficient roadways. Rural intersections may be illuminated with a single luminaire that highlights the intersection location and calls the driver's attention to it. Urban streets may be illuminated with nearly continuous lighting to improve drivers' and pedestrians' advance awareness of each other and to deter street crime. Interchanges and ramp areas are frequently lighted with high-mast illumination to create large-area, low-level illumination that provides drivers with a panoramic perspective of the roadway environment similar to daylight conditions and increases traffic flow efficiency during night hours beyond that expected without illumination. In roadway tunnels, lighting is designed to minimize any difficulties a driver may have in making the transition from a very dark roadway environment to a lighter environment. Thus, while failures in street and roadway illumination may not be catastrophic, burned-out luminaires or lighting not working as intended in the original design reduces the safety and efficiency benefit intended. Illumination should receive a service and repair priority in accordance with the importance of its intended benefit. Because the power consumption associated with street and highway illumination may be a significant budget item, some agencies contract illumination maintenance as a part of the payment for power to operate illumination. If illumination maintenance is contracted out of the agency, then the agency maintenance division should still be sufficiently familiar with the maintenance requirements for street and highway illumination to conduct inspections that verify the quality of the contracted maintenance.

Roadway Illumination: For maintenance purposes, street and highway illumination can be divided into categories according to the type of lamp used.

Gas Discharge: Fluorescent, low-pressure sodium, mercury, metal-halide, and high-pressure sodium lamps are types of gas-discharge lamps. Light is produced by the excitation of gas molecules or metal vapors in the lamp's arc tube.

Incandescent: Light in incandescent lamps is produced when an electrical current heats a filament and causes it to glow. Since these are less efficient and generally require more frequent lamp replacement than gas-discharge lamps, they tend to be found only in areas with old illumination systems where the benefits of changing to gas-discharge technology have not proven to outweigh the annualized capital cost of a changeover.

Other: An example of the final category is a gas flame lamp. Generally these lamps are only found in historic districts where architectural considerations require a form of illumination no longer in common use.

Special Maintenance Considerations: Luminaires have the following features that require special maintenance considerations:
- Poles or standards upon which the lighting equipment is mounted.
- Mast arms and hangers that attach the lamp units to the poles or standards.
- Lamp housing or fixture.
- Wiring.
- Service control.
- Ballast to control voltage, wave form, and current for proper operation of the lamp.
- Lamps.
- Leveling indicator.
- Starting devices, such as photoelectric cells.

General Considerations: Personnel conducting maintenance on lighting equipment should be properly trained in the requirements and procedures to be applied according to the manufacturer's recommendations. High-mast lighting and linear lighting patterns with high mounting heights should be relamped when 50% or more of the lamps in a grouping or area of concentration are not working. If a burned-out individual lamp significantly alters the light distribution at the roadway surface, lamps should be replaced as they burn out in addition to any program of large-scale replacement. Typical cycles for group replacement of lamps and for checking of all components of the lighting installation is every 2 years for fluorescent lamp lighting and halide lamp lighting, about every 5 years for sodium vapor lamp lighting and mercury vapor lamp lighting, and at about 90% of rated lamp life for other types of lamps.

Luminaires with mercury vapor lamps are typically cleaned when a lamp is replaced or at about the rated midlife of the lamp. Luminaires that do not have mercury vapor lamps are typically cleaned when they are relamped. Environmental factors can require adjusting the cleaning cycle to maintain proper illumination levels.

- More frequent cleaning may be required in tunnels or other roadway situations where vehicles can readily splash water on the light fixture.
- Sealed luminaire units can extend the cleaning cycle time.
- Unsealed luminaire units may require more frequent cleaning in areas with concentrated insect populations.
- Areas with high levels of particulate air pollution (dust storms, industrial contaminants, etc.) can shorten the cleaning cycle required to maintain lumen output.

Photoelectric cells controlling lighting should be checked for proper on and off operation about every 6 months unless the manufacturer's quality assurance program provides a longer inspection cycle. Lighting circuits underground and in conduits should be subjected to a megaohm test once per year.

Factors to be considered in developing an adequate lighting maintenance program are as follows:

- Level of reduced light output caused by accumulated dirt and dust on lamps, reflectors, and glass elements.
- Level of reduced lumen output due to normal aging of the lamp components.
- Lamp outages caused by lamp or circuit component failure, accidental breakage, and vandalism.
- Voltage drop in the circuit or reduced current at the lamp owing to abnormal line losses or transformer loading.
- Shading or masking of the illumination pattern on the roadway as a result of vegetation and foliage being allowed to grow into the zone of maximum light.
- Alignment and leveling of the luminaire.

Special consideration needs to be given to safety during the maintenance of lighting systems. Normal work safety procedures must be observed by the operating personnel. Special attention needs to be given to work zone traffic control because of the exposure of personnel while maintaining lighting systems, especially on bridges and in tunnels. Other special precautions are needed because of unique potential hazards in maintaining lighting systems.

- When servicing any lighting system and especially high-pressure sodium lamps, the electrical circuit should be turned off and locked to avoid the possibility of lamps switching on. Starting voltages of some lamps can exceed 4,000 volts.
- A disposal program should exist for old mercury vapor lamps in accordance with state and federal environmental guidelines.

Maintaining roadway illumination

Cleaning: Accumulated dirt on both the outside and the inside of luminaires absorbs the light generated by the lamp and reduces effective illumination on the roadway. Smoke and fumes in the area can increase the need for cleaning. Dirt on the glass refractor surface can create abnormal heating patterns and cause cracking of the glass. Reflecting surfaces can become stained from the prolonged presence of air pollutants or "bug juice," reducing their effectiveness. Allowing dirt and contaminant material to remain on glass surfaces for extended periods can cause a baked film (because of the lamp heat) that is difficult to remove by ordinary cleaning. In this situation, the luminaire may have to be prematurely replaced. Fluorescent lighting used in tunnels and bridge-rail-mounted illumination will typically have a large translucent cover over the light tubes that needs to be maintained with an airtight seal, keeping dirt away from the tubes. Low-mounted lights may require frequent washing of their cover surfaces to remove vehicle wheel splash and wheel spray dirt. Cleaning of all glass and metal surfaces should always be done with cleansers that do not have any abrasive materials; the luminaire is a precisely designed optical system, and even fine scratches can reduce its illumination effectiveness. Use only soft brushes or sponges to remove crusted material; no acid or alkaline cleaning agents should be used on luminaire surfaces fabricated from aluminum.

Lamp Replacement: An efficient, effective lamp replacement program is important to maintain desired illumination levels on a roadway. Lacking a planned replacement program, traffic accident potential may increase, street crime may be encouraged in high-risk urban areas, public perception of the transportation agency's quality of maintenance may decline, and inefficient special trips may be required to replace single lamps at critical locations.

Spot Replacement: Spot replacement is the policy and practice of only replacing a lamp after it has burned out. Such a practice makes obvious which lamps to replace but does not consider the decline in lamp effectiveness as the lamp weakens. Lamps that have declined in their lumen output from the designed illumination level do not provide the intended benefit to vehicular and pedestrian traffic. Since there is a time lag between a lamp burning out and its replacement, a preferred policy and practice is group replacement.

Group Replacement: Group replacement programs replace all of the lamps in a given area or on a particular circuit at one time and repeat the replacement process at a regular interval (usually at a specified percent of the rated lamp life). Group replacement has three main advantages over spot replacement:

(1) It greatly reduces the number of lamp failures between replacements.
(2) It maintains the illumination level more nearly at the intended design level when it is combined with a regular cleaning program.
(3) An added personnel safety advantage is that crews can work in daylight hours when the circuits are not energized. It also reduces the unit labor cost per lamp replaced.

Mechanical Inspection: Periodic inspection of luminaires is important to maintaining effective illumination systems. Typically, an inspection is conducted when lamps are replaced, especially if group replacement is practiced. If the persons who replace the lamps are not qualified or equipped to make repairs on the luminaire, they should be trained how to properly inspect and report the findings so that any needed repair can be scheduled.

- Inspect gaskets to see if they are loose, worn, or missing; schedule repair or replacement as necessary. Poor-quality gaskets will decrease luminaire effectiveness and lead to increased maintenance costs.
- Transparent enclosures should be inspected for cracked or broken glassware. Determine the cause of the damage (thermal shock or impact), if possible, and replace if necessary. A thermal shock break is caused by sudden chilling of the glass, primarily from sudden cold or driving rainstorms, when dirt accumulation has caused the luminaire to become excessively heated. Thermal shock breaks usually show a clean, single break with no obvious origin and no chipping. Impact breaks resulting from vandalism or vehicle-generated projectiles will appear to start from a chipped or bruised spot.
- Hardware items that must be removed during routine maintenance should be refastened with appropriate thread lubricant. Check the tension of all spring latches and ensure that all set screws are secure.

- Insulators should be inspected for cracks or breakage. Broken or cracked insulators should be scheduled for replacement or for corrective action and repair.
- Mounting mechanisms should be inspected to ensure that the luminaire is held rigidly in its proper position. Adjustments should be made as needed.
- Sockets and receptacles should be inspected for burned parts carrying electrical current and for broken insulation. Damaged parts should be scheduled for replacement as necessary. Arcing from loose sockets or improperly seated connections will significantly reduce lamp life and increase overall maintenance costs.
- Wiring should be inspected for abrasions that could develop shorts or grounds, and repairs should be scheduled as needed.

Circuits and Controls: Inspection for maintenance needs is necessary to ensure that the lighting system operates as intended.
- Constant current transformers tend to drift to increase the amount of current delivered to a series circuit; therefore, they should be inspected annually. They should be cleaned, adjusted, and regulated to correct any deviation from the design current value. The mechanical operation should be checked. In oil-insulated units, check for moisture or sludge development and replace the oil, if necessary.
- Oil switches should be inspected; check the electrical leads for abrasions, the connections to ensure that they are tight, and the bushing for cracks. Clean the bushings. Inspect the gaskets and replace any in poor condition. Check the oil level and the oil's dielectric strength; filter the oil, if needed; or replace the oil, if necessary.
- Relays should be inspected to identify any excessively burned contacts or badly pitted contact faces that may require the relay circuit to be disconnected and resurfaced with a burnishing tool. The relay magnet should be inspected for quiet operation because magnet noise indicates that the magnet may need to be cleaned or have some corrosion removed.
- Photoelectric controllers generally do not require much maintenance other than cleaning of the cover (the window through which ambient light level is sensed) and recalibration to the intended light level for on and off operation. Replace any controller that has failed in the on position. If a timing clock is part of the controller, annually inspect the mechanical parts for proper condition and operation.
- Series film cutouts can cause excessive trouble and expense if they are not properly selected, installed, and maintained. It is important that properly sized cutouts are used. Visual inspection of new film cutouts before installation can eliminate defective units. Such inspection should include checking for side parallel, broken insulation, moisture absorption, crimping, and centering of disc.

Poles and Brackets: Steel poles and brackets that are not galvanized should be repainted as frequently as local environmental conditions dictate. All hand-hole and access covers should be checked to ensure that they are secured. Generally, poles fabricated from aluminum, galvanized steel, and wood require very little maintenance. However, maintenance patrols should always be on the alert for poles that have been damaged by vehicle impacts or vandalism.

Tunnel Illumination: Because of sudden visibility contrasts when drivers enter and exit tunnels, lighting in tunnels is a special maintenance concern. During daylight hours, the contrast creates the greatest visibility difficulty for drivers. Therefore, good tunnel lighting is more critical to maintaining safe, efficient traffic flow during the day than during the night. Burned-out lamps in tunnels should be replaced immediately, and group replacement should be scheduled before 90% of the rated lamp life has elapsed. Rated life of tunnel lighting is usually based on the lamp's continuous burn use. Lighting at the entrance and exit needs to be maintained at the same level as the interior lighting because the illumination design within the tunnel is based on the expected level of illumination at the tunnel's entrance and exit.

Maintenance engineers and managers should be involved in the review of the design of any roadway tunnel facilities, ensuring the structure's maintainability. It is especially critical that they provide feedback on the tunnel's lighting design. A relatively new tunnel in Phoenix has three rows of lamps mounted in the ceiling of the tunnel as opposed to the more typical sidewall-mounted luminaire [94]. The type of lamp selected in the design was strongly influenced by a desire to have long-life lamps, minimizing the need to replace them. However, it is not clear that the operational impact of changing lamps in the tunnel (traffic safety, maintenance personnel safety, or traffic delays) was a significant factor in the tunnel's design analysis.

Bridge Lighting: Bridge lighting generally illuminates the roadway carried on the bridge superstructure. However, as aesthetic considerations become a significant factor in urban highway development, architectural lighting of bridges has developed as a design concept. Miami has a striking example of an architecturally lighted bridge [95]. While roadway and tunnel illumination should have a high priority in maintenance response, care should be taken to avoid including architectural lighting in any high-priority response procedure unless it also has a dual safety purpose.

Maintaining bridge lighting frequently requires dealing with problems not often encountered in maintaining other roadway illumination systems [96]. Vandalism damage seems to be more frequent on bridges. Damage to lighting components from boats and vehicular traffic occurs on large bridges. Bridge vibration can degrade lamp life; one agency tried using lamps designed for carnival lighting because they must withstand frequent vibration loads. In coastal areas, heat and corrosion damage to power cables and lines is a common problem for which the use of solar power as a backup, especially for navigational lights on large bridges, has shown promise, reducing maintenance expense. (See chapter 3 for additional details.)

Other Considerations: Some factors to consider in developing a maintenance program for illumination systems are as follows:

Vandalism: Since vandalism is a perpetual maintenance problem in all areas, document any type of vandalism damage that seems to be repetitive and seek out repair or replacement materials that are resistant to that type of vandalism. If the luminaires are being damaged, changing the poles to a higher mounting height may eliminate the problem. However, mounting height should never be changed until qualified design engineers have evaluated the impact of any proposed mounting height change on the illumination level of the roadway.

Trees: Low-hanging branches on trees frequently obstruct the intended spread of light on urban streets and highways. In some urban areas, if trees near light sources are not kept pruned, the trees will completely grow around the light pole and the streetlight will provide little positive illumination.

Inspection Checklist: A maintenance checklist helps ensure that inspection and maintenance of illumination systems is comprehensive and complete. This is especially helpful for personnel doing routine cleaning and relamping. Separate areas of a checklist for optical, electrical, and mechanical components are helpful. Notes detailing the condition and need for repair or replacement will help agency personnel provide proper maintenance. This checklist can be used to monitor the quality of contract maintenance for illumination systems and to defend against any litigation claiming that an accident was caused by deficient maintenance of illumination systems.

2.1.12 Roadways on Bridges and in Tunnels

Removing flood debris from a bridge

Roadways on Bridges: Maintenance of bridges is covered in detail in the chapter on bridge maintenance and management. However, it is appropriate to outline here some roadway maintenance considerations pertinent to the roadway carried over bridges. Actual delineation of the responsibility to maintain the roadway carried over a bridge varies from one agency to another. Regardless of which agency unit is responsible for the actual maintenance work, the following points should be considered in developing a program to maintain roadways across bridges:

Inspection: The roadway approaching a bridge and the roadway across a bridge should be regularly inspected as part of any roadway inspection program. Personnel conducting routine inspections of the roadway for maintenance needs should not halt their roadway inspection at the bridge and then resume it once the bridge is crossed.

Condition of Approaches: Inspect approaches including paving, slopes, and drainage. Give special attention to the structure's pavement areas. Make repairs as soon as possible where the sagging approaches or deficiencies such as potholing, scaling, or spalling are noted. If the pavement is eroding or sinking, determine the cause and make corrections. All repair efforts should be coordinated between personnel responsible for roadway maintenance and personnel responsible for bridge maintenance, ensuring proper response.

Condition of Any Asphalt Wearing Surface: Asphalt wearing surfaces on bridges should be kept at the established level of service intended by agency policy. Any condition indicating a surface quality less than that intended by agency policy should be scheduled for treatment or repair.

Condition of Concrete Roadway Deck: Scaled, cracked, and spalled areas that expose reinforcing steel should be scheduled for repair as necessary. Concrete decks should be repaired with concrete if at all possible. Bituminous patches should be considered temporary and used where situations dictate. Bonding compounds and curing accelerators should be only those approved by the agency. Decks should be treated or maintained for protection against chlorides. All repair efforts should be coordinated between personnel responsible for roadway maintenance and personnel responsible for bridge maintenance, ensuring proper response.

Condition of Steel Grid Deck Roadways: Areas where welds are broken should be rewelded. Broken welds or clips should be repaired as soon as noticeable rattling or deck movement from live loads is noticed. Traffic should be observed and accident rates on slippery grid decks should be studied to determine if corrective action to a traffic or accident problem is necessary. If a traffic signal is added or traffic conditions change at a bridge with grid decks, special attention should be paid to skidding problems and corrective action should be initiated if an engineering study suggests it is needed.

Condition of Concrete Curbs and Rails: Curbs or rails where reinforcing steel is exposed should be scheduled for repair as necessary. Curbs or rails where spalling is deep should be repaired with approved materials only. Curbs and rails showing checking, small cracks, and very minor spalling should be treated with an approved sealer. If curbs or rails have damage that presents a potential hazard to pedestrian traffic, if the bridge contains a sidewalk, or if there is a possibility of vehicular traffic hitting the curb or rail, then repair or treatment should be scheduled as soon as possible. Otherwise, repairs and treatment can be coordinated with other bridge maintenance activities.

Condition of Paint on Metal Rails: Rust or pitted areas on steel rails should be brush cleaned and spot-painted. Sections of steel rails on which rust pitting appears to be extensive in depth should be replaced. Treatment and repair should be coordinated with any other bridge maintenance activity.

Condition of Expansion Devices: Expansion joints and devices should be checked to ensure that they are working. Steel expansion plates should be kept clean and free of accumulated debris and dirt. Steel expansion devices that are loose and banging, jammed shut, or showing signs of failure should be repaired as soon as possible. Before repairs are made, an engineer knowledgeable about the bridge's design and construction should be consulted to determine what type of supports or anchorage the expansion device has and if there are any special features to observe during concrete drilling or removal during repair. Joints sealed with elastomeric sealants should be checked for watertightness and failures. Repairs should be made to damaged sections following the recommendations of an engineer knowledgeable about the bridge's design and construction. It is important that the repaired joint configuration be correct to ensure that it will function properly.

Preformed compression seals may fail because of one or more of the following reasons: the joint was not designed and constructed to the correct dimension for the seal or the seal was the wrong width; the wrong shape of seal was used; the seal was installed incorrectly; the seal material has age hardened; incompressible material has entered the joint; or the joint has not been properly cleaned. The cause of failure in any compression joint needs to be determined and the recommendation of an engineer knowledgeable in the bridge's design and construction obtained before any repair is made.

Condition of Waterway: While a detailed inspection for waterway scour will be associated with a bridge inspection, roadway maintenance personnel should be alert for any new erosion. This condition might need to be checked in more detail. Roadway maintenance personnel should be especially alert for conditions that might result in logjams or ice jams at bridges, endangering the bridge's stability. Logjams and other debris piled up against piers, bulkheads, or pilings should be promptly removed. Bridges with piers or bulkheads experiencing floating debris problems should be checked during and after each flood condition. It may be necessary to remove debris lodged against piers or bridge superstructures on an emergency basis to protect the bridge from damage. Bridges over streams with ice flows should be watched for ice jams because the jams may have to be removed on an emergency basis to protect the bridge.

General Conditions: Each structure should be inspected regularly for dirt and debris accumulations on the roadway carried across the bridge, ensuring that the drainage openings in the deck are providing proper roadway drainage and that debris has not accumulated under the bridge, creating a fire hazard. Before removing any debris from under a bridge, check with the agency's coordinator for wildlife conservation activities to ensure that the debris is not part of a wildlife migration path effort. Accumulated dirt, grit, and debris should be removed from the roadway across the bridge by washing, brooming, or some other equally effective means, usually in the spring.

Movable Bridges: Special note needs to be taken of some unique maintenance characteristics of roadways across movable bridges. Common types of movable bridges include swing spans, bascule bridges, and vertical lift bridges. Important aspects of movable bridges that need to be inspected to ensure continuity of roadway maintenance across the bridge are as follows:

- Check span position alignments to ensure that the roadway surface is suitable for all forms of traffic expected when the bridge deck is positioned for traffic use and that the roadway surface is as serviceable to traffic as a stationary bridge deck roadway would be.
- Check any traffic signal controls that regulate the traffic using the bridge to ensure that they are fully operational and are properly timed for safe traffic operations.

Special Maintenance Inspections: Certain situations arise that create special concerns about bridge serviceability. While roadway maintenance personnel may not be involved in these special inspections if a separate group handles bridge maintenance, they should be aware of these situations and why they are special. If knowledgeable, roadway maintenance personnel can assist in the alert for special maintenance needs that may result.

- Excessively high water or a rapid flow of water may affect bridge piers, particularly if ice or debris can collect against the piers.
- Severe storms, tornadoes, and hurricanes may cause bridge damage, including bottom scour around the substructure.
- When roadway or marine traffic, including trucks or trailers that exceed the vertical clearance, collide with a bridge or its piers, damaging fascia and steel floor beams, the bridge must be evaluated, determining if the roadway can safely carry traffic across the bridge.
- Following an earthquake of sufficient magnitude to cause bridge damage, the bridges in that area need to be evaluated, determining if their roadways can safely carry traffic across the bridge.

Marine Navigation Lights: The lights on bridges having marine navigation lights, aircraft warning lights, or both, should be frequently inspected for burnout. Double lamping navigation lights with an indicator when one of the lamps burns out is desirable. If a bridge spans navigable waters, the maintenance supervisor responsible for the bridge must be familiar with all applicable U. S. Coast Guard requirements for the maintenance of navigation lights, ensuring that the maintenance practices are in compliance.

Roadways in Tunnels: Motor vehicle tunnels are a highly restrictive roadway environment in which to conduct maintenance. As tunnel length and the traffic using the tunnel increases, so does the number of safety features and the amount of equipment requiring maintenance. Tunnels that operate as high-volume critical links in a metropolitan area or provide a critical surface transportation link across a regional barrier require backup personnel and backup equipment systems to ensure that tunnel ventilation is not disrupted; that drainage pumps are always available, as needed; that emergency power is available in the event of general power failure; that traffic control and traffic surveillance equipment is active; and that smoke detectors, fire detectors, emergency alarms, emergency telephones, and emergency aid equipment is always available. Plans must be made in advance for disasters such as a crash or fire in the tunnel or a hazardous material spill even if hazardous materials are not supposed to be transported through the tunnel. Regular inspections of the tunnel and roadway are usually the responsibility of the maintenance division of the owning agency. Inspections should include, but may not be limited to, the following items:

- Inspect tunnel walls and ceilings periodically for new cracks and water leaks. Water seepage is not unusual, but new sources of water or new cracks in the tunnel should be reported, triggering an engineering evaluation.
- If any timber supports exist, they should be inspected annually for any evidence of decay, insect damage, or other deterioration.
- Tunnel lighting is especially important to safe, efficient traffic operation. Daily note should be made of burned-out lamps, and a group replacement program for all lamps should be in place, minimizing lamp burnout.
- Ceramic-tile-lined tunnels use tile reflectivity to increase lighting visibility. Therefore, tunnels lined with ceramic tile should be washed and rinsed at least once per year. However, high-traffic-volume tunnels may need more frequent cleaning. Damaged ceramic tiles should be replaced annually, unless an area of extensive tile damage has developed. In that case, the entire area should be repaired as soon as practicable.
- Ventilation systems should be inspected monthly for proper operation. Inspect air shafts for water or ice formation. Airflow meters should be used to determine if the equipment is performing as designed.
- All traffic control and safety devices associated with the operation of a vehicular tunnel should be high-priority items for maintenance service.
- Inspect the areas around a tunnel portal for loose rock or debris and ice-packed or heavy snow deposits. Any similar surface or geological condition that might contribute to blocking a tunnel portal needs to be removed as soon as practicable. Water dripping from the roof of a tunnel portal needs to be diverted away from the roadway surface, preventing the formation of any slick spots at the entrance or exit. Tunnel portal paint and marking should be kept as clean and clear as practicable, aiding good driver communication.
- Monitor any increase in truck use of a vehicular tunnel and the possible increase in damage resulting from the greater number of heavy loads.

2.1.13 Environmental Aspects of Roadway Maintenance

Current Assessment of Issues: A Transportation Research Board conference report on environmental issues in transportation identified 13 categories of critical concern [97]. While the conference report did not contain a proposed environmental research project directed toward maintenance in each category, it is appropriate to examine each category of environmental concern with respect to the performance of roadway maintenance. Environmental issue categories identified in the report include the following:

Aesthetics and Visual Quality: The general public has a quality-of-life interest in the appearance of the street and highway landscape. The conference group identified four possible research projects, but the focus of the problem statements was directed at the planning and design process. The planning and design of streets and highways has major aesthetic and visual quality impacts; however, maintenance has a much longer term effect on the visual environment. Wildflower fields in the right of way, areas of native prairie grasses, variety in tree plantings native to an area, roadside wetland areas providing bird nesting, well-landscaped and neatly groomed rest areas, relatively litter-free rights of way, and a host of other regular and routine maintenance practices contribute immensely to maintaining and improving the visual quality of the travel experience. Even if these maintenance efforts are not deemed worthy of research for expansion and improvement, it is in the agency's and especially the maintenance division's best interest to use public relations opportunities to inform and enlighten the public about maintenance activities that have a positive impact on the visual and aesthetic environment.

Air Quality: Air quality issues are primarily associated with urban areas. Consequently, most of the report's proposed research projects are directed toward seeking ways of improving individual classes of vehicle performance to reduce internal combustion pollution. If funded and successfully completed, one project titled Heavy Duty Vehicle Emissions and Activity Levels might profoundly affect maintenance by generating new truck operation guidelines that require significant changes in the way trucks are used in maintenance activity. Another proposed project titled Particulate Matter Source Apportionment and Control Strategy Synthesis suggests that de-icing sand and salt mixtures be examined as a source of particulate matter in the mix of air pollutants. In any case, maintenance organizations must ensure that research results on the effects of reduced salting and sanding such as Kallberg [68] be included in the analysis of any project such as the one proposed. A total system analysis must be applied to maintenance practices if the true environmental cost and benefit are to be assessed. Furthermore, while a maintenance agency is willing to share the strides it has made to improve air quality through the reduction of volatile organic compounds in traffic markings [98] with the rest of the maintenance community, this same type of information needs to be broadly disseminated to the general public through media sources.

Cultural Resources: Cultural resources are usually associated with antiquities and efforts to preserve historic buildings. The proposed projects identified in the report are directed toward facility planning and design. While maintenance is not typically associated with preserving or impacting cultural resources, it may be advantageous for maintenance engineers and maintenance managers to be aware of any maintenance activity that supports or reinforces the preservation of cultural resources. For example, if access routes to a cultural festival are given any special maintenance effort in support of a particular festival, that activity might be a source of positive public relations for the maintenance agency. Maintenance of access routes to recreational and historic areas might be promoted as a source of positive public relations at the beginning of the tourist season. Granite block paving of streets in urban areas, or any similar historic architectural treatment that requires special maintenance, might become the basis of a public relations effort that communicates that the maintenance agency supports historic preservation of cultural resources. While maintenance is not a major player in preserving cultural resources, because strong public sympathy often exists for such activities, a maintenance agency should consider what image and status benefit might be derived from association with preservation of cultural resources.

Energy Conservation, Alternative Fuels, and Climate Change: Global warming and fuel consumption are the primary focus of the projects proposed in this section of the report. While it is appropriate for maintenance agencies to be interested in alternative fuels, such fuels for maintenance processes should be evaluated from an economic basis and as alternatives in times of shortage, rather than from a global-warming impact perspective, because the scientific community cannot agree about the validity and extent to which any global-warming effect is caused by human activity. Energy conservation efforts and alternative fuel investigations with respect to maintenance practices and processes need to focus on universally accepted criteria.

Environmental Review Process: Either directly or indirectly, all maintenance activity is subject to environmental review; consequently, maintenance agencies have an interest and a stake in changes that might develop in the environmental regulatory process. However, the projects proposed under this topic in the report are mostly directed at industrial processes related to transportation development.

Hazardous Materials Transportation: Since maintenance forces are usually involved in some aspect of the cleanup of any hazardous materials spill or incident that occurs on a street or highway right of way, this issue has direct bearing on maintenance divisions. One proposed project titled Integration of Design, Operation and Maintenance of Hazardous Materials Transportation Facilities and Equipment, if successfully completed, might yield some revised maintenance management strategies that more effectively deal with hazardous material spills on streets and highways.

Hazardous Waste: Many of the materials used in street and highway maintenance are hazardous to some degree at some time in their materials life. Increasing the general knowledge of the degree to which a material in the waste stage is hazardous would be useful to maintenance agencies. Several projects proposed in the report have a direct linkage to the interests of the street and highway community. Among those projects are Evaluating Constituent Leachability, Migration and Fate Issues Associated With Incorporation of Contaminated or Reusable Materials in Highway Construction (although directed to construction, any findings will have some applicability to maintenance repair processes); Minimizing Transportation Agencies' Liability Associated With Use of Contaminated Property (although directed toward planning, design, and construction of new projects, the findings might be useful for evaluating the use of maintenance yards known to have site contamination); Interactive Right-to-Know, Health and Safety and Waste Management Awareness Training for Transportation Agency Employees (since maintenance division employees are frequently exposed to hazardous materials, the results of this project could make training more efficient and effective); and Evaluate the Use of Universal Wastes and Other Transportation-Generated Wastes as a Replacement for Virgin Materials in Highway Construction (although directed to the construction process, the results would apply to maintenance repair and rehabilitation processes). If any of these projects are initiated, the maintenance community should monitor the progress and the results for any technology transfer deemed to be advantageous to the maintenance engineering and management process.

Noise: Transportation noise is an emotional, highly sensitive issue with the general public. An extensive list of proposed projects are given in the report to address what was considered to be a variety of current issues. Two of the proposed projects might have an indirect impact on maintenance activity. A project titled Investigation/Validation of Testing Procedures for Sound-Absorbing Barrier Materials might, if successfully completed, result in guidelines regarding the effectiveness of various highway noise barrier materials. As the highway noise barriers now in place continue to deteriorate, the repair process might shift to a replacement process in which materials rated as more effective at reducing highway noise are used. A project titled Investigation of Sound Propagation Over Irregular Terrain might, if successfully completed, result in guidelines for regrading foreslopes and backslopes, reducing noise propagation in suburban growth areas where sensitivity to noise has increased since the original highway design and construction took place.

Operations and Maintenance: The conference participants recognized the strong relationship between maintenance and environmental issues by reporting on five proposed projects directly associated with maintenance and operations. Proposed project titles included Analysis and Improvement of BMP Methodology, Environmental and Safety Impacts of Snow and Ice Control, Environmentally Sensitive Design for Highway Facilities, Roadside Vegetation Management: Ecological/Economic Solutions, and Environmental Protocol Development for Product and Waste. Should any of these proposed projects be successfully completed, the maintenance community could benefit from the results. These topics also indicate the range of issues for which there is direct environmental concern in roadway maintenance.

Social and Economic Impacts: While the performance of roadway maintenance (or conversely, the lack of quality maintenance being performed) has a substantial impact on the social and economic value of travel, the projects proposed in this area were oriented toward transportation policies in ways that were largely tangential to maintenance.

Water Quality and Hydrology: Seven projects were proposed in this environmental category; four have some relationship to maintenance environmental issues. A proposed project titled Method to Assess Effects of Highway Runoff on Aquatic Life and Receiving Waters would help the maintenance community evaluate criticisms related to chemical use. However, the maintenance community already has studied this area [36, 65]. One proposed project titled Impacts of Air Emissions on Highway Runoff could benefit the maintenance community in the same fashion as the previously cited project but to a lesser degree. Two other proposed projects, if successfully completed, may potentially encourage modifications of roadway storm water controls during the process of maintaining storm water drainage.

Wetlands: Five projects were proposed in this environmental issue category. Because the process of maintaining roadside drainage is related to any adjacent wetlands activity, all of these proposed projects have a potential bearing on roadway maintenance.

Wildlife and Ecosystems: Five projects were proposed in this environmental issue category. While the major thrust of all of these projects is the planning and design process, maintaining the roadside can have a significant impact on wildlife habitat in some regions. Thus, the maintenance community has an interest in monitoring the results of these projects, if any are successfully completed.

Maintenance Actions: The maintenance community is a very proactive environmental influence through the performance of their activities. Many agencies are quite effective at keeping the public informed of their herbicide use; such agencies find that a knowledgeable public complains only when there are real problems requiring action [35]. Maintenance agencies actively seek better ways to store and handle materials that can potentially degrade the environment if not used carefully (e.g., salt for snow and ice control [62]). The maintenance community has long recognized that training personnel in the proper way to handle and apply environmentally hazardous materials is the right thing to do for both worker safety and environmental safety [36]. Maintenance agencies are constantly evaluating materials and processes to seek more environmentally sensitive approaches to their work [70, 72]. Maintenance engineers and maintenance managers have continued to improve roadside maintenance and reduce erosion, knowing that reducing erosion reduces maintenance costs and negative impacts on surface water quality [26, 38]. Many agencies have a long history of attempting to develop native vegetative beauty along the roadsides, recognizing that the public enjoys the enhanced visual quality and that maintenance costs are simultaneously lowered [30]. However, agencies that have undertaken the application of materials, methods, procedures, etc. that are environmentally sensitive need to conduct objective evaluations of these efforts and to share their experiences with the maintenance community. In the area of environmental issues, a little knowledge can be a dangerous thing, but fully shared knowledge almost always has positive benefits. In Synthesis of Highway Practice 196 [104], the maintenance community has developed a concise but comprehensive guide to the procedures that should be implemented when responding to hazardous materials. Synthesis of Highway Practice 196 covers the education and training of personnel, equipment needs, response procedures, containment, and cleanup processes. Maintenance engineers and maintenance managers who are not familiar with the range of activities that can be associated with participation in a hazardous material incident response should review the contents of this synthesis. Many agencies have had to deal with the petroleum contamination of maintenance yards, and Synthesis of Highway Practice 226 documents the various processes by which these agencies have remediated such sites [106].

Routine maintenance activities that produce environmentally sensitive materials include sweeping streets and cleaning catch basins. In many areas, the materials swept from the streets and the sludge removed from catch basins have to be sampled for heavy metals and oils that may require disposal as an environmentally controlled material (in a landfill or disposal site authorized to accept such materials). Frequent, regular street sweeping helps avoid such problems because the crankcase drippings, antifreeze drippings, etc. have less opportunity to reach a concentration level requiring special disposal. Likewise, frequent flushing of catch basins limits sludge development in the bottom of the basins, reducing the disposal problem.

2.1.14 Maintenance Work Zone Traffic Control

The work zone, especially when traffic is routed through it, is potentially a very hazardous area, either to a motorist or to a maintenance worker. A work area and a traffic environment that are as safe as reasonable application of engineering principles can produce necessitates that proper work zone traffic control be a high priority. Poorly placed, nonserviceable, leaning, fallen, or dirty warning signs are detrimental to safe, convenient highway travel. Work crews should ensure that they have proper warning devices on their trucks and equipment when they leave the garage or yard to begin their assigned activity. If the assigned maintenance activity is so complex that warning devices on trucks and equipment will not provide reasonable safety for the maintenance personnel and equipment or the traffic that will pass through the work zone, then the supervisor needs to ensure that the crew implements additional warning and traffic control measures including barricades, warning signs, flaggers, flashers, temporary regulatory signs, etc. All work zone traffic control procedures and devices should be consistent with the guidance of the *Manual on Uniform Traffic Control Devices* (MUTCD), Part VI, or the state-adopted equivalent of it. (Also see chapter 3.)

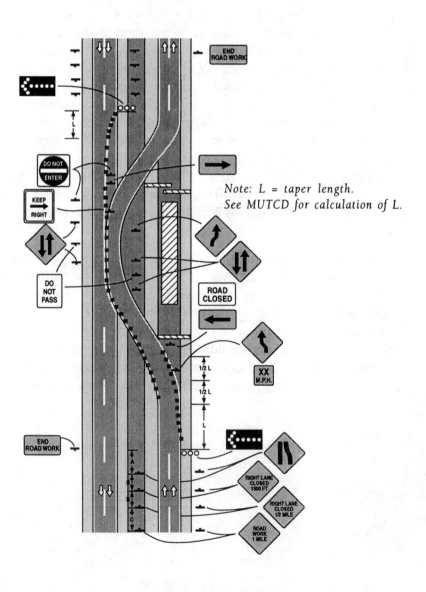

Note: L = taper length.
See MUTCD for calculation of L.

MUTCD Example Layout: Median Crossover on Freeway

Principles of Good Practice: Part VI of the MUTCD outlines the principles that guide good work zone traffic control [99]. Maintenance engineers and maintenance managers should be familiar with part VI of the MUTCD, and maintenance supervisors should have enough training in applying work zone traffic control that they can recognize when the principles in the manual or specifications should be followed to create a safer traffic control environment. Part VI of the MUTCD contains many typical illustrations of good practices in applying traffic control to work zones. However, these illustrations should always be taken as just that, and persons responsible for work zone traffic control should always use a traffic control scheme that will provide the safest traffic environment possible. Persons responsible for work zone traffic control should remember that an effective traffic control device must meet the following five basic requirements:

   (1) Fulfill a need.
   (2) Command attention.
   (3) Convey a clear, simple meaning.
   (4) Command respect from road users.
   (5) Give adequate time for proper response.

A traffic control plan and suggested traffic control device layouts are developed with these requirements in mind. However, if in the field a supervisor finds that something is not working as intended, then any changes that could improve work zone traffic control with the resouces immediately available should be considered. If the resources to make an improvement are not available, then the problem and any possible solution should be reported up the chain of command.

Work zone traffic control for urban roadway repairs

Work zone traffic control for rural roadway repairs

Chapter 2 Roadway Maintenance and Management

The five considerations for using traffic control devices in work zones to ensure that the above five basic requirements are met are as follows:

- Device *design* needs to be consistent with the guidance of the applicable manual. A traffic control device that is not consistent with the uniform examples and practices that drivers have been taught to expect in formal driver training or public information campaigns can add an element of confusion to what is already a confusing driving environment. Human factors studies of persons' response to signs indicate that familiarity with a sign shape, color, and message can increase the speed with which a person can remember the meaning. If a driver has to study a sign or message to figure out the meaning, then the sign is less helpful.

- The device should be *placed* so that it is within about 5 to 10 degrees of the driver's line of sight when he or she is looking straight ahead. This is the general range where a driver sees most clearly. In the work zone, it is not always possible to place a control device in that lateral location; thus, if the driver is going to have a high probability of truly seeing the device in a way that its intent is recognized, it may be necessary to increase the attention factor (e.g., by using flashers on low barricades).

- Device *operation or application* should meet the traffic requirements at a given location. Simply put, where the traffic control devices are located and how they are arranged should make sense with respect to what the driver needs to do to negotiate the work zone properly.

- *Device maintenance* should be of a high standard. It is not helpful to do everything else right but to not clean signs that are so dirty the message is difficult to read (especially at night), to not repair knocked-down vertical panels adjacent to a low shoulder, to not repair or replace a large advance warning sign when one-third of it is broken off by a wide load, etc.

- Traffic control devices in work zones should be *used uniformly*. It is desirable for traffic control device application in work zones to be uniform across all jurisdictions, but unique needs and circumstance create the need for some variability. However, it is important that within any given project and within a given jurisdiction, traffic control devices be applied as uniformly as practicable to aid driver response and traffic enforcement.

A recent study of truck drivers' perceptions of the adequacy of work zone traffic control provides an indication of just how important these principles are in actual practice [100]. Among 930 semitrailer truck drivers surveyed in Illinois, about 90% considered driving through a work zone to be significantly more dangerous than driving on the road outside of a work zone. However, most of the drivers thought work zones were clearly marked and were not confusing (about 14% disagreed), and about 20% thought some additional signs and markings would be helpful. The truck drivers agreed that they wanted advance warning signed far ahead of the actual work zone (about 50% want to be warned at least 5 to 8 km [3 to 5 miles] ahead). Most of the drivers surveyed thought arrow boards were too brightly illuminated. Overall, surveyed truck drivers appreciated the general uniformity and predictability of the traffic control. Understanding the work zone traffic control appears to be related to safety because almost none of the surveyed drivers had been involved in a work zone crash even though Illinois is reporting about 10,000 work zone crashes per year. Perhaps if roadway maintenance agencies could get auto drivers to understand work zone traffic control as well as this sample of truck drivers, a significant increase in maintenance personnel safety could be achieved.

Some Practice Notes: Warning signs should be erected on the appropriate side or sides of the work zone at a sufficient distance to warn motorists so they can easily bring their vehicles to a stop before coming to the first piece of equipment or the first maintenance worker on the roadway or in the work zone. Of course, this presumes the driver understands the signs and heeds the message. Recognition of drivers failing to respond properly has encouraged the use of attenuator devices on "shadow vehicles" for maintenance work zones. Details and guidance in proper placement and location of these signs is contained in MUTCD, part VI, [99] or in the appropriate state signing manual.

At the end of the working day or upon completion of the work, all temporary signs for worker protection must be removed or in some way changed to avoid presenting a misleading message to drivers. All signs warning traffic of hazards must be removed as soon as the hazard has been eliminated.

No state-owned vehicle should be used in work on roadway pavement unless proper warning signs have been placed on the right of way in advance warning traffic that the equipment is in the roadway. This is also applicable to the placing of flaggers, where such flaggers are necessary. These practices also should apply to the activities of any utility agency vehicles and any organization conducting contract maintenance for the street or highway agency.

Conditions arise that make it desirable or necessary to conduct maintenance operations at night (emergency repairs, daytime traffic demand is too great, daytime temperatures are adverse to material or process quality, etc.). Extra care must be taken in work zone traffic control under nighttime conditions, because many drivers will not expect night maintenance activity. When concern for both worker and motorist safety indicates that a work zone needs to be illuminated, the guidelines presented in reference [101] should be followed if the agency does not have its own policy or practice guidelines. All conditions should be taken into account when planning to conduct night maintenance operations. Extra precautions should be taken to protect both motorists and workers adequately. The work area should be outlined with properly illuminated and positioned traffic control devices.

When hazardous road conditions occur that endanger the traveling public, such as a fallen tree, washouts, or material on the highway, then advance signs and barricades with flasher lights should also be used to encompass the hazardous section. It is helpful if warning can be provided far enough in advance to encourage traffic to take alternate routes around these types of maintenance situations.

Flaggers are assigned the primary job of protecting their fellow workers from dangerous traffic conditions. Their secondary job is to assist in guiding traffic through the work zone. Because flaggers are exposed to the hazards presented by any errant vehicle, if an adequate level of work zone traffic control can be provided without using them, that type of work zone traffic control should be pursued. Because flaggers are such a prominent and important part of work zone traffic control, select and train persons with the following minimum qualifications to be flaggers:
- A sense of responsibility for the safety of the public and the workers.
- Training in safe traffic control practices.
- Average intelligence.
- Good physical condition, including sight and hearing.
- Mental alertness and the ability to react in an emergency.

- Courteous but firm manner.
- Neat appearance.

Flaggers should be attired in high-visibility clothing, preferably in fluorescent colors for daytime flagging and highly reflective materials at night. Reflective clothing must have reflective material on enough of the person's body that it will be obvious to all drivers that the material covers a person, not some inanimate object, and that hand motions and signals will be clear in the reflective pattern. Hand-signaling devices and procedures used by flaggers and the station or position taken up by the flagger in the work zone should all be as recommended by the MUTCD, part VI, or the applicable state traffic control manual. All flaggers need to be trained in the proper procedures and not just thrust into the position without any safety training, both for their own protection and the protection of the workers and motorists in the work zone.

Speed limits in work zones continue to be an issue. Setting a speed limit optimal for the safety of workers and for safe, efficient traffic flow through the work zone requires some flexibility. Maintenance engineers and maintenance managers involved in formulating an agency policy and a practice guide should review both the MUTCD [99] and the results of NCHRP Project 3-41 [102] to explore the range of alternatives that may be appropriate.

2.1.15 General Maintenance Worker Safety Notes

The agency maintenance division should have a well-documented safety policy and safety training program. All organizations contracting to perform maintenance activities for the agency should be required to demonstrate that they provide safety training for their own employees and that they have a well-documented safety policy and program meeting the same standards of care that the agency does. Safety training is just as important for long-time, experienced employees as it is for the new or inexperienced employee. Safe operations are the result of building safe work habits. Safe operations are also efficient operations. Nothing is as inefficient as downtime caused by a collision or personnel accident, in addition to the material and equipment that is often lost in such incidents. Safety training is related to training personnel to use the right tool for the job, the right equipment for the job, and wearing and using the right personal safety equipment. Maintenance supervisors are responsible for workers knowing the correct, safe procedures for conducting a maintenance activity and for ensuring that the proper tools and resources are available to that worker to do the job safely. However, it is always the individual worker's responsibility to act on that knowledge and those resources to be safe. Maintenance employees at all levels are responsible for being aware of and studying the agency's safety rules and procedures as they apply to their jobs and those under their supervision.

2.1.16 Developing Issues

Developing issues are issues or problems that are not presently being studied for alternatives or solutions, but these issues or problems are also not so far into the future that a person must be a technological forecaster to anticipate them.

Use of Indirect Waste Products: There will continue to be pressure on roadway maintenance activities and processes to absorb more of the ever-growing volume of waste products from our industrial society within the materials used in maintaining roadways. Some products, such as worn-out tires, are politically attractive since tire consumption is related to the amount of roadway use. However, the use of any nontraditional material in a maintenance process must be subjected to unbiased research and evaluation to estimate life-cycle costs and benefits before adopting a material substitution on a large scale. It is bad engineering, bad economics, and bad environmental policy to trade one problem for another of equal or greater magnitude. For example, shredded rubber in unbound granular bases has been proposed as a means of recycling some used tires, but research has shown that such a material substitution policy would result in lower pavement strength and not be as cost-effective as traditional dense-graded aggregate base courses [103]. It is not necessary to enumerate the many proposals of alternative materials flowing from the industrial waste stream, but only to emphasize that when considering the use of these materials in maintenance practices, it is important to

- search the transportation material literature databases (such as those maintained by the Transportation Research Board) for the results of research on the material in the maintenance application under consideration;
- seek technology transfer information about using the material in your type of process or a similar process (such as might reside in the information flow among the network of the various state technology transfer centers); and
- be willing to participate in an experimental application of the alternative material in a designed experiment so that, if successful, statistically valid results can be obtained.

Synthetic Materials With Designed Characteristics: Historically, maintenance divisions have been willing to experiment with industrial synthetics that approximate some naturally occurring material. The use of epoxy-based patching materials and polymer concretes for repairs is an example. Today, maintenance divisions are looking beyond synthetic materials to substitute for naturally occurring materials or for other traditional maintenance materials. For example, the growing awareness of some potential undesirable health effects from sand being pulverized under tires on dry pavement after it has been applied to the roadway for snow and ice control is creating interest in developing a synthetic sand that would have fewer adverse health effects in this application. Materials science and engineering is now recognized as a separate applied science and engineering field of technology distinct from physics and physical chemistry, from which it originated. To use the creativity of people engaged in this field to enhance roadway maintenance engineering and management, it is necessary to begin opening up the perspective of materials:

(1) Begin by enumerating what is wrong with any material in its application in a specific maintenance process.
(2) Add the characteristics that would be desired in a material for ideal application in the specified process.
(3) Define the geographical and economic extent of the potential to apply an ideal material in the specified process.
(4) Initiate a dialogue with professionals from the materials science and engineering profession to investigate the potential to "design" a material to meet the specific needs defined.

There is the same potential to advance the state of the art of maintenance materials as structural ceramics were advanced as a consequence of the U.S. space exploration program.

Equipment Systems Evolution to Fit Work Force: Historically, maintenance organizations have always been interested in newer, better equipment that is more productive and efficient. In that regard, an agency maintenance division is not so very different from a private contracting organization. Thus, maintenance equipment research is examining machine vision, robotic controls, computer control of processes (much like numerical control of manufacturing processes), real-time quality sensors with near instantaneous feedback, and similar applications of high technology. A collected body of people-oriented dimensions needs to be incorporated into the search for better equipment to conduct maintenance practices.

Computational Literacy: While much publicity has been developing about the increasing quality of mathematical skill and knowledge of persons graduating from high school, sadly most of this advancement is associated with college-bound young people. They are the high school students who take the ACT and SAT tests. The majority of young people, who include potential maintenance employees in the near future and those who are already employees, often lack the mathematical skills to reliably conduct routine calculations necessary for quality control analysis. These young people are not prepared to make any numerical interpretation of electronic output displays for quality control efforts. In the retail trade business, this fundamental weakness in the work force (clerical sales personnel could not reliably count and make change) was compensated for 10 years ago with computerized cash registers, electronic check readers, bar code scanners, and credit cards. Maintenance agencies are likewise either going to have to compensate for this fundamental weakness in the work force through equipment and communications advancements or implement employment standards precluding persons not capable of the required sophistication.

Continual Residual Drug Effect: Much is being made of drug testing for prospective employees, and some occupations with widely recognized public safety concerns are administering random drug tests. What is not as widely recognized is that as many persons who are over the age of 50 in 1997 retire in the next 10 years, the work force will become largely populated by people who regard illegal substances as recreational and often consume them. Many of these people who reached the age of consent in 1960 or later think of marijuana, cocaine, and other such drugs as having no more effect on their ability to function than alcohol did on the generations that preceded them. Alcohol does significantly affect worker safety and effectiveness, but because it is a water-soluble drug, employees who are willing to end alcohol abuse can return to effective functioning quite quickly. Unfortunately, many of the chemical compounds in other drugs have fat-soluble components that the body cannot cleanse quickly from itself and that can have lingering detrimental effects on cognitive and body motor functions. The potential for diminished worker capability because of residual drug effects should be considered when sophisticated equipment controls and safety features are developed.

Non-English Multilanguage Work Force: Labeling of manufactured equipment has been moving to the use of international symbols because of a desire to have units manufactured at one site be marketed directly in all countries where the company has an economic presence without having to alter the originally manufactured unit. Because of the ever-increasing multicultural makeup of the maintenance work force in the United States, this same philosophy needs to be developed for all equipment operating instructions, especially safety-related instructions. The use of symbols and multiple-language instructions should not supplant efforts to assist personnel to become competent enough in English to be safe, productive employees. It is, however, necessary for maintenance engineers and maintenance managers to think about how practices, procedures, and equipment systems need to be modified to assist people who are in cultural transition be a safe, effective work force.

Traffic Volume Demand Exceeding Work Zone Capacity: In the past, when a new route was planned, designed, and built, it always had a traffic capacity that significantly exceeded the existing route being replaced. Once the interstate highway system was completed, however, the concept of incremental expansion to the highway network vanished. Consequently, as we begin to focus almost exclusively on maintenance and rehabilitation of the national highway system, it is inevitable that we encounter some projects in which using even a small portion of the traveled roadway to conduct maintenance and rehabilitation reduces the capacity of the roadway such that traffic demand exceeds capacity. In this case, traffic queues (backed-up waiting lines) will grow until the traffic demand voluntarily goes away or we remove the work zone. Twenty to 30 years ago, it was possible to use a detour to move traffic around a work zone that created a massive bottleneck to traffic flow. That is still possible today unless the bottleneck is on the interstate system. Creating a detour to an interstate route grossly congested because of maintenance and rehabilitation is nearly impossible or prohibitively expensive. Doing some creative thinking today in advance planning of maintenance operations may yield great dividends in the future as traffic volumes and maintenance collide in their claim to use our highest classification of streets and highways.

2.1.17 REFERENCES

1. O'Brien, Louis G. *Evolution and benefits of preventive maintenance strategies.* Synthesis of Highway Practice 153. Washington, D.C.: National Cooperative Highway Research Program, Transportation Research Board, 1989.

2. Federal Highway Administration. Pavement preventive maintenance: An idea whose time has come. *Focus.* Strategic Highway Research Program Implementation, United States Department of Transportation-Federal Highway Administration, August 1995.

3. Burns, E. Nels. *Managing urban freeway maintenance.* Synthesis of Highway Practice 170. Washington, D.C.: National Cooperative Highway Research Program, Transportation Research Board, 1990.

4. Ceran, T., and R. B. Newman. *Maintenance considerations in highway design.* National Cooperative Highway Research Program Report 349. Washington, D.C.: Transportation Research Board, 1992.

5. Byrd, L. Gary. *Short-term responsive maintenance systems.* Synthesis of Highway Practice 173. Washington, D.C.: National Cooperative Highway Research Program, Transportation Research Board, 1991.

6. Transportation Research Board. *Relationship between safety and key highway features: A synthesis of prior research.* State of the Art Report 6. Washington, D.C.: Transportation Research Board, 1987.

7. Wikelius, Mark R. *Driving customer quality into highway maintenance.* Workshop on Performing Highway Maintenance Using Total Quality Management, Transportation Research Board, Whitefish, MT, May 21–23, 1995.

8. SHRP binder specification validated. *Focus.* Strategic Highway Research Program Implementation, United States Department of Transportation-Federal Highway Administration, March 1993.

9. Quality materials key to better pothole patches. *Focus.* Strategic Highway Research Program Implementation, United States Department of Transportation-Federal Highway Administration, December 1993.

10. Strategic Highway Research Program. *Asphalt pavement repair manuals of practice.* SHRP-H-348. Washington, D.C.: Strategic Highway Research Program, 1993.

11. County turns to Superpave for solution to low-temperature cracking. *Focus.* Strategic Highway Research Program Implementation, United States Department of Transportation-Federal Highway Administration, June 1994.

12. Geoffroy, Donald N. *Cost-effective preventive pavement maintenance.* Synthesis of Highway Practice 223. Washington, D.C.: National Cooperative Highway Research Program, Transportation Research Board, 1996.

13. Strategic Highway Research Program. *Concrete pavement repair manuals of practice.* SHRP-H-349. Washington, D.C.: Strategic Highway Research Program, 1993.

14. Iowa Department of Transportation. *Maintaining granular surfaced roads.* Iowa Highway Research Project HR-223. Ames, IA: Iowa Department of Transportation, November 1980.

15. National Association of County Engineers. *Blading aggregate surfaces.* Training Guide Series. Washington, D.C.: National Association of County Engineers, 1986.

16. Federal Highway Administration. *Maintenance of drainage features for safety: A guide for street and highway maintenance personnel.* FHWA-RT-90-005. Washington, D.C.: Federal Highway Administration, 1990.

17. Brown, Dan. Highway drainage systems. *Roads and Bridges.* Des Plaines, IL, February 1996, pp. 34–ff.

18. Stein, Edward G., Jr. Erosion control a priority on Maryland road projects. *Roads and Bridges.* Des Plaines, IL, May 1988, pp. 85–87.

19. Drainage channel maintenance helps Texas county survive massive flood. *Roads and Bridges.* Des Plaines, IL, September 1995, pp. 56–ff.

20. Cave-in avoided by relining pipe. *Better Roads.* Park Ridge, IL, March 1988, p. 36.

21. Coated culvert pipe solves problems. *Better Roads.* Park Ridge, IL, March 1991, p. 32.

22. Hassan, Hosam F., Thomas D. White, and David Andrewski. Indiana switches to edge drains. *Better Roads.* Park Ridge, IL, August 1996, pp. 24–27.

23. Edge drain inspections eased with equipment. *Roads and Bridges.* Des Plaines, IL, September 1988, p. 66.

24. Geocomposite drains "edge out" skeptics. *Roads and Bridges.* Des Plaines, IL, September 1996, pp. 50–54.

25. Good drainage extends pavement life. *Better Roads.* Park Ridge, IL, April 1989, pp. 21–23.

26. Erosion control: A basic of roadside drainage. *Better Roads.* Park Ridge, IL, May 1995, p. 33.

27. Road and Hydraulic Engineering Division of the Directorate General for Public Works and Water Management. *Nature across motorways.* Delft, Netherlands, 1995.

28. Federal Highway Administration. *Vegetation control for safety: A guide for street and highway maintenance personnel.* FHWA-RT-90-003. Washington, D.C.: Federal Highway Administration, 1990.

29. Roadside planting that reduces costs. *Better Roads.* Park Ridge, IL, May 1990, pp. 33–35.

30. Johnson, William D. How to establish wildflowers. *Better Roads.* Park Ridge, IL, September 1995, pp. 27–28.

31. Heine, Martha. Charge cards help fund wildflowers in Oklahoma. *Roads and Bridges.* Des Plaines, IL, September 1990, p. 64.

32. Caution critical in mowing rights of way. *Roads and Bridges.* Des Plaines, IL, February 1991, pp. 37–38.

33. Sixteen roadside mowing safety tips. *Better Roads.* Park Ridge, IL, February 1991, p. 17.

34. Herbicides cut costs for Jersey turnpike. *Roads and Bridges.* Des Plaines, IL, September 1990, pp. 60–63.

35. Informing the public calms concern about herbicide use. *Better Roads.* Park Ridge, IL, January 1988, p. 38.

36. McWilliams, Denise A. Ground water quality threatens herbicide use. *Roads and Bridges.* Des Plaines, IL, February 1991, pp. 40–41.

37. Training modules target U.S. herbicide users. *Roads and Bridges.* Des Plaines, IL, February 1991, pp. 30–32.

38. Geosynthetic controls erosion at ODOT. *Better Roads.* Park Ridge, IL, April 1996, p. 39.

39. Federal Highway Administration. *W-beam guardrail repair and maintenance: A guide for street and highway maintenance personnel.* FHWA-RT-90-001. Washington, D.C.: Federal Highway Administration, 1990.

40. Keep, Dale, and Dick Parker. Tests clear snow, path for use of liquid anti-icing in Northwest. *Roads and Bridges.* Des Plaines, IL, August 1995, pp. 50–52.

41. SHRP's role in roadway anti-icing. *Better Roads.* Park Ridge, IL, June 1997, pp. 22–25.

42. Dye, David L., Harry O. Krug, Dale Keep, and Raymond Willard. *Experiments with anti-icing in Washington State.* Transportation Research Record 1533. Washington, D.C.: Transportation Research Board, 1996, pp. 21–26.

43. Gerstlinger, Lee. The blizzard of '96. *Roads and Bridges.* Des Plaines, IL, March 1996, pp. 48, 50, 52–54.

44. Dickinson, Jerry L. The blizzard of '96: Midwest edition. *Roads and Bridges.* March 1996, pp. 55–56, 58.

45. Kuemmel, David E. *Managing roadway snow and ice control operations.* Synthesis of Highway Practice 207. Washington, D.C.; National Cooperative Highway Research Program, Transportation Research Board, 1994.

46. Early preparation key in snow fighting nationwide. *Roads and Bridges*. Des Plaines, IL, June 1990, pp. 40–41.

47. Maintenance management. In *Proceedings of the seventh maintenance management conference, July 18–21*. Washington, D.C.: Transportation Research Board, 1994.

48. Miner, Max. Simulations show ways to boost plow productivity 90%. *Better Roads*. Park Ridge, IL, April 1997, pp. 26–30.

49. Winter maintenance tips. *Better Roads*. Park Ridge, IL, June 1989, pp. 37–38.

50. Winter equipment maintenance: Tips you need to know. *Better Roads*. Park Ridge, IL, June 1993, pp. 18, 20–21.

51. Seventeen winter maintenance ideas that work. *Better Roads*. Park Ridge, IL, June 1997, pp. 18–19.

52. Wallace, Susan. Iowa DOT wins snow battle. *Better Roads*. Park Ridge, IL, June 1997, pp. 20–21.

53. Strategic investment in winter maintenance. *Roads and Bridges*. December 1996, Des Plaines, IL, pp. 20–21, 39.

54. Wisconsin expands weather system to counties. *Better Roads*. Park Ridge, IL, June 1990, pp. 20–22.

55. Hearn, Daryl L. New technology aids winter maintenance. *Better Roads*. Park Ridge, IL, June 1990, p. 25.

56. DOTs applaud winter weather systems. *Better Roads*. Park Ridge, IL, February 1991, p. 29.

57. How thermal mapping works. *Better Roads*. Park Ridge, IL, February 1991, pp. 30–31.

58. Mickes, Joseph. Managing winter weather. *Roads and Bridges*. Des Plaines, IL, December 1996, pp. 22–23.

59. Crosby, John D. Visibility technology improves RWIS. *Better Roads*. Park Ridge, IL, May 1997, pp. 37–40.

60. Boselly, S. Edward III. *Benefit-cost assessment of the utility of road weather information systems for snow and ice control*. Transportation Research Record 1352. Washington, D.C.: Transportation Research Board, 1992, pp. 75–82.

61. Hoover, Thomas R. Toledo's staged snow plan maximizes men, equipment. *Roads and Bridges*. Des Plaines, IL, June 1990, pp. 42, 44–45.

62. A better shed for storing salt? *Better Roads*. Park Ridge, IL, January 1997, pp. 25–26.

63. Heine, Martha. Use trucks efficiently for best snow removal. *Roads and Bridges*. Des Plaines, IL, June 1990, pp. 49–50, 55, 57.

64. GPS latest ally in war on winter snow. *Roads and Bridges*. Des Plaines, IL, June 1996, p. 67.

65. Granato, Gregory E. *Deicing chemicals as source of constituents of highway runoff*. Transportation Research Record 1533. Washington., D.C.: Transportation Research Board, 1996, pp. 50–58.

66. Boselly, S. Edward III, John E. Thornes, and Cyrus Ulberg. *Road weather information systems, Vol. I: Research report*. SHRP-H-350. Washington., D.C.: Strategic Highway Research Program, 1993.

67. Reiter, Elmar R., David K. Doyle, and Luis Teixeira. *Intelligent and localized weather predictions*. SHRP-H-333. Washington., D.C.: Strategic Highway Research Program, 1993.

68. Kallberg, Veli-Pekka. *Experiment with reduced salting of rural main roads in Finland*. Transportation Research Record 1533. Washington., D.C.: Transportation Research Board, 1996, pp. 32–37.

69. Kallberg, Veli-Pekka, Heikki Kanner, Tapani Mäkinen, and Matti Roine. *Estimation of effects of reduced salting and decreased use of studded tires on road accidents in winter.* Transportation Research Record 1533. Washington, D.C.: Transportation Research Board, 1996, pp. 38–43.

70. McCrum, Ron L. CMA and salt mix cuts corrosion. *Better Roads*. Park Ridge, IL, June 1988, p. 32.

71. Gall, Jim. Liquid calcium chloride helps keep Michigan's roads to the slopes open. *Roads and Bridges*. Des Plaines, IL, June 1996, pp. 58, 60.

72. Beazley, Scott, Bill Sader, and Paul Brown. Liquid calcium chloride hard on ice, easy on corrosion. *Roads and Bridges*. Des Plaines, IL, June 1996, pp. 61–63.

73. Lawson, Milan W. Smart salting: Field techniques to use. *Better Roads*. Park Ridge, IL, September 1995, pp. 24, 26.

74. Oklahoma's plans for RWIS and anti-icing technologies get boost from Nevada's experiences. *Focus*. Strategic Highway Research Program, June 1997.

75. Equipment and systems that help control snow. *Better Roads*. Park Ridge, IL, November 1989, pp. 41–42, 45.

76. Snow-winging tips. *Better Roads*. Park Ridge, IL, November 1989, p. 44.

77. End gate eliminates snowplow windrows. *Better Roads*. Park Ridge, IL, June 1997, p. 28.

78. Smithson, Leland D. DOTs push for better snow control vehicles. *Better Roads*. Park Ridge, IL, June 1997, pp. 27–29.

79. National Cooperative Highway Research Program, Transportation Research Board. *Winter maintenance technology and practices—Learning from abroad*. Research Results Digest 204. Washington, D.C., January 1995.

80. Lessons from blizzard of '96 heeded in winter of '97. *Focus*. Strategic Highway Research Program, November 1996.

81. Indiana DOT bolsters snow and ice control using high technology. *Roads and Bridges*. Des Plaines, IL, June 1995, pp. 38–39.

82. New technologies keep snow and ice—and winter maintenance expenses—under control. *Focus*. Strategic Highway Research Program, March 1997.

83. Ring, Stanley L. *Wind-tunnel analysis of the effect of plantings on snowdrift control.* Transportation Research Record 766. Washington, D.C.: Transportation Research Board, 1980, pp. 8–12.

84. Tabler, Ronald D. *Snow fence guide.* SHRP-W/FR-91-106. Washington, D.C.: Strategic Highway Research Program, 1991.

85. Federal Highway Administration. *Maintenance of small traffic signs: A guide for street and highway maintenance personnel.* FHWA-RT-90-002. Washington, D.C.: Federal Highway Administration, 1990.

86. Collins, Thomas J., and Michael J. Garlick. Sign structures under watch. *Roads and Bridges.* Des Plaines, IL, July 1997, pp. 38, 40–44.

87. Federal Highway Administration. *Manual on uniform traffic control devices for streets and highways* (including *Revision 3 to part IV* dated 1993). Washington, D.C.: Federal Highway Administration, 1988.

88. Burnham, Archie C., Jr. *Sign policies, procedures, practices, and fees for logo and tourist-oriented directional signing.* Synthesis of Highway Practice 162. Washington, D.C.: National Cooperative Highway Research Program, Transportation Research Board, 1990.

89. Dudek, Conrad L. *Changeable message signs.* Synthesis of Highway Practice 237. Washington, D.C.: National Cooperative Highway Research Program, Transportation Research Board, 1997.

90. Cunard, Richard A. *Maintenance management of street and highway signs.* Synthesis of Highway Practice 157. Washington, D.C.: National Highway Cooperative Research Program, Transportation Research Board, 1990.

91. Traffic Operations Branch, Wyoming State Highway Department. *Pavement marking manual.* Laramie, WY, 1985.

92. Yauch, Peter J. *Traffic signal control: State of the art.* Synthesis of Highway Practice 166. Washington, D.C.: National Cooperative Highway Research Program, Transportation Research Board, 1990.

93. Giblin, James M. *Traffic signal installation and maintenance manual.* Englewood Cliffs, NJ: Prentice-Hall, 1989.

94. Lighted tunnel design prevents city split in Phoenix. *Better Roads.* Park Ridge, IL, February 1991, pp. 40–41.

95. Bridge lighting brightens Miami skyline. *Roads and Bridges.* Des Plaines, IL, June 1997, pp. 40–42.

96. How to overcome bridge lighting problems. *Better Roads.* Park Ridge, IL, May 1995, pp. 13–15.

97. Transportation Research Board. *Environmental research needs in transportation.* Transportation Research Circular 469. Washington, D.C.: Transportation Research Board, March 1997.

98. Mississippi road agency changes with the times. *Roads and Bridges.* Des Plaines, IL, July 1997, pp. 48–50.

99. Federal Highway Administration. Part VI of the *Manual on uniform traffic control devices*. 3d rev. Washington, D.C.: Federal Highway Administration, September 3, 1993.

100. Benekohal, Rahim F., E. Shim, and P. T. V. Resende. *Truck drivers' concerns in work zones: Travel characteristics*. Transportation Research Record 1509. Washington, D.C.: Transportation Research Board, 1995, pp. 55–64.

101. National Cooperative Highway Research Program, Transportation Research Board. *Illumination guidelines for nighttime highway work*. Research Results Digest 216. Washington, D.C., December 1996.

102. National Cooperative Highway Research Program, Transportation Research Board. *Procedure for determining work zone speed limits*. Research Results Digest 192. Washington, D.C., September 1996.

103. Speir, Richard H., and Matthew W. Witczak. *Use of shredded rubber in unbound granular flexible pavement layer*. Transportation Research Record 1547. Washington, D.C.: Transportation Research Board, 1996, pp. 96–106.

104. Russell, Eugene R., Sr. *Highway maintenance procedures dealing with hazardous material incidents*. Synthesis of Highway Practice 196. Washington, D.C.: National Cooperative Highway Research Program, Transportation Research Board, 1994.

105. McGee, Kenneth H. *Design, construction, and maintenance of PCC pavement joints*. Synthesis of Highway Practice 211. Washington, D.C.: National Cooperative Highway Research Program, Transportation Research Board, 1995.

106. Friend, David J. *Remediation of petroleum-contaminated soils*. Synthesis of Highway Practice 226. Washington, D.C.: National Cooperative Highway Research Program, Transportation Research Board, 1996.

2.2 ROADWAY MANAGEMENT
2.2.1 Introduction
General agreement has existed since the late 1960s that maintenance management requires a systematic, quantified approach to
- develop work programs,
- budget and allocate resources,
- schedule work, and
- report and evaluate performance and cost.

This process has become recognized as fundamental to all levels of maintenance agencies [1, 2] and has been the driving force behind continued refinements in the organization and processing of databases that support these maintenance management activities. Hyman et al. [3] examined a wide variety of technologies that improved data collection in managing maintenance; a number of these technologies have been field-tested and evaluated for their effectiveness in improving the data collection process [4]. Agencies having lengthy experience with maintenance management usually upgrade their maintenance management capabilities, which does not interfere with their ongoing management process [5]. This upgrading often includes, but is not limited to, the following:
- Gaining a user's perspective on the effectiveness of maintenance activities [6].
- Training personnel to do the present job correctly and effectively and to advance in their capabilities.
- Broadening the scope of the maintenance management system's technical capabilities through the adoption and transfer of technologies appropriate to the agency's needs.
- Establishing a regular program of reviewing, and where justified, updating of computer hardware and software.
- Recognizing and responding to changes in the type of maintenance management support personnel needed to retain and enhance the management system's effectiveness.

Unlike roadway maintenance management, roadway management is not as clearly defined. It is an evolving management process seeking to organize and coordinate roadway assets and thus optimize the total roadway function. Roadway maintenance is a resource available to manage the total roadway. In a similar fashion, traffic operational methods that control traffic flow, reduce delays, increase safety, and improve efficiency are a resource to manage the roadway. Thus, roadway management is evolving into a growing effort to coordinate the maintenance, rehabilitation, and control of pavements, shoulders, roadsides, traffic access, drainage, lighting, etc. to maximize the safety and efficiency of traffic movement while minimizing the cost, environmental impact, and social impact of using the roadway.

2.2.2 Interaction With Other Highway System Management Systems
Models have been proposed and developed that work toward an optimal allocation of maintenance activity, such as the mathematical programming approach outlined in reference [7]. According to systems analysis principles, optimization of a subset of an organization's structure rarely produces an organizational optimum. The next step in advancing the optimization process is to link models optimizing maintenance (and other components of the roadway system such as pavement management systems) to the financial planning database system. Several case studies of efforts in this direction are available in the literature [8–11]. What is still lacking is a comprehensive structure that analyzes, evaluates, and interprets the effect that a change in one system management component has on the other interrelated system management components. Initial efforts are underway to bring this capability, accomplishing large-scale optimization and system management integration, to agencies [12, 13, 14]. What is not yet fully understood is how to bring the evaluation and interpretation

of assembled and produced information to bear on the desire to more fully optimize maintenance while optimizing all resources applied to the roadway. Perhaps the management processes used in large, integrated technology-based corporations may facilitate development of the next step in this ongoing evolutionary process. An early effort at defining an interpretation and evaluation process within the maintenance process is documented in chapter 3 of Butler et al. [15]. These concepts are still of merit in working toward a more global management of roadway systems. Synthesis of Highway Practice 238 outlines transportation performance measures that will likely become a part of the management system optimization process by widening the scope of roadway maintenance optimization [16]. The general principle behind the changed focus of performance measures is "outcome assessment" measures, which in turn suggests that data collected in the future will have to be more outcomes oriented than production oriented. This revised orientation suggests that roadway maintenance will focus more on roadway condition and serviceability, which may reduce the priority on maintenance and preservation of the roadway. Both the short-term and the long-term possible consequences of pursuing alternative strategies presented by revised management system models need to be considered. All strategies evolving from a wider scope management system process must be subject to fiscal resource limitations for maintenance; hence, prospective maintenance strategies should conform to general budget guidelines [17]:

- Strategies being compared should be consistent, comprehensive, and flexible.
- Each strategy needs to be credible within the maintenance division as well as credible to external maintenance forces.
- Each strategy should build on successful maintenance and successful past budgets (when uncertainty through unknown results begins to exceed 10% of the effort, a high probability of failure generally exists).
- Budget strategies need to adapt to changes that may threaten the success of the activity being budgeted or present opportunities for sudden positive adjustments.

2.2.3 Roadway Inspection and Condition Inventory Process

Roadway inspection and condition inventory is a critical element in the development of effective global maintenance management systems. Response maintenance activities are likely to continue to depend on the firsthand observations of maintenance patrols and the routine notations of maintenance engineers, maintenance managers, and maintenance supervisors during the course of their normal duties. However, roadway environmental condition sensors, low-level infrared thermography from satellites, digital video imaging transmitted through satellite telemetry for image processing, and computer-coded location markers in the roadway may possibly permit some response maintenance decisions to be based upon remotely observed data. The roadway inspection and condition inventory process is changing most rapidly in the programmed maintenance activities for preventive maintenance, preservation maintenance, environmental maintenance, and aesthetic maintenance. Traditional photo logging is giving way to digital video logging. Photo log files are giving way to digital video disc files and geographic information systems files. As object-oriented programming systems become more widely used and maintenance activity programming initiation thresholds become defined in graphic image criteria, then the same machine and pattern recognition analysis that drives industrial robot processes may be used to automate certain maintenance areas. Until such technology processes are readily available to maintenance agencies, inspection and condition inventory processes need to be evaluated and the collection of any nonessential data ended. Longitudinal time data also need to be maintained for benchmarking quality controls.

2.2.4 Environmental and Nonroadway Issues

As long as roadside maintenance involves elimination of intrusive vegetation, which requires chemical applications, surface water and groundwater quality as well as worker and wildlife exposure to chemicals will be an issue [18, 19]. Maintenance agencies will need to continue to create knowledgeable personnel in environmental, health, safety, and hazardous-materials-handling regulations [20]. Creating knowledgeable personnel requires continued investment in and attention to training [21, 22]. Finally, since solid waste management is becoming an increasingly sensitive issue far beyond roadway maintenance, public policy will continue to apply pressure on roadway maintenance agencies to increase efforts that reduce the volume of roadside litter [23]. Since nonroadway issues are extremely difficult to quantify and it is even harder to define their economic value, they do not lend themselves to incorporation into the rational management models forming the evolving global maintenance management systems. For the foreseeable future, programming responses to environmental and nonroadway issues in maintenance will continue to be guided by legislative edicts, social priorities adopted by upper agency management, and heuristic (intuitive) evaluations by maintenance engineers and maintenance managers.

2.2.5 Developing Issues

The contracting of maintenance is an existing issue, both past and present, but it is also a developing issue. Continued efforts to evaluate what should be contracted out, what is feasible to contract, and how contracting efforts should proceed in the best interest of both public and private entities should be based on the information compiled in NCHRP Report 344 [24]. An account of British Columbia's experience in massive privatization of roadway maintenance in NCHRP Report 344 is especially useful. The report also discusses the various elements of contracting maintenance that are useful to know before an agency proceeds to revise or modify its contracting efforts for maintenance services. The State of Washington has studied a project cost-evaluation method for assessing maintenance privatization [25] that expands upon the knowledge compiled in NCHRP Report 344. The State of Wisconsin has developed pavement warranties [26], an approach with merit for all aspects of contracting for maintenance services, especially if the agency's management system is moving to "outcomes assessment" criteria in program review and evaluation.

The proper professional development of maintenance engineers and maintenance managers is a current, evolving issue. NCHRP Report 360 provides a concise yet definitive examination of a prototype structure that develops the professional capabilities of maintenance engineers and managers [27]. The report examined the need for professional development at three levels (with the lowest level being the resident or area maintenance engineer or manager) in terms of 24 different activity or subject areas in which a maintenance organization should review personnel capabilities. The general consensus of NCHRP Report 360 is that maintenance engineers and maintenance managers should

- have the opportunity to participate in traditional higher education where collegiate offerings will enhance personnel knowledge in areas necessary for one or more of the 24 activity areas;
- participate in professional continuing education programs that address needed knowledge areas;
- have the opportunity to participate in training programs that develop skills needed in one or more of the 24 activity areas; and
- take short-course programs that incorporate training and continuing education modules supporting needed knowledge and skill areas.

Only for maintenance training do most agencies have sufficient resources to provide regular professional development to maintenance engineers and managers. To ensure that they provide the breadth of professional development needed, agencies should foster cooperative education and training programs with educational and training organizations. The application of advanced, and advancing, technology to roadway maintenance and management is a continuing, developing issue. As maintenance management systems evolve into increasing levels of sophistication, more data analysis could be relegated to expert systems. Expert system computer programs also could potentially structure and organize the practical knowledge of retiring maintenance engineers and managers, accelerating the transfer of this knowledge to young engineers and maintenance managers entering maintenance management. Synthesis of Highway Practice 183 summarizes the application of expert systems to transportation; many aspects of such systems may potentially be applied to maintenance engineering and management [28]. New mathematical concepts are being applied in systems analysis of networks. Neural network analysis is being applied to the prediction of bridge maintenance cycles and the management analysis of highway maintenance strategies [29]. Once procedures are created that can be applied to highway maintenance activities with neural network analysis, it may not be necessary for maintenance engineers and managers to be skilled in neural network computations. However, to be aware of a technique's strengths and limitations, maintenance engineers and managers need to have a basic understanding of such concepts. Internet home pages are developing into an advanced mechanism by which construction projects are being managed when the constructor, the owner/client, the various design professionals, and the manager/financier are housed in dispersed locations [30]. While this new project management technology was developed for construction projects, it has some attributes that may prove useful in the management of wide-area maintenance activity. A system using portable computers, digital cameras, voice and fax transmission computer boards, and high-grade communications links among the parties connected through the home page can replace cellular telephones, fax machines, and plan/specification files and potentially shorten the time required to authorize maintenance action for roadway defects or suspected conditions that an expert needs to review to determine how to best repair or rehabilitate.

Maintenance agencies are just learning how to be effective messengers of their work on behalf of the public. For decades, it was assumed that if maintenance forces did a good job, everyone would understand and support their work. For maintenance engineers and maintenance managers who do not have a background, by education or experience, in public affairs communication, NCHRP Report 364 provides a concise discussion of public affairs communication strategies and tools [31].

All public organizations are struggling with change. Some of the pressures to change are external to the organization and some arise from within. Maintenance engineers and maintenance managers, especially at the local level, may suffer from the syndrome "it's hard to remember your objective was to drain the swamp when you are up to your rear end in alligators" because of the immediate, urgent needs of accomplishing response maintenance activities. When the direction for change comes to an agency's maintenance division, a review of NCHRP Report 371 can help a person understand how such changes may be an integral part of the overall development of the total organization [32]. When maintenance engineers and maintenance managers understand how organizational changes have originated, they are better equipped to support proposed changes that will have a positive impact on maintenance and to suggest alternatives to proposed changes that will have a deterimental impact on maintenance.

2.2.6 REFERENCES

1. Jorgenson, Roy E. *Summary remarks: Maintenance management.* Special Report 100. Washington, D.C.: Highway Research Board, 1968, pp. 7, 8.

2. National Association of County Engineers. *Maintenance management.* NACE Action Guide Volume I-5. National Association of County Engineers, 1992.

3. Hyman, William F., A. D. Horn, O. Jennings, F. Hejl, and T. Alexander. *Improvements in data acquisition technology for maintenance management systems.* Report 334. Washington, D.C.: National Cooperative Highway Research Program, Transportation Research Board, 1990.

4. Hyman, William A., and Roemer M. Alfelor. Field testing and evaluation of innovative technologies for maintenance data collection. Maintenance Management, in *Proceedings of the Seventh Maintenance Management Conference.* Washington, D.C.: Transportation Research Board, 1995, pp. 9–17.

5. File, Dennis H. How to keep your maintenance management system from growing old. Maintenance Management, in *Proceedings of the Seventh Maintenance Management Conference.* Washington, D.C.: Transportation Research Board, 1995, pp. 50–54.

6. Miller, Jerry. Maintenance management from the customer's viewpoint. Maintenance Management, in *Proceedings of the Seventh Maintenance Management Conference.* Washington, D.C.: Transportation Research Board, 1995, pp. 3–8.

7. Sinha, Kumares C., and Tien F. Fwa. *Framework for systematic decision making in highway maintenance management.* Transportation Research Record 1409. Washington, D.C.: Transportation Research Board, 1993, pp. 3–11.

8. George, K. P., W. Uddin, P. J. Ferguson, A. B. Crawly, and A. R. Shekharan. *Maintenance planning methodology for statewide pavement management.* Transportation Research Record 1455. Washington, D.C.: Transportation Research Board, 1994, pp. 123–131.

9. DeCabooter, Philip, K. Weiss, S. Shober, and B. Duckert. *Wisconsin's pavement management decision support system.* Transportation Research Record 1455. Washington, D.C.: Transportation Research Board, 1994, pp. 76–81.

10. Mijuskovic, Vera, Dragan Banjevic, and Goran Mladenovic. *Impact of different economic criteria on priorities in pavement management systems.* Transportation Research Record 1455. Washington, D.C.: Transportation Research Board, 1994, pp. 178–187.

11. Humplick, Frannie, and Asif Faiz. Strategies for managing public expenditures for road maintenance. Maintenance Management, in *Proceedings of the Seventh Maintenance Management Conference.* Washington, D.C.: Transportation Research Board, 1995, pp. 38–49.

12. Markow, Michael J., and William A. Hyman. Highway maintenance and integrated management systems. Maintenance Management, in *Proceedings of the Seventh Maintenance Management Conference.* Washington, D.C.: Transportation Research Board, 1995, pp. 31–37.

13. Evans, L. D., A. R. Romine, A. J. Patel, and A. G. Mojab. *Concrete pavement repair manuals of practice.* SHRP-H-349. Strategic Highway Research Program. Washington, D.C.: Transportation Research Board, 1993.

14. Cumberledge, Gaylord, Charles A. Wilson, and Gary L. Hoffman. *Integration of management systems for maintenance activities*. Maintenance Management, in *Proceedings of the Seventh Maintenance Management Conference*. Washington, D.C.: Transportation Research Board, 1995, pp. 26–30.

15. Butler, Bertell C., Jr., R. F. Carmichael III, P. Flanagan, and F. N. Finn. *Evaluation of alternative maintenance strategies*. Report 285. Washington, D.C.: National Cooperative Highway Research Program, Transportation Research Board, 1986.

16. Poister, Theodore H. *Performance Measurement in State Departments of Transportation*. Synthesis of Highway Practice 238. Washington, D.C.: National Cooperative Highway Research Program, Transportation Research Board, 1997.

17. Reno, A. T., William A. Hyman, and M. E. Shaw. *Guidelines for effective maintenance-budgeting strategies*. Report 366. Washington, D.C.: National Cooperative Highway Research Program, Transportation Research Board, 1994.

18. Dickens, Ray. *Herbicide fate and worker exposure*. Maintenance Management, in *Proceedings of the Seventh Maintenance Management Conference*. Washington, D.C.: Transportation Research Board, 1995, pp. 57–59.

19. Conrad, John F., and Doug Pierce. *Storm water management strategies to meet national pollution discharge elimination systems requirements*. Maintenance Management, in *Proceedings of the Seventh Maintenance Management Conference*. Washington, D.C.: Transportation Research Board, 1995, pp. 156–159.

20. Tarrer, A. R., Gregory T. Whetstone, and James W. Boylan. *Impacts of environmental, health, and safety regulations on highway maintenance*. Maintenance Management, in *Proceedings of the Seventh Maintenance Management Conference*. Washington, D.C.: Transportation Research Board, 1995, pp. 144–151.

21. Swearingen, Keith, and Robert E. Tatman. *Environmental training for hazardous materials management*. Maintenance Management, in *Proceedings of the Seventh Maintenance Management Conference*. Washington, D.C.: Transportation Research Board, 1995, pp. 152–155.

22. Holt, Harvey A. *Applicator training materials on use of chemicals for vegetation management*. Transportation Research Record 1409. Washington, D.C.: Transportation Research Board, 1993.

23. Andres, Dorothy, and Chester J. Andres. Roadside litter and current maintenance waste management practices: Are we making any progress? Maintenance Management, in *Proceedings of the Seventh Maintenance Management Conference*. Washington, D.C.: Transportation Research Board, 1995, pp. 135–143.

24. Newman, Robert B., Jeffrey E. Garmong, and Harry P. Hatry. *Maintenance contracting*. Report 344. Washington, D.C.: National Cooperative Highway Research Program, Transportation Research Board, 1991.

25. Conrad, John F., Paul Nelson, and Kelly Jones. *Project cost evaluation methodology approach to privatization in the Washington State Department of Transportation*. Transportation Research Record 1409. Washington, D.C.: Transportation Research Board, 1993, pp. 12–22.

26. Shober, Stephen F., Gary C. Whited, and Kevin W. McMullen. *Wisconsin Department of Transportation's asphaltic pavement warranties.* Transportation Research Record 1543. Washington, D.C.: Transportation Research Board, 1996, pp. 113–119.

27. Carter, Everett C., M. Ed Shaw, and Jeffrey E. Garmong. *Professional development of maintenance engineers and managers.* Report 360. Washington, D.C.: National Cooperative Highway Research Program, Transportation Research Board, 1994.

28. Cohn, Louis F., and Rosewell A. Harris. *Knowledge based expert systems in transportation.* Synthesis of Highway Practice 183. Washington, D.C.: National Cooperative Highway Research Program, Transportation Research Board, 1992.

29. Randolph, Dennis A. *Application of neural network technology to highway maintenance.* Transportation Research Record 1533. Washington, D.C.: Transportation Research Board, 1996, pp. 3–10.

30. Avney, Jon. *A day in the life of a project manager: Using the Internet.* NSPE Professional Edge Seminar 3343. Rapid City, SD: National Society of Professional Engineers, July 17, 1997.

31. Frank Wilson & Associates. *Public outreach handbook for departments of transportation.* Report 364. Washington, D.C.: National Cooperative Highway Research Program, Transportation Research Board, 1994.

32. National Academy of Public Administration. *State departments of transportation: Strategies for change.* Report 371. Washington, D.C.: National Cooperative Highway Research Program, Transportation Research Board, 1995.

3.0 BRIDGE MAINTENANCE AND MANAGEMENT

3.1 BRIDGE MAINTENANCE

Bridge maintenance has been defined as work performed to keep a facility in its current condition [1]. However, bridge maintenance has a broader scope because maintenance includes all activity in a facility's life that does not require a redesign and development project; thus, some agencies properly include work often classified as bridge rehabilitation (intended to upgrade the bridge to a condition better than its existing condition) within the context of bridge maintenance. Some agencies have not chosen to develop specialized skills or to provide the specialized equipment systems needed to perform some bridge maintenance activities. Instead, they contract for some part of their bridge maintenance. The fact that certain work activities may be a contract activity, however, does not remove them from the realm of bridge maintenance.

Functionally, bridges are simply a special portion of the total roadway in the transportation network. However, they are a critical link in the highway network, and therefore are considered separately from maintenance of the roadways approaching them. Since bridges are also a large capital investment per unit length of roadway, a high level of sophistication can be justified in maintaining them.

3.1.1 Introduction
3.1.1.1 Load-Carrying Capacity

Each bridge has an estimated capability to carry a certain total load limit. A bridge is designed for a specific load, and there is an estimated load at which the bridge is expected to fail (ultimate strength load). The ratio of the two loads is the factor of safety estimated to exist in the bridge.

$$\text{Factor of Safety} = \frac{\text{(Estimated Failure Load)}}{\text{(Design Load)}}$$

The "capacity" of a bridge is the sum total of the various loads that a bridge can safely carry in its existing condition or state. The various loads on a bridge may include the following:

Dead Load: This is defined as the load from the weight of the beams and deck and all of the structure above the piers and abutments upon which the vehicular and pedestrian traffic weight is supported. Some elements of the bridge included in the dead load are as follows:
- Wearing courses.
- Structural decks.
- Structural members.
- Curbs, sidewalks, railings, and fencing.
- Utility pipes, conduits, lighting masts, and traffic signal hardware.

The dead load may be significantly altered by the removal or addition of such elements during the bridge maintenance process, thereby reducing (or increasing) the bridge capacity to carry a specified live load. When considering the foundation support of the bridge, the piers (and the footings, if piers are supported by footings rather than piles) must be included in the dead load. That is, at whatever level within a bridge structure the capacity is estimated, all elements of the structure permanently in place are included in the dead load.

Live Load: The weight of vehicles, pedestrians, and other traffic. Highway loadings used in establishing load-carrying capacity will be the standard AASHTO vehicle loadings or the maximum legal loads of the state.

Impact Load: Dynamic load resulting from vertical acceleration of vehicles and persons while moving. It is estimated as a percentage of the live load according to the approved bridge design and analysis process.

Wind Load: Pressure on the beams, trusses, and other parts of the bridge exposed to steady, buffeting wind gusts. Sometimes this loading is significant for tall bridges and may require a detailed engineering analysis to determine its effect. The total effect of wind load depends upon the expected maximum wind velocity, the vertical area of all bridge members in the side profile, and the torsional resistance of the bridge structure to lateral loading.

Longitudinal Forces: Through acceleration and braking, traffic moving across a bridge can generate forces parallel to the centerline of the bridge. Such forces are often estimated to be 5% of the estimated live load.

Thermal Forces: Changes in temperatures create stresses and strains due to the thermal expansion and contraction of bridge materials.

Earth Pressures: Soil pressure on abutments and other components of a bridge against which any significant depth of earthen fill rests. Additional pressure can build up if weep holes (drainage openings) become clogged and do not drain the hydrostatic pressure.

Stream Forces: Bridge piers must resist horizontal loads caused by water flowing around them. During flood flow, this pressure can significantly increase. Allowing debris to build up against a pier during flooding greatly increases this pressure.

Ice Impacts and Pressures: Streams that ice over may release large ice floes that can collide against piers with massive force, damaging the structure. Ice jams can develop that act as a partial dam to the stream at the bridge with the potential to wash the bridge out in extreme conditions.

Earthquake Forces: Bridges in earthquake zones are subject to potential vibration loadings from any direction that the geologic earth crust plates may move during an earthquake tremor.

Pavement or Deck Joint Pressures: Joints in the pavement approaches to a bridge and joints in the bridge deck can become filled with debris, can become corroded, or may fail to function properly because of some other cause. Fouled joints can cause compressive thermal stresses to build up.

Bridge Capacity Defines the Capability of the Bridge to Carry the Load From These Various Forces: The engineering mechanics principles by which these forces are resisted or transmitted throughout the bridge structure are limited by the type of structure. *Axial forces* are those of compression or tension only. Beams may be subjected to axial forces, but trusses and suspension members are designed to carry only axial forces. *Bending forces* result when beams carry loads with compressive stresses in part of the beam and tensile stresses in the opposite portion of the beam cross section. Bending forces result from moment force couples and shearing forces vertically throughout the structural member.

3.1.1.2 Structural Systems

Bridges incorporate three basic types of structural systems to provide capacity to carry the various loads and forces resulting from those loads.

Beam-type Bridges: A subdivision of a beam bridge is a bridge incorporating simple spans. A special category of a simple span beam is the cantilever beam. A second subdivision of beam bridges is bridges having continuous beams over at least one pier. Vertical loads on simply supported beams produce only compression in the upper fibers of the cross section and tension in the lower fibers of the cross section. Continuous beams have positive moments (compression on top and tension on bottom) between the piers and have negative moments over the piers (tension on top and compression on bottom of the beam cross section). Material failure theories used in the design of beam bridges originate in the analysis of beams of one material, but bridge beams are frequently designed to be fabricated with distinct portions of the beam cross section produced from two or more construction materials. Such bridges are therefore identified as "composite beam" construction. The most common example of this type of bridge construction is a structure of rolled steel beams with shear studs attached to the upper flange and a reinforced-concrete deck placed on top of the beams that then acts as an integral structural unit when the studs transmit forces from the deck to the beams.

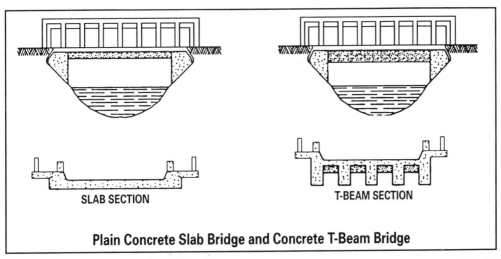

Plain Concrete Slab Bridge and Concrete T-Beam Bridge

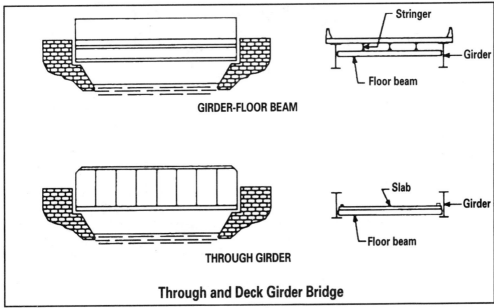

Through and Deck Girder Bridge

Chapter 3 Bridge Maintenance and Management

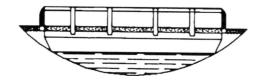 

I-Beam Bridge

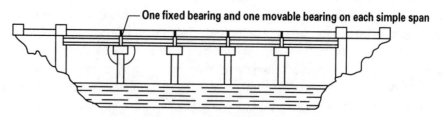

Multispan Bridge With Simple Spans

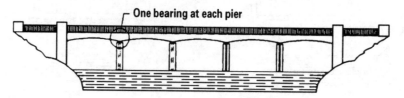

Continuous Girder Bridge

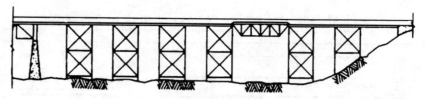

Steel Viaduct Bridge

Arch Bridges: Structural arches have been incorporated into bridge design to permit the loads on the bridge to be carried entirely by compression against the bridge abutment. In early bridge-building technology, arch bridges were required to permit the construction of bridges with masonry materials and other materials that could not carry tension or be fastened to carry tension. Modern arch bridges may be designed to be constructed from steel or reinforced-concrete members that are so delicate in their dimensions that the tension forces they are capable of resisting are negligible in comparison to the compression forces they resist.

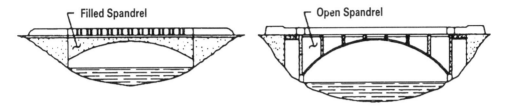

Filled- and Open-Spandrel Arch Bridge

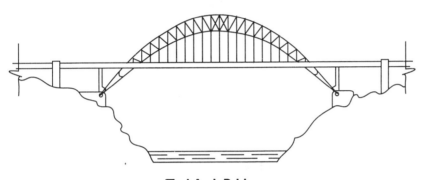

Tied-Arch Bridge

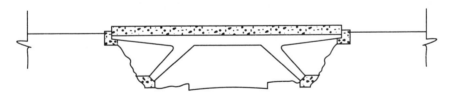

Steel Rigid-Frame Bridge

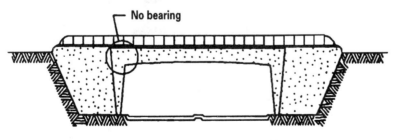

Concrete Rigid-Frame Bridge

Chapter 3 Bridge Maintenance and Management

Cable-Supported Bridges: Cable-supported bridges are usually classified as either cable suspension bridges or cable-stayed bridges. Typically, suspension bridges have two tall towers near the ends of the cable-supported span from which the cables on each side of the suspended roadway are hung. In contrast, a typical cable-stayed bridge may have one or more tall towers from which cables run to the adjacent suspended roadway in a radiating pattern of individual cables. Cable-supported bridges are often used to cross long spans where high clearances above the water are needed.

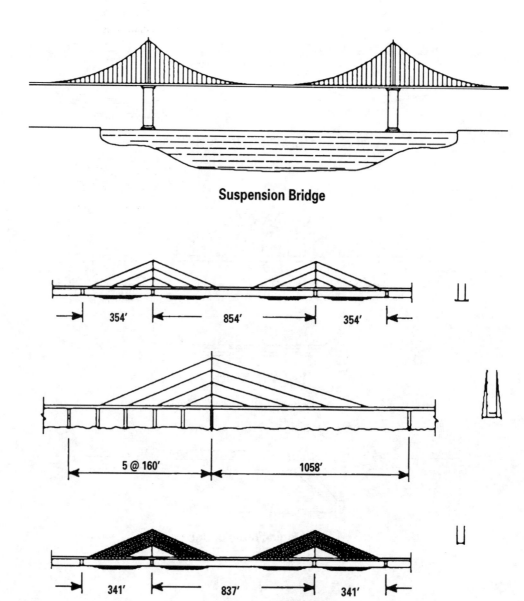

Suspension Bridge

Cable-Stayed Bridge

Truss Bridges: A truss is a system of structural members joined at their ends to form a stable framework. A bridge may be designed such that the truss acts as a beam with the same function as reinforced-concrete beams or structural steel beams, or an arch may be created from a series of curved trusses. The unique characteristic of trusses is that all loads are transformed into axial (either tension or compression) loads in the individual members of a truss, even though the truss itself may be resisting shear and moment loads.

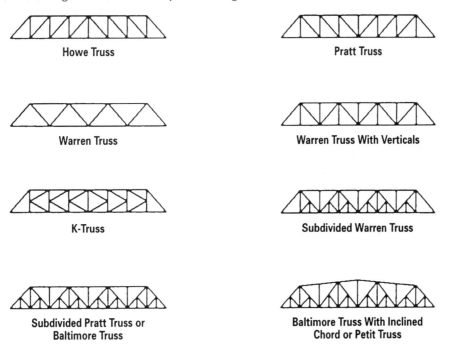

Conventional Bridge Truss Types

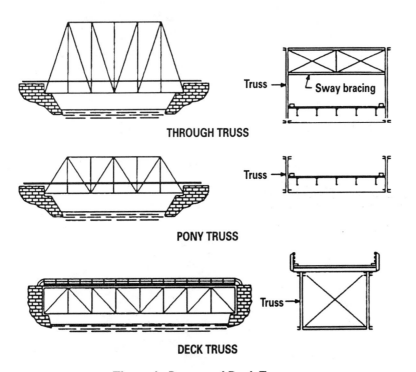

Through, Pony, and Deck Trusses

Chapter 3 Bridge Maintenance and Management **3-7**

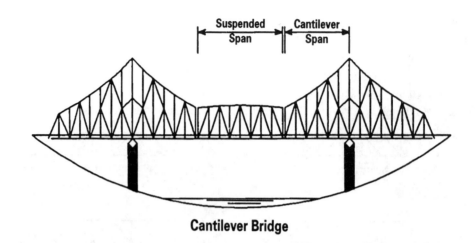

Cantilever Bridge

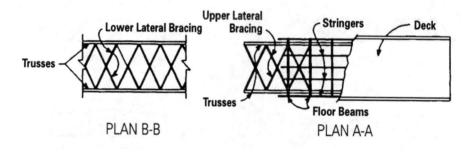

PLAN B-B PLAN A-A

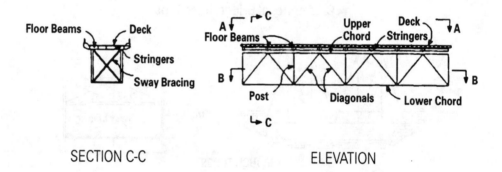

SECTION C-C ELEVATION

Identification of Main Truss Members

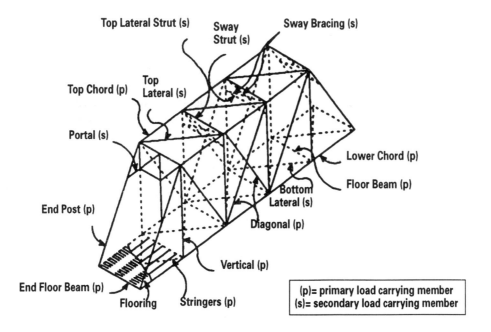

Primary and Secondary Members in a Truss

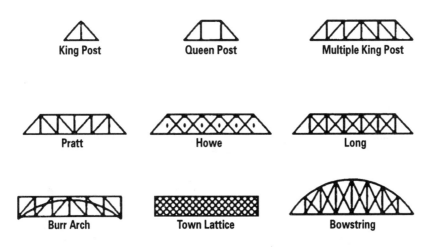

Types of Timber Trusses

Materials: Bridges are commonly constructed of the following three basic materials to create the structural system: timber, reinforced concrete, and steel.

Timber (wood) has a low cost per pound of material; it has a high strength-to-weight ratio; and it is durable when protected from changes in moisture content. However, it is subject to attack by insects and fungi, it cannot provide the total structural capacity to cross long spans that other materials can, and it is not as homogeneous as other materials.

Reinforced concrete usually provides a high strength-to-cost ratio, it can be designed to provide a particular capability required by the structure, and an architecturally and aesthetically pleasing bridge can be created though concrete formwork patterns. Some disadvantages of concrete include its susceptibility to structural deterioration when exposed to chlorides, the large weight of material required to achieve a high structural strength, and the need for special consideration in protecting the reinforcing steel from chlorides.

Chapter 3 Bridge Maintenance and Management **3-9**

Structural steel can develop very high-strength structures, can be used in a wide variety of structural forms, and can be provided as a quite homogenous material permitting good predictability of the structural performance of the bridge. Disadvantages of structural steel include the risk of corrosion from environmental influences; the need to coat or paint the steel to inhibit corrosion, with attendant maintenance costs; potential loss of structural strength through fatigue failure; and temperature-induced stresses when large temperature changes occur.

Some bridges have been designed and fabricated from aluminum in order to have a bridge with high strength and low dead-load weight. However, because of the high cost of aluminum in comparison to other common structural materials and the special skill required to fabricate structures with it, aluminum has not been a common material. When closing part of the bridge to traffic to perform maintenance and rehabilitation is an unacceptable cost, composite structural materials made of synthetic fibers may become widely used in bridge construction because of the long expected life of their structural components and because of their modular part replacement capabilities.

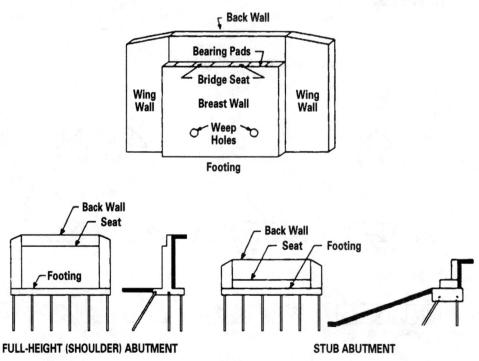

Abutment Types and Components

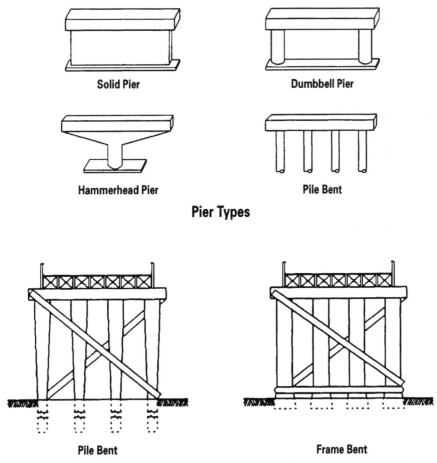

Pier Types

Pile Bent and Frame Bent Pier

3.1.1.3 Bridge Maintenance Concepts

With respect to bridge maintenance, maintenance managers distinguish between deferred maintenance and programmed maintenance. It is important to keep in mind that bridge maintenance may be deferred for one of two reasons:

(1) There may be an intentional programmatic decision to delay (or to omit) performance of a bridge maintenance activity to save money or to make the conduct of the maintenance activity more efficient.

(2) Some maintenance activities are not performed when needed because the budgeting process is ineffective, necessary funds are not allocated, or inspection and reporting procedures are defective. Maintenance deferred for these reasons is to be avoided.

Bridges may be identified as deficient for either one or both of the following reasons: structural deficiency and functional deficiency. Bridges can become structurally deficient because of corrosion or concrete deterioration associated with wear and environmental degradation, the effects of which can be reduced through good bridge maintenance. A bridge may also be structurally deficient because vehicle loadings are exceeding the bridge's design capacity, a problem for which maintenance can do nothing except post and monitor load limits.

Bridges become functionally deficient when some aspect of the design or structure type is no longer appropriate to handle the traffic because of dimensional or geometric problems. Bridge maintenance activities cannot address functional deficiencies.

Preventive bridge maintenance is directed at performing activities that will preserve bridge components in their present (or intended) condition, forestalling development of a structural deficiency. Preventive maintenance activities can be classified into two groups: scheduled and response.

(1) Scheduled (programmed at intervals): Typical activities that are conducted on a scheduled interval basis include:
- cleaning decks, seats, caps, and salt splash zones;
- cleaning bridge drainage systems;
- cleaning and lubricating expansion-bearing assemblies; and
- sealing concrete decks or substructure elements.

(2) Response (done as needed and as identified through the inspection process): Typical activities that are performed on an as-needed basis include:
- resealing expansion joints;
- painting structural steel members;
- removing debris from waterway channels;
- replacing wearing surfaces;
- extending or enlarging deck drains; and
- repairing damage from a vehicle hitting the structure.

The concept of preventive bridge maintenance suggests that many relatively small repairs and activities are performed to keep the bridge in good condition and avoid large expenses in major bridge rehabilitation or bridge replacement.

3.1.2 Traveled Surface

The wearing surface of the bridge deck should provide the same smooth riding surface as the street or roadway approaching the bridge as much as practicable. The layer or wearing course of material applied to or integral with the structural deck also protects the deck from traffic deterioration, environmental degradation from weather effects, and de-icing chemicals.

Concrete wearing surfaces may be placed with the concrete structural slab (i.e., a monolithic deck) or may be cast as a separate wearing surface on top of a previously cast slab. Maintenance forces typically need to be alert for scaling, spalling, and cracking in these wearing surfaces.

Asphalt wearing surfaces over structural deck slabs of reinforced concrete should be placed with a waterproofing membrane between the asphalt course and the concrete deck if the deck will remain in service for an extended period of time. In general, asphalt wearing surfaces should not be used in regions where de-icing chemicals are applied, especially if there is no waterproof membrane.

Agencies will not normally be responsible for stone and brick wearing surfaces except in areas where historical or architectural aesthetic concerns have retained or reintroduced such surfaces. Where these surfaces exist for aesthetic reasons, their maintenance may be a high-unit-cost activity and as much as possible should be supported by organizations interested in maintaining the aesthetic values.

Steel grating (or steel grid floors), both open grate and filled grid, are not usually found as a separate wearing surface. Grate floors are normally designed to be an integral part of the structural deck.

Timber plank and wood block surfaces may be encountered on older bridges with low traffic volumes. Localities using discarded railroad flatcars as the structural system for low-volume road bridges are an example where timber plank surfaces will be encountered.

3.1.2.1 Concrete Surfaces

Maintenance is expected to treat common problems with concrete wearing surfaces such as scaling, spalling, cracking, and pop outs.

Scaling: Scaling results when mortar and aggregate are gradually and continually lost over an area of the deck surface. If the surface is scaled to a depth less than about 6 mm (0.25 inch), this is usually considered to be light scaling; a depth of about 6 mm to 13 mm (0.25 inch to 0.5 inch) is usually considered medium scaling; a depth of about 13 mm to 25 mm (0.5 inch to 1 inch) is usually considered heavy scaling; and a depth exceeding about 25 mm (1 inch) is usually considered severe scaling. Severe scaling is often the result of improper concrete construction methods rather than negligent preventive maintenance. Light scaling may be corrected by applying a concrete sealing membrane or other approved concrete surface sealant. Sealing the surface initially may be helpful in preventing the scaling condition from arising in the first place. For any scaling condition exceeding a light scale, concrete repair procedures similar to those applicable to concrete pavement must be performed. When removing the deteriorated concrete, use care to avoid damaging any reinforcing steel. If the scaling results from chlorides penetrating the deck wearing surface, the deck needs to be monitored for further deterioration from chloride intrusion producing delamination of the structural concrete over the reinforcing steel.

Spalling: A spall appears as a depression resulting from a chunk of concrete, usually in a circular or oval shape, breaking loose from the concrete deck surface. Corrosion of the underlying reinforcing steel or deteriorated aggregate is usually the cause. If the spall is about 25 mm (1 inch) or less in depth or about 15 cm (6 inches) in diameter or less, it is considered a small spall. Dimensions exceeding this are usually considered large spalls, and their size begins to affect vehicular travel over the bridge. A hollow-sounding area (a hollow sound produced when struck with a hammer or steel bar or when swept with a drag chain) indicates a fracture plane (or delamination) below the deck surface. When any spalling or delamination is suspected or evident, the entire deck area should be surveyed to determine the extent of spalling and delamination before beginning repairs. Since surveying a bridge deck interferes with traffic flow, exposes maintenance personnel to traffic hazards, and is generally time-consuming and expensive, all aspects of the deck condition should be examined, if practicable, when conducting the survey. Survey aspects to be considered include the delamination survey, reinforcing cover survey, chloride content survey, and corrosion potential survey. The delamination survey covers rod sounding, hammer sounding, drag chain sounding, ultrasonic delamination detecting, etc. The reinforcing cover survey uses a magnetic field detector to estimate the depth of concrete over the reinforcing steel. The chloride content survey analyzes samples of concrete powder produced by drilling holes in the deck concrete over the reinforcing steel. The corrosion potential survey consists of electrical resistivity measurements with a half-cell probe; however, this is not an effective procedure when the deck contains epoxy-coated reinforcing steel or contains galvanized coated steel, or if the deck surface has been treated with a dielectric material.

Once the extent and severity of the root cause of the spalled deck surface condition has been determined, an appropriate corrective action can be scheduled on the basis of the agency's guidelines and policies regarding whether the deck should be repaired with a maintenance treatment or scheduled for an overlay or replacement. If the deck is overlaid, use of low-slump dense concrete should be considered to improve resistance to moisture penetration.

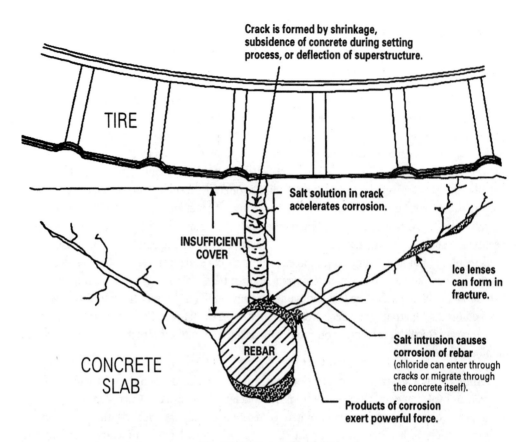

The Evolution of a Spall

Cracking: A crack is a linear fracture in the concrete wearing surface that may only extend partway through the deck or it may occur as a failure plane running completely through the concrete structural element. The five types of cracks are as follows: longitudinal, transverse, diagonal, pattern or map, and random.

Longitudinal cracks are reasonably straight cracks running parallel to the centerline of the roadway. These cracks are usually caused by shrinkage, settlement, differential deflection of adjacent beams or girders, voids in the slab, or corrosion of reinforcing steel.

Cracks classified as transverse cracks will appear in patterns roughly perpendicular to the centerline of the roadway. These cracks are usually caused by shrinkage, settlement, corrosion of the reinforcing steel, or deflection of the superstructure.

Diagonal cracks are similar to longitudinal and transverse cracks, but they tend to run at an angle to the centerline of the roadway. Frequently a bridge constructed on a skew angle to the centerline will exhibit diagonal cracks.

When an interconnected network of cracks appears, similar to the cracks that occur in dried mud flats, it is classified as pattern or map cracking. This normally results from improper curing of the concrete or a weakness in the concrete mix design.

If cracks are meandering, irregular, and have no particular form or direction, they are classified as random.

Isolated longitudinal, transverse, or diagonal cracks may be sealed. Wearing surfaces with severe cracking sometimes are overlaid with concrete in an attempt to seal the deck.

3.1.2.2 Asphalt Surfaces

The original bridge design may have included an asphalt wearing surface, or maintenance or rehabilitation may have required an overlay asphalt wearing surface during the life of the bridge.

Cracking: Cracks in an asphalt wearing surface may take different forms from those appearing in concrete wearing surfaces. They include alligator or map cracks, edge cracks, lane joint cracks, reflection cracks, shrinkage cracks, and slippage cracks.

Alligator or map cracking is defined as interconnected cracks forming a series of small blocks resembling an alligator skin or chicken wire. This type of cracking is generally caused by the asphalt material drying out, but could also be caused by excessive deck deflection.

Edge cracks are longitudinal cracks near the edge of the deck. These are usually caused by a lack or loss of lateral support under the wearing course, by the asphalt drying out, or by deterioration of the underlying concrete deck.

Lane joint cracks can develop along the seam between two paving lanes in a longitudinal pattern. This type of cracking is usually caused by a weak seam between two adjacent passes of the asphalt paving machine.

Reflection cracks can develop in the asphalt wearing course that reflect a crack pattern in the underlying concrete deck.

Shrinkage cracks are interconnected cracks forming a series of large blocks in the surface. It may be difficult to determine if the cracks are caused by volume change in the asphalt. Shrinkage cracks can be repaired by filling the cracks with an asphalt emulsion slurry followed by a surface treatment or slurry seal over the entire surface.

Slippage cracks are crescent-shaped cracks caused by a lack of bond between the asphalt surface course and the concrete deck beneath it.

Alligator and slippage cracks are repaired by removing the asphalt wearing course down to the concrete deck surface and in all directions out to sound-wearing surface material. If the overall wearing surface is deteriorated, it may be necessary to mill the entire wearing course off and reapply it as a new overlay. If upon removing the asphalt wearing course, the concrete deck material beneath is found to be deteriorated, it must be repaired first. Regular asphalt paving and patching procedures should be used. If the original surface was treated with a sealant, also seal the newly patched area.

Lane joint cracks, edge cracks, and reflection cracks are usually repaired by cleaning the cracks by brooming and by using a compressed air lance, then filling them with an emulsion slurry or liquid asphalt mixed with sand. After the filled area is cured, it can be sealed with liquid asphalt and the surface blotted with sawdust, sand, or absorbent paper to reduce vehicle tracking.

Preventive maintenance of cracking involves having an inspection process to detect cracks and minor defects early and responding quickly to early distress so relatively minor surface treatment will arrest the rate of deterioration.

Distortion: Surface distortion is any change of the surface from its original shape. It can take a number of forms: channels (ruts), corrugations (shoving), and grade depressions. Channelized depressions may develop in the wheel tracks of an asphalt surface. They may result from consolidation of the asphalt under the pressure of wheel loads; from lateral movement of the surface material because of wheel loads; or from erosion of the surface by studded tires in snow and ice regions permitting the use of such tires.

Corrugations result from plastic movement of surface materials that lack internal friction and stability, producing ripples across the asphalt surface. Corrugations may also appear in a crescent shape with the curved part pointing in the direction of traffic movement, or they may develop into a noticeable bump in the direction of travel.

Depressions are characterized as localized low areas of relatively small size that may or may not display cracking around the slightly sunken area. After a rainstorm, a small quantity of water will pond, known as a "birdbath," and in freeze-thaw climate zones, these small depressions can produce ice patches that drivers have difficulty detecting before they lose traction.

If spot repairs are to be made to channels or grade depressions, the surface should be tested with a straightedge to define the necessary limits of repair. A typical maintenance treatment will be to fill the depressed areas with a thin overlay. Some surface planing may also be required before filling. Corrugations and shoving should be repaired in a manner similar to alligator-cracked areas.

Disintegration: An area that disintegrates will typically exhibit raveling and potholes where fragments and chunks of the aggregate and asphalt become dislodged. If these areas are not treated, the damage will typically continue to grow. Potholes in the wearing course of a bridge deck should be repaired in the same fashion as potholes in an asphalt pavement. Raveling is usually a symptom of poor mix design and defective construction methods and may require milling the surface and relaying the asphalt surface course. If the raveling occurs only at the top part of the surface course, the area may be swept clean, a fog seal applied and allowed to cure, and then the area treated with an asphalt pavement surface.

3.1.2.3 Stone and Brick Surfaces

These surfaces typically suffer from the effects of open joints and rutting. Open joints appear when joints between the bricks fail or are not sealed, permitting de-icing chemicals to penetrate into the underlying deck in snow and ice climates. To correct an open-joint situation, any failed joint sealant must be removed, the joints cleaned with a compressed air lance, and proper joint sealant applied. Preventive maintenance to prevent open joints on stone and brick surfaces involves keeping joints sealed and sweeping the surfaces to preclude dirt and debris from abrading the joint surfaces under tires. Rutting in stone and brick surfaces results from structural failure beneath the stones and bricks, allowing them to displace. Repair of rutting requires removing the stone or brick surface to repair the structural failure in the deck, then replacing the stones or bricks on an appropriate sand cushion, and finally sealing all joints.

3.1.2.4 Steel Grate Surfaces
Open-grate or open-grid bridge deck floors tend to become slippery when wet or frost covered. Additional traction can be provided to vehicles by spot-welding small-diameter studs (about 9 mm [3/8 inch]) to the intersection points of the cross members of the steel floor. Rutting can develop in grid or grate floors that have been filled with concrete if studded tires are permitted in the region during the snow and ice season. Studded tire damage can be corrected by overlaying the area with an asphaltic concrete surface or a portland cement concrete surface if the bridge capacity will permit the additional dead load.

3.1.2.5 Wood Surfaces
Wood wearing courses or deck surfaces are subject to decay problems, mechanical wear problems, and surface planks becoming loose or displaced.

Decay: Both treated and untreated timber floors are subject to decay when moisture penetrates the flooring material around nails, bolt holes, notches, cracks, and in end-grain areas. Timbers showing evidence of decay should be replaced with treated timbers and sealant should be applied around all nails, bolts, notches, and saw cuts. Preventive maintenance usually takes the form of using treated lumber in the first place, sealing every place where moisture might be able to enter the wood with a waterproofing compound, and checking surface drainage to ensure that water will not pond or stand on the bridge surface.

Mechanical Wear: Tire abrasion wears down the wood fibers in the decking. This wear is accelerated by dirt and debris on the deck as well as by any use of studded tires during the snow and ice season. Maintenance repair involves replacing worn planks with new treated lumber planks as needed. Preventive maintenance is generally limited to keeping the plank surface free of debris and dirt.

Loose, Displaced Deck Planks: This condition usually results from broken deck planking fasteners or failures developing in supporting members. Maintenance repair involves replacing failed fasteners and any failed supporting structural members. Preventive maintenance consists of replacing any deteriorated fasteners with galvanized fasteners and keeping the wood wearing surface clean, limiting water accumulation and moisture retention on the surface area.

3.1.3 Structural Decks
3.1.3.1 Introduction
Bridge decks carry traffic and protect the underlying structure. In this dual role, bridge decks are subjected to the effects of mechanical wear and weather-induced deterioration. As a result, bridge decks typically require more maintenance than other elements of the structure. Concrete bridge decks must resist wear and weather well, provide good traction and a smooth ride, and are often an integral part of the bridge's structural support system. Design features that reduce the level of maintenance required include providing optimum depth of cover over the reinforcing steel, creating a surface that provides good deck drainage, and generally developing deep-thickness structural slabs. Materials and construction practices that reduce the level of maintenance required include maintaining the design depth of cover over the reinforcing steel, using air-entrained concrete, holding the water-to-cement ratio as low as possible to still achieve a workable mix, developing good consolidation of the placed concrete, and controlling the environment that may cause excessive shrinkage [1].

In snow and ice climate areas, de-icing chemicals applied to bridges provide one of the main sources of damage. Most of these chemicals gradually penetrate the deck and then may attack the reinforcing steel; in some cases, they can even attack the aggregate in the concrete. Using high-quality aggregates in the original construction, using low water-to-cement ratio concrete, using reinforcing bars with corrosion-resistant coating, sealing the deck surface, or installing cathodic protection systems are some of the ways to counteract the corrosive effects of de-icing chemicals.

Proper repair of deteriorated decks requires the most cost-effective treatment within the constraints of the fiscal budget, the personnel available, the work zone traffic control required, the likely weather conditions, and the availability of equipment and materials. A good decision on treatment cannot be made if good information on the extent and nature of the deterioration is not available. A visual inspection by a skilled, experienced person may be sufficient to identify the location and description of spalling, scaling, and cracking on both the top and bottom of a deck, but more intensive testing and analysis will be required for other information to guide the maintenance and rehabilitation process.

3.1.3.2 Preventive Maintenance of Concrete Decks

Preventive maintenance of concrete bridge decks is mostly aimed at controlling salt and moisture penetration (from de-icing chemicals in the snow-and-ice belt and from marine saltwater exposure in coastal areas) to prevent or retard corrosion of reinforcing steel and deterioration of the cement-aggregate matrix of the concrete.

- Keep the deck clean. Maintain the original surface drainage of the deck. Keep the drains from the deck open.
- Monitor deck condition by testing for chloride penetration, delaminations, and active corrosion on a regular interval schedule.
- Seal or overlay the surface to prevent or reduce salt and moisture penetration.
- Seal cracks to prevent or reduce reinforcing corrosion.
- Remove and replace deteriorated concrete. If some concrete areas are highly chloride contaminated, they may have to be removed and replaced.
- Apply cathodic protection when appropriate. Cathodic protection is growing in popularity as a means of preserving and maintaining concrete decks that are experiencing (or may experience) deterioration from chloride-induced corrosion of reinforcing steel. The process is based on creating an electrical current flow that reverses the ion exchange that occurs at the surface of the reinforcing steel when water containing chloride ions reaches the steel. Thus, it is very difficult to apply cathodic protection to a deck in which the reinforcing steel has an epoxy coating. Monitoring of some decks with cathodic protection systems has indicated that reinforcing steel corrosion can be halted with the proper electrochemical design of the application.

3.1.3.3 Concrete Deck Sealing

Sealer materials require penetration since the surface film is worn away by traffic abrasion. Sealer materials include silanes, siloxanes, silicone, epoxies, and methyl methacrylates. Good sealing requires a clean deck, a dry deck, and warm temperatures. Maintenance procedures should include the following steps [1] (NCHRP Synthesis of Highway Practice 220 provides a comprehensive assessment of available waterproofing membranes and the range of practices in applying them [10]):

- Clean the deck by brooming, washing, air-blasting, or a combination thereof. For some sealants, sandblasting or shot-blasting may be recommended by the manufacturer.
- Check the environmental conditions with respect to the agency's and/or the manufacturer's specifications for sealant application. Typical items to be checked are as follows:
  → Correct temperature of air and deck.
  → Temperature range of materials as recommended.
  → Wind velocity low enough to prevent air drift.
  → No falling temperature during or immediately after application.
  → Clean deck with no moisture or oil on it.
  → Other conditions that could prevent satisfactory application.
- Mark off or measure areas to be sealed, ensuring the correct rate of application.
- Prepare sealant as recommended by manufacturer.
- Apply sealant by a recommended method, such as by spraying using pump tanks or mechanical spray equipment, or by using notched or unnotched squeegees, rollers, or distributors.
- The freshly sealed surface may have to be treated to retain acceptable skid resistance. Sealants that do not penetrate the deck surface usually require an immediate application of sand or other grit material that will embed into the sealant. Sealants that penetrate the deck surface may have to be blotted with sand until the sealant has been absorbed into the deck, temporarily increasing skid resistance.
- Estimate cure and deck absorption times for the application method used and match lane closures and traffic control to these estimates.
- Monitor procedures to maintain maintenance personnel safety in the handling and application of sealant chemicals, and to ensure that the natural environment beyond the bridge deck is not contaminated with any hazardous materials.

3.1.3.4 Concrete Deck Patching

Deck deterioration requiring patching is usually caused by corrosion of the reinforcing steel. Patching is a temporary repair unless all of the chloride-contaminated concrete is removed before the deck is patched. If only the spalled and delaminated concrete is removed, the corrosion process continues and additional spalling will soon appear. Studies of deck-sealing practices suggest that while sealing or overlaying chloride-contaminated concrete will not stop the concrete's corrosion and deterioration, it can slow the process, which may be acceptable, depending upon the schedule of future major rehabilitation efforts, such as entire deck replacement.

Potholes in a deck sometimes require a "temporary" patch, while at other times a "permanent" patch may be best. They should not be allowed to remain without some action if they are severe enough to adversely affect the rideability of the deck or the safety of vehicles operating at normal speeds. In addition, the vertical acceleration of vehicles hitting potholes increases the impact loading on a bridge and can increase damage to other parts of the bridge structure.

Procedures to patch concrete decks should include the following steps:
- Evaluate surface areas to be patched, using hammers or a drag chain for hollow-sounding delaminated areas (or using instruments to detect unsound areas).
- Outline the area to be patched with spray paint (or lumber crayon). A concrete saw is best for a rectangular area with square corners. Mark the area about 150 mm (6 inches) beyond the detected delamination to ensure coverage of all damage.
- The saw cut is usually about 20 mm (3/5 inch) deep around the edge of the patch to provide a good vertical edge face. Monitor sawing to ensure that no reinforcing steel is cut. Saw operators should not cut beyond the patch area in the corners or a future spall will form there. Patch areas that are within 600 mm (2 feet) of each other should be combined into a single larger patch.
- Workers should use hand tools or pneumatic hammers weighing 15 kg (30 pounds) or less (7 kg [15 pounds]) (at reinforcing steel level) at an angle of 45° to 60° to the deck, removing the concrete within the patch area. The patch area should be periodically sounded to ensure that the area and depth are correct and that all deteriorated concrete has been removed. If fracture lines are observed over a reinforcing steel bar, this indicates an area that will soon spall and should thus be removed.
- The patch area should be thoroughly cleaned by sandblasting or water-blasting to remove loose concrete, rust, oil, or other materials that will prevent good concrete bonding.
- Reinforcing steel will probably be deteriorated from any corrosion action. As a general rule, if the reinforcing steel bar has lost over 20% of its original cross section, new steel bars should be added by lapping, welding, or mechanically connecting them to the deteriorated bars.
- Patching is often classified according to the depth of the patch required to make the necessary repairs.
    - Type A Patches only above the top layer of reinforcing steel. This type of patching may require special aggregate since the largest diameter of aggregate particles can not exceed the depth of the patch. If the patch depth is too thin for effective concrete patching and, simultaneously, too thick for epoxy mortar patching, then the patch depth may need to be increased to create an effective concrete patch.
    - Type B Patches from the deck surface to at least 25 mm (1 inch) below the top mat of the reinforcing steel. Type B patching will require removing enough concrete from under the reinforcing steel bars to permit fresh concrete to flow under the bars and to ensure that no voids exist.
    - Type C Patches to the full depth of the deck. If Type C patching is suspected to be required, ensure that preparations include gaining access to the deck's underside. Type C patching will require formwork to support the bottom of the hole. When any Type C patch exceeds a 1.3 m × 1.3 m (4 feet × 4 feet) area, an engineer qualified to assess the structural implications should be consulted, determining if patching or some other rehabilitation action should be pursued.
- When the fresh concrete is placed in the patch hole, the surface of the concrete to be patched should be damp but free of standing water. The patch concrete should be finished with a straightedge or float to produce a patch surface that is not more than 3 mm (1/8 inch) above or below the surrounding concrete surface.
- If a patched area must be opened to vehicular traffic flow quickly, rapid-setting patch materials can be used. However, maintenance personnel must follow the manufacturer's or agency's specifications exactly while applying these materials because the quality of the patch is very sensitive to application condition control limits.

- Any patching concrete mixed from scratch (not arriving at the job site in premixed sacks or containers) should be mixed in accordance with a design developed by a qualified materials engineer and with the agency's specifications. If large areas of a deck are being patched, a qualified bridge engineer should be consulted to determine if the patching process will adversely affect the bridge's structural capacity.
- Proper curing of the patch is important. If the cement being used is supposed to have a wet cure, ensure that the surface remains wet until it is opened to traffic. Membrane sealing cures may be appropriate.

3.1.3.5 Epoxy Deck Patching

Epoxy is not a commonly used patching material. Its main advantage is that it does not require a sawed hole. The maximum depth of the hole is generally limited to about 19 mm to 25 mm (3/4 inch to 1 inch) owing to high shrinkage of the material during the curing process. If neither an agency specification nor any manufacturer's recommendation is available for a mix, experience has shown that about one part epoxy to four to seven parts of well-graded dry sand is a good epoxy-patching mix [1]. Maintenance personnel should remember to add the sand to the epoxy, rather than adding the epoxy to the sand.

3.1.3.6 Asphaltic Concrete Patching

Asphaltic concrete is a porous material that will permit moisture (and any associated chloride ions) to reach the concrete deck and the reinforcing steel unless a waterproofing membrane is applied to the cleaned, exposed concrete before the asphalt is placed in the patch area. Patching a deteriorated area in a reinforced-concrete deck with asphalt should be considered a temporary patch until a better time to apply permanent repairs becomes available.

3.1.3.7 Emergency Full-Depth Patching

A full-depth hole in the deck will often require emergency repair. Maintenance personnel must be cautious to keep traffic from moving across a bridge with a full-depth deck hole unless it is clear (after consulting with a qualified engineer if necessary) that the hole is not a symptom of some greater structural weakness that may require closing the bridge for major rehabilitation and repair. If it is appropriate to make emergency repairs to the hole, it may be possible to hang plywood underneath the deck, allowing the hole to be filled with one of several proprietary materials that will develop high strength quickly, even in low temperatures. Sometimes steel plates can be bolted over a full-depth hole to provide an emergency cover. Steel plates have the disadvantage of creating a bump and a slick spot on the deck surface. In either type of emergency repair, the materials needed typically cannot be obtained on emergency notice. Therefore, advance planning is required so the necessary materials are available when needed. This type of emergency repair should be replaced by permanent repair or rehabilitation as soon as is practicable.

3.1.3.8 Crack Sealing

Cracks in the concrete deck that reach the reinforcing steel and have widths greater than about 0.18 mm (0.007 inch or the thickness of two sheets of paper) can allow moisture and associated chlorides to initiate and support corrosion of the reinforcing steel. Some experiences suggest that even epoxy-coated reinforcing steel will develop corrosion over an extended time if concrete cracks are left unsealed. Some common causes of cracking in a concrete deck are as follows:

- Thermal stresses or drying shrinkage stresses.
- Defective aggregates that cause shrinkage or expansion of the concrete matrix.
- Too much water in the concrete mix.
- Insufficient or improper curing of the concrete.
- Bending and flexure stress in the deck.
- Movement between beams and girders supporting the deck.
- Bridge foundation settlement.

If small cracks exist over large areas, application of a liquid sealer to the entire deck surface may be an effective treatment to seal out moisture.

Large, open cracks that are no longer expanding may be treated by injecting a sealant (e.g., an epoxy- or polyurethane-based material). If the crack has passed completely through the deck, it will be necessary to seal off the lower surface of the crack before injecting sealant into the cracked area from above. To seal large cracks successfully by injection, a clean crack (vacuumed, air-blasted, water-blasted, chemically washed, and flushed) is important.

If a crack is open and appears to be a moving crack, if a crack has been sealed before and has failed again, and if evidence of recent recracking exists, the crack is a working crack and may be sealed with a flexible crack sealant material. This usually requires routing a groove over the crack to hold the sealant and any recommended backer material.

In all crack-sealing operations for bridge deck maintenance, observe the following cautions:

- The cause of the cracking condition should be determined and corrected (if practical) at the same time.
- The crack surfaces must be clean and dry before sealing the crack.
- The depth of the seal should be less than or equal to the width of the seal.

Laboratory quality control tests suggest that gravity-fill polymer crack sealers effectively seal cracks in concrete bridge decks [1]. With the wide variety of sealers available, a methodology has been developed to predict concrete sealer service life so that a reapplication interval can be estimated [12].

3.1.3.9 Overlays to Bridge Decks

Overlays to bridge decks may be applied as part of a preventive maintenance program or as part of the deck repair process. In some cases, chloride-contaminated deck concrete may be removed before the overlay is placed. However, if a waterproofing sealer is applied with the overlay and chloride-contaminated concrete is allowed to remain, the combined process can add to the remaining deck service life because even though the corrosion process will continue at the level of deck-reinforcing steel, it will be retarded.

Cement-Based Overlays: These include concrete, latex-modified concrete, silica fume concrete, and fiber-reinforced concrete. These overlays are usually about 25 mm to 50 mm (1 inch to 2 inches) thick with small aggregate (maximum 13 mm [0.5 inch]) and a low water-cement ratio, producing a very dense mix that resists water penetration when it is properly cured. Because proper curing is so important with these thin overlays, special care should be taken to support good curing conditions.

The existing deck should be prewetted so it will not draw moisture from the mix. However, it is best if the surface is dry at the time the concrete overlay is placed. Avoid placing the concrete on very hot or windy days. Such conditions can contribute to cracking caused by rapid evaporation of water from the concrete surface. To reduce cracking and permeability, the finishing process should be completed as quickly as possible, and the curing should begin as quickly as possible after the finishing is completed.

Asphaltic Concrete With a Waterproof Membrane: Bituminous concrete is sometimes used with a waterproof membrane to prevent intrusion of moisture and chlorides into the deck. The asphaltic concrete acts as an armor coat to protect the membrane from wear and damage by vehicular traffic. Membranes are either applied in place or preformed. Membranes applied in place are usually squeegeed across the surface of the deck, while preformed membranes are delivered from the manufacturer in rolls of thin membrane material that are applied directly to the deck after priming it. "Outgasing" of moisture and air in the deck can result in bubbles and pinholes in the membranes. The deck can also be sealed with a liquid sealant material or an epoxy-resin-based sealer prior to placing the bituminous concrete surface overlay. Regardless of the waterproofing membrane or sealing process used, one of the keys to successful treatment is carefully laying the asphaltic concrete to avoid damaging the waterproofing film. Since a bituminous concrete overlay is usually less durable than a cementatious concrete overlay, it is important to plan for removal and replacement of the bituminous overlay periodically. If the bridge normally carries relatively low traffic volumes, sometimes a chip seal may be sufficient to protect the waterproofing membrane.

Polymer Overlays: A polymer overlay results when a polymer waterproofing material is applied to the concrete deck and covered with sand or a similar fine granular material (to improve the polymer's skid resistance). The application process can be repeated to build up the thickness, generally limited to three applications. Some epoxy overlays have exhibited brittleness, some have lost excessive sand under the abrasive pressures of vehicular traffic, and some have lost their bond to the concrete in large areas. Polymer concrete has a different coefficient of expansion than deck concrete, which can introduce large thermal stresses when the overlay is exposed to large temperature variations. These thermal stresses contribute to the polymer concrete bond breaking loose from the deck concrete, and where the bond is stronger than the deck concrete, it can even fracture the deck concrete just below the polymer overlay. Polymer concrete materials applied to bridge decks should have test application data from environments comparable to those in which the agency is applying them. Tests on improved polymer concrete overlays have shown a satisfactory service life of 5 to 7 years [13], while other studies suggest that a 10- to 20-year life might be possible [14].

Steel-Fiber-Reinforced Concrete: This is an experimental overlay that may prove advantageous in the future as more experience is gained using it [15].

3.1.3.10 Concrete Deck Replacement

Many agencies do not undertake entire or partial replacement of concrete decks because of the large commitment of personnel, materials, and equipment required; instead the work is contracted out. If an agency does undertake deck replacement with maintenance forces, it is usually best to have a crew that specializes in this work. When undertaking concrete deck replacement, other aspects of the bridge deficiency need to be assessed for possible upgrading, rehabilitation, or repair at the same time. Consider the extent to which a deck replacement is a significant proportion of the total cost or function of the bridge; replacing the deck may place the agency in a program obligation or liability obligation to upgrade the bridge's other functional deficiencies. If the bridge design permits, it may be appropriate to replace sections of the bridge deck by cutting them out and inserting precast panel sections. If future bridge designs are developed with more flexibility to replace decks one panel at a time, the maintenance of bridge decks will gain an additional degree of flexibility. A methodology has been developed that allows maintenance engineers and managers to consider a variety of factors if they choose to apply a rational planning approach to programming concrete deck replacement [16].

3.1.3.11 Timber Bridge Decks

Timber decks are usually the easiest type to repair. Broken, worn, or decayed planks can generally be replaced with little difficulty. When replacing them, nail the planks securely because loose planks increase future surface wear and create traction difficulties for vehicles. Plank or laminated timber decks can be fastened to steel I-beams, using metal fasteners or wood spiking pieces. A timber deck can be replaced by maintenance personnel relatively easily. Consequently, if a significant number of individual planks need to be replaced as part of the maintenance effort, the total life-cycle cost of material and labor may be reduced by replacing the entire deck, especially if an inspection indicates that further deterioration of individual planks will require more replacements in the near future. If an asphalt wearing surface is placed over the timber deck to improve surface traction and to improve the surface's ride quality, it must be frequently checked for needed maintenance. A deteriorated asphalt wearing surface over a timber deck may hinder more than help the deck surface's quality. Typical steps in replacing a timber deck are as follows [1]:

(1) Remove and store deck-mounted curbs, parapets, wheel guards, and railings.
(2) Remove existing deteriorated decking that can be repaired or replaced in the same work day.
(3) Clean and paint top flanges of stringers.
(4) Place deck planks on the stringers parallel to the abutment. Place the lumber with the smallest cross-sectional dimension bearing on the stringer. Make certain that each plank rests on at least three stringers and that all end joints of planks are staggered across the deck.
(5) Spike each plank in place with spikes spaced 300 mm (12 inches) apart, and alternate adjacent spikes with one being about 50 mm (2 inches) from the top of the plank and the next spike being about 50 mm (2 inches) from the bottom.
(6) Place anchor plates on each stringer spaced about 300 mm (12 inches) apart on alternating flange edges.
(7) Apply wood preservative to the ends of any plank that is field cut. (Timber deck maintenance and replacement should always use treated lumber.)
(8) Replace all deck-mounted appurtenances (curbs, wheel guards, railings, etc.) after any wearing surface has been applied.

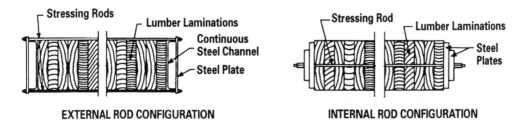

Stress Laminated Deck

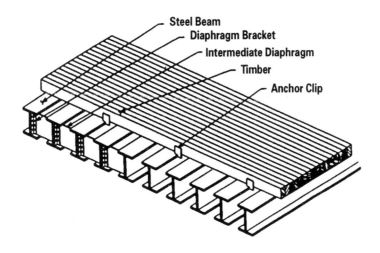

Timber Deck Replacement

3.1.3.12 Steel Grid Decks

Steel grid decks are usually the bridge designer's choice when a major design criterion is to have a bridge deck of the lightest possible weight capable of providing load capacity for heavy live loads. Thus, steel grid decks are common on movable lift span bridges. Some common concerns in maintaining steel grid decks include the following:

- Welds or rivets that join the grids or that hold them down may break. When inspection indicates this problem exists, these fastenings should be repaired or replaced since a loose grid can be a serious safety problem. Maintenance and repair may require reinforcing or replacing a damaged grid plate. If several grid plates have been damaged or displaced, it may be best to cut out an entire portion of the deck and replace it.
- Elements below open-grid decks should be cleaned regularly to flush out chlorides since the open grid does not offer any protection to those elements.
- Steel grid decks generate high levels of traffic noise. When noise complaints (especially in urban areas) are a problem, and if the bridge design will permit the additional dead load, filling the grid with lightweight concrete may reduce traffic noise generation.
- Moisture, especially frost, reduces any skid resistance of a steel grid deck. Welding studs (such as 8 mm [5/16 inch] in diameter and 10 mm [3/8 inch] high) to the intersections of the cross members may increase the grip of vehicle tires to the deck surface. One agency has used a scabbler device to roughen the surface of the steel grid deck, increasing the traction of rubber tires on the deck surface.

3.1.3.13 Maintaining Deck Joints

Some deck joints are sealed and some are not. Damaged joints or leaking seals need to be repaired or replaced for the bridge to function as the original design intended. Leaving a damaged deck joint or breaking seal unattended also increases exposure of the bridge structural elements under the deck to increased damage from debris and contaminating materials from the deck surface. No bridge deck joint is perfect, but if a bridge deck joint was properly installed at the original construction and is well maintained, it will contribute to a bridge's long, effective service life. Ideally, a joint should be watertight, allow deck movement in expansion and contraction, last at least as long as the adjacent deck materials, and never require any maintenance. However, maintenance personnel will have to deal with less than ideal joints.

Joint Types: Joints in bridge decks can be classed into four general types: construction, contraction, nonexpansion, and expansion.

Construction Joints: Those joints are boundaries separating sections of a concrete deck placed at different times. Generally, these joints will be watertight because of the chemical bonding of the newer concrete section to the previous section when the concrete placement was resumed.

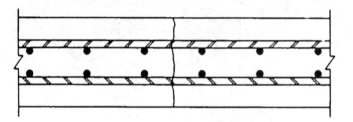

Construction Joint

Contraction Joints: These joints control the cracking effect of concrete shrinking as it cures. Generally these joints have a water stop incorporated into them, and they are typically found in wing walls and retaining walls where seeping water needs to be controlled.

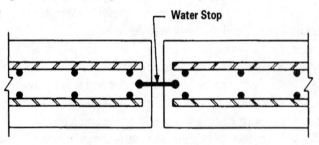

Contraction Joint

Nonexpansion Joints: These joints allow rotation of a beam end but do not permit expansion. They are generally used at the abutment supporting a fixed bearing. These joints are typically smaller than expansion joints.

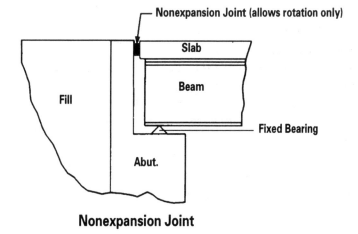

Nonexpansion Joint

Expansion Joints: These joints are used if both rotation and expansion need to be allowed. In long span bridges, such as suspension bridges, multiple expansion joints may be needed to allow up to several meters of total movement.

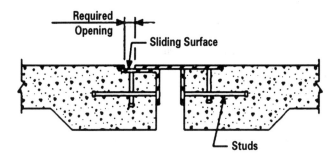

Expansion Joint (Sliding Plate Type)

Joint Functions: Bridge deck expansion joints accommodate movement of the superstructure as live loads move across the bridge, as environmental conditions change, and as the bridge materials themselves change over time. Live-load deflection and substructure settlement can cause rotational movement. Skewed bridges can develop significant maintenance problems because of the difficulty of allowing transverse movement through the joints. Movement is usually allowed by providing a space between rigid sections of the superstructure equal to or greater than the movement expected. This space requires a break in the deck surface that adversely affects ride quality and provides an opportunity for water and other contaminating materials to reach the structural elements below the deck.

Joint Classification: The two general types are open joints and closed joints.

Open Joints: In open joints, water and contaminating materials are allowed to pass through the joint and onto elements beneath or are collected in a drainage trough to move them away from sensitive areas of the bridge structure. This type of joint is not used as much now as in the past because the drainage trough must be well maintained and because it allows water and material to move directly into the deck. Some types of open joints are butt, sliding plate, and finger.

Butt Joints: Butt joints may or may not be reinforced at the joint face. They are often used where the movement is limited to 25 mm (1 inch) or less and may be commonly found in a bridge at a point where only rotation movement is expected with some minor thermal expansion or contraction. It is difficult to protect the joint from corrosion. If the deck receives an asphalt overlay and the joint has a reinforced face, then the reinforced face should be extended through the overlay. Normal maintenance includes periodically clearing the opening of roadway debris, painting, and repairing the roadway surface adjacent to the reinforcement (armor plate).

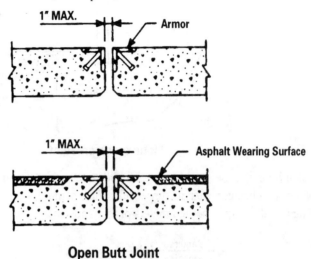

Open Butt Joint

Sliding Plate Joints: Sliding plate (plate dam) joints are typically used when movement between 25 mm (1 inch) and 75 mm (3 inches) is expected. This was the predominant type of bridge expansion joint used before neoprene joints were developed. Plate joints tend to be difficult to maintain. The plates clatter and bang when hit by a vehicle; they sometimes come loose from the bridge, presenting a roadway safety problem. Maintenance personnel need to clean the joint periodically to prevent excessive buildup of debris in the joint, to check for deterioration of the concrete in the vicinity of the joint, to check for deterioration of the anchor bolts holding the plate in place, and to keep all exposed metal parts painted and free of corrosion.

Finger Joints: When movement is expected to be more than 75 mm (3 inches), a finger or cantilever joint is typically used. These joints require the same type of maintenance care as sliding plate joints, in addition to maintenance of the drainage trough typically installed under the finger joint.

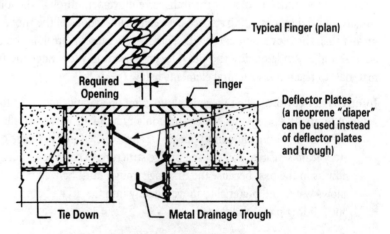

Finger Joint With Drainage Trough

Closed Joints: Closed joints are designed and intended to be waterproof. Some types of closed joints are filled butt, neoprene compression seal, membrane seal, cushion seal, and modular dam.

Filled Butt Joints: This joint is very similar to an open butt joint and is intended to be applied to the same general movement conditions, except that a premolded joint material is usually attached to one face of the joint or supported from beneath by an offset in the vertical slab of the deck. Sealant material is placed into the joint at the roadway surface, sealing the opening and preventing water from entering the joint. Maintenance typically includes periodic cleaning of the point, replacing the sealant at the roadway surface, replacing the filler material when needed, and repairing the roadway surface adjacent to the joint when needed. Generally, a poured-in-place seal performs best when the expected movement is less than about 13 mm (about 0.5 inch). Under the best conditions, the seal can be expected to remain watertight for a maximum of 2 years. If the joint is not maintained to retain a watertight seal, the filler material below the seal will deteriorate and incompressible fine material can enter the joint, preventing it from relieving the stress of deck expansion.

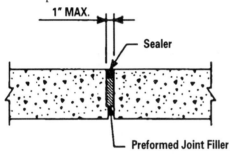

Filled Joint

Neoprene Compression Seals: These seals can be used where deck movement up to about 65 mm (about 2.5 inches) is expected. Successful joint operation requires that the initial installation provide a properly sized opening so that under low temperatures tensile stresses will not separate the seal from the deck face and under high temperatures compressive stresses will not crush or damage the seal. Maintenance includes sweeping or flushing the deck to prevent incompressible fine materials from building up in the joint area and periodically inspecting the seal for cracking and deterioration from weathering effects.

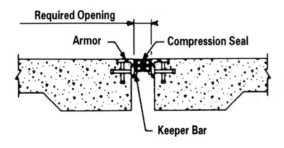

Compression Seal

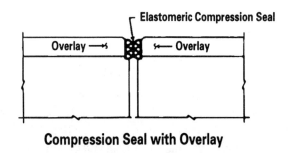

Compression Seal with Overlay

Membrane Seal: A membrane (strip) seal is a flexible sheet of neoprene rigidly attached to two metal facings at the joint. The installed seal has a downward curved shape and flexes ("stretches") with the bridge deck's movement. Membrane seals can be applied where movement up to about 100 mm (about 4 inches) is expected. Installations sometimes leave breaks or cracks in the seal at gutter lines and at points where a change in the deck cross section occurs. Debris and fine material in the joint space can tear the seal or cause it to come out of the metal facing under the downward force of vehicle wheels. Maintenance, therefore, should include periodically removing debris and reattaching or replacing defective membranes.

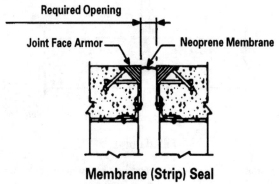

Membrane (Strip) Seal

Cushion Seal Joint: In a cushion seal (elastomeric) joint, a reinforced neoprene pad is rigidly attached to each side of the joint. The neoprene pad has elastic characteristics that allow it to stretch when the joint opens and shrink when the joint closes, while the reinforcing materials in the pad provide sufficient strength to span the joint gap in a durable manner. Cushion seal joints are normally used to span joints with movement up to about 100 mm (about 4 inches). These joints need regular inspection to ensure that the cushion remains firmly anchored, especially when the bridge contracts. Joint maintenance at a curb line can require frequent attention also. A cap to seal the anchor area can be "glued" down with adhesive and aid in maintaining a watertight seal, but frequent inspection is needed because the adhesives tend to deteriorate, resulting in cap loss and leaking joint seals. These joints should be periodically cleaned, the anchoring devices should be inspected and replaced when needed, and the seal repaired when needed.

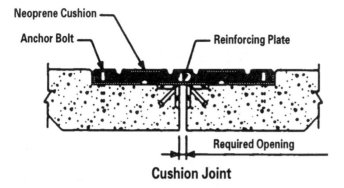

Cushion Joint

Modular Dam Joint: These joints are typically fabricated to accommodate joint movements in excess of about 100 mm (about 4 inches). These joint seals are special designs incorporating strip or compression seals separated by beams and supported by a series of bars. They are designed to allow some individual sections and components to be replaced as needed. These joint seals should be inspected periodically, especially after snowplowing season in snow and ice climates because snowplow blades can damage them. Impact from heavy traffic is also a source of damage, necessitating regular inspections. Like strip seals, these joints need to be periodically cleaned for the same reasons.

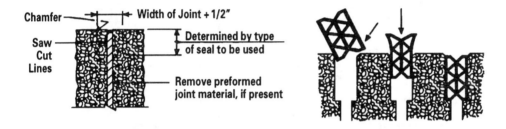

Upgrading Joint to Accommodate Compression Seal

Chapter 3 Bridge Maintenance and Management

Joint Preventive Maintenance: For maximum effectiveness and highest return on resources expended, preventive maintenance of bridge deck joints should start when the bridge is new and continue throughout its life. This is especially important so that maintenance personnel may identify any problems that arise early for which the source is improper design or construction, such as improper forming of the joint opening, wrong joint opening (too wide or too narrow), wrong seal size for the joint movement, placement of a seal too high, construction damage to the joint edge, weak bonding of a seal to the adjacent concrete, or failure to install a bond breaker in the lower part of the joint opening. Early initiation of preventive maintenance will help detect damage problems early such as those arising from later damage to the edge of a joint, raveling of the wearing surface over a joint, joint plates working loose, joint fillers deteriorating, and anchorage failures developing in cushion joints. Failure to continuously and properly maintain bridge deck joints can result in damaged end diaphragms, damaged beam ends, damaged bridge bearings, damaged seats and caps, accelerated deterioration of the bridge substructure, and embankment erosion problems at the abutments. Where deck joints are a significant source of maintenance problems, the possibility of upgrading the joint seal to a compression seal should be investigated for the potential to decrease the frequency and intensity of maintenance required.

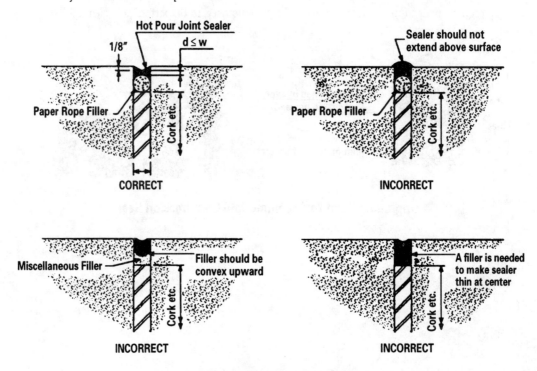

Hot Pour Joint Seal Installation

3.1.3.14 Deck Drainage Systems

Proper maintenance of deck drainage systems to ensure the flow of water off the deck and away from the structure is just as important as keeping deck joints sealed. It does little good to keep deck joints watertight if the water and any associated contaminants cannot flow away from the bridge structure. Furthermore, if water ponds on a deck, it can result in reduced traction for vehicles crossing the bridge under certain conditions. If the deck drainage is not efficient and effective, dirt and debris may accumulate on the deck and in the joints, leading to increased maintenance requirements for other bridge elements.

Scuppers (Drop Through and Piped): Small-diameter pipes, long downspouts, and any horizontal runs with relatively flat slopes or sharp directional changes all contribute to clogged drains. Short drops and through pipes that drain directly under the bridge may corrode structural steel, deteriorate concrete piers, and contribute to erosion of abutment earth slopes. Dirt and debris should be removed from scuppers and downspouts by high-pressure water jets or with metal probes (taking care not to puncture any pipe if metal probes are used). Preventive maintenance requires regular, frequent inspection; cleaning to limit accumulation of dirt and debris; application of coatings to piers and structural steel as needed to prevent corrosion from water passing through the drainage system; and if deck drainage threatens abutment earth slopes, protection of them by paving the slope or using ditch liners. Deck drains should be designed and constructed to be readily accessible to maintenance personnel for cleaning. If this is not the case, maintenance engineers and managers need to ensure that design engineers are made aware of maintenance personnel needs. Deck drains designed and constructed to drain through bridge structural elements, such as columns, must be inspected and cleaned as needed in freeze-thaw climates before the onset of freezing weather, ensuring that water will not stand in the drain and damage the structural column.

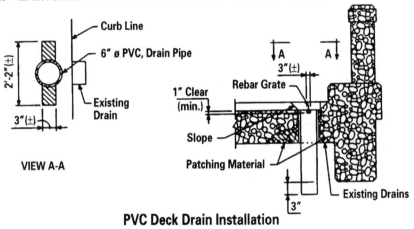

PVC Deck Drain Installation

Chapter 3 Bridge Maintenance and Management

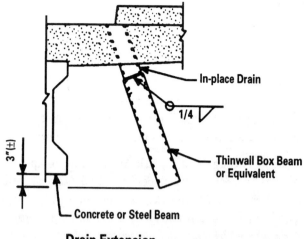

Drain Extension

Gratings: Open steel grate decks or open-grid deck floors usually provide good drainage of the deck floor but tend to allow dirt and corrosive elements to accumulate on the structural members below the deck. Required maintenance activity includes removing dirt and debris from beams, girders, and pier caps on a regular basis. Regular and frequent inspection with cleaning, as needed, is necessary to prevent dirt and debris from accumulating. Applying protective coatings, covers, or shielding plates to structural members below the grating helps reduce the corrosive impact of water passing through the gratings.

Open Joints With Troughs: Troughs under open joints are also subject to clogging and dirt accumulation that then produces a drainage backup that can contribute to steel corrosion, concrete deterioration, and earth erosion at abutments. Maintenance activity should provide regular, frequent inspection and cleaning of troughs under open joints. Protective coatings applied to piers and abutments in the vicinity of open joints with troughs will help slow any corrosive effect of the water passing through the open joint. Earth slopes under the discharge pipes should be protected by concrete paving or rock riprap. Long-term maintenance expenditures may be less with riprap-stabilized earth slopes.

3.1.3.15 Curbs, Sidewalks, and Railings

Curbs: While curbs may be constructed of metal, stone, asphalt, or wood as a raised barrier parallel to the intended traffic movement, generally the curb will be of the same material as the deck. Curbs with a high enough face will act as a limited traffic barrier and aid in guiding a vehicle in the correct path. If deck surface drainage is through inlets and downspouts, the curbs channel surface water flow at the edge of the deck. If curbs are intended to protect pedestrians on a sidewalk on the bridge deck to the maximum degree practicable, the curb should be integral with a "New Jersey–type" barrier between the vehicle-traveled lane and the sidewalk; however, in climates with snow and ice, this will increase snow removal maintenance.

Metal Curbs: Metal curbs that are intended to channel water are often difficult to keep sealed at the joint with the deck surface (as compared to curbs intended to allow water to pass through). Metal curbs are also subject to high levels of corrosion as well as misalignment and damage from vehicles hitting them and loosening anchor bolts. If metal curbs are to channel water, they should be set in a noncorrosive sealant material. Any curbs with structural sections significantly reduced by corrosion should be replaced, and decks with metal curbs should be flushed frequently to reduce corrosion activity. Metal curb corrosion can be reduced by applying protective coatings early in the life of the curb. Misaligned (or damaged) curbs should be replaced and brought back to the original design and construction alignment.

Stone Curbs: Stone curbs are usually associated with bridge decks where architectural and aesthetic values have to be maintained in an historic district. The joints of stone sections are normally sealed and keyed with mortar. When stone curb sections become loosened because of mortar deterioration but the curb is not displaced, the deteriorated mortar should be removed down to sound mortar and replaced with new mortar. If any curb sections are displaced from their proper position, the stone sections should be removed and reset in new mortar. Mortar deterioration can be retarded by using mortar with additives to reduce permeability to moisture and by applying a waterproofing sealant to the mortar surface.

Concrete Curbs: Problems with concrete curbs include freeze-thaw deterioration, deterioration from de-icing chemicals, spalling at expansion joints, cracking, and deterioration from corrosion of reinforcing steel or anchor bolts. Deteriorated concrete should be removed down to good concrete (even if the entire curb section must be removed), the area thoroughly cleaned (sandblasted if needed), and a patch applied. Spalling is often associated with expansion joint movement, so in the repair process additional clearance for longitudinal movement should be provided. Cracks often appear as transverse cracks in adjacent deck paving or adjacent sidewalk paving. Minor cracks can be sealed with a waterproofing sealant material. Where reinforcing steel and anchor bolts are encountered in curb repair, if corrosion has caused the bar or bolt cross section to be reduced, they should be replaced with new bars or bolts in the concrete repair process. Preventive maintenance includes regular deck flushing, treatment of curbs with waterproofing sealants in snow and ice climate areas, and good deck joint maintenance.

Asphalt Curbs: Asphalt curbs are primarily used to provide visual tracking delineation to drivers and to aid in channeling surface water on the deck since this type of curb is easily damaged by vehicles and often loses its bond to the deck. Any sections that have been damaged by vehicles or lost bond with the deck should be removed and replaced. Small sections of an asphalt curb can be repaired by hand shaping, but if any significant length of curb is to be replaced, it is usually more efficient to use a curb-extruding machine. A loss of bond between the curb and the deck can be minimized by ensuring that (1) the deck surface is dry, clean, and properly primed before installing an asphalt curb; (2) the asphaltic concrete is between 225°F and 300°F; and (3) deck surface temperatures are above 40°F to 50°F.

Timber Curbs: Timber curbs should exist only on low-volume road bridges with timber decks. They are subject to the same problems as deck timbers (splitting, warping, weather checking, decay, and corrosion of anchor bolts and fasteners). Any sections that have deteriorated or are misaligned and likely to snag a vehicle tire should be replaced as soon as practicable. Preventive maintenance measures include using treated lumber, using galvanized bolts and fasteners, and periodically checking for loose sections.

Sidewalks: Since sidewalks facilitate the safe movement of pedestrians over a bridge, the surface needs to be maintained so that people do not encounter any surface hazards. In concrete sidewalks, cracks should be sealed; spalls or potholes need to be filled; scaled areas need to be resurfaced for a smooth, well-draining surface; joints need to be kept sealed to minimize any tripping potential; and surface drainage needs to minimize any surface ponding of water (which may form ice patches in freeze-thaw climates). While sidewalk deterioration can be reduced by applying a surface sealant, any sealant should be applied in such a way that the surface retains good traction for pedestrians during wet weather. Where steel plates or steel grating is used for sidewalks, regularly check for any loose fasteners, ensuring that the surface is not slippery for pedestrians when wet (a sand, aluminum oxide, or other abrasive coating may be added to improve skid resistance). Asphaltic concrete sidewalk surfaces are often overlaid on portland cement concrete, and as such, need to be maintained so that raveled spots, potholes, and cracks do not present pedestrians with difficulties or hazards. Timber deck sidewalks need to be inspected for cracks, warped areas, rotted areas, or knothole openings that can create pedestrian difficulties; when such defects are observed, the deteriorated planking needs to be replaced as soon as practicable. In urban areas, police agencies often inform maintenance agencies of any areas where vandalism associated with pedestrians throwing objects from bridges has been reported. To reduce such vandalism, maintenance agencies can consider adding chain-link fencing to sidewalks and bridges.

Railings: Railings include any barrier or parapet that runs parallel with the traffic on either side of the bridge. Avoid leaving the end of a bridge rail or a parapet wall exposed to traffic flow. If a bridge rail or barrier has been identified as substandard, at the first rehabilitation opportunity practicable consider upgrading it to the current applicable standard as part of the rehabilitation activity. Repair collision damage to railings as soon as possible. Since major collision damage to a section of railing may make complete replacement of the entire run of railing more cost-effective than repair, any substandard railing should be considered for an upgrade replacement in such a situation.

Concrete Railings: Like concrete decks, concrete railings are susceptible to cracks and spalls from de-icing chemical damage. Sealing cracks and applying surface sealant as is done to decks is helpful in minimizing needed maintenance. If a sealant is applied to a concrete railing soon after construction is completed, as a preventive measure use of a sealant containing a light-reflecting material should be considered to enhance driver awareness of the bridge railing. When one or more sections of a concrete railing have been broken (assuming a replacement with concrete will be done), remove the damaged sections of post and rail with a jackhammer and acetylene torch; straighten or position existing reinforcing steel as needed; splice and replace reinforcing steel as needed; form new concrete sections to conform with rail dimensions as shown on the original construction plans; place and properly cure the concrete; finish the concrete as needed with a rubbing stone to match the existing rail and seal; and clean up the job site to remove any traffic hazard debris. Maintenance inspections need to pay particular attention to the joints of any electrical conduit in a concrete railing because differential expansion can create conduit breaks.

Steel Pipe and Tubular Railings: Steel pipe and tubular railings are susceptible to collision damage, loose anchor bolts and connections, corrosion damage, and weaknesses resulting from original design deficiencies. Corrosion damage is minimized by using galvanized materials in the original installation (or in repair replacement) and a zinc-rich paint to coat any parts not galvanized. Repair and replacement typically consists of straightening any collision-damaged components for which replacement is not deemed necessary; installing a temporary railing with traffic warning devices if the repair and replacement process cannot be completed quickly; tightening all loose anchor bolts and connections; painting any corroded areas with zinc-rich paint; cleaning and painting any rusted areas where posts or anchors enter concrete; replacing any railing components for which corrosion has significantly reduced the structural cross section; repairing or replacing all damaged anchor bolts; welding any anchor bolt extension at a point below the finished surface of the concrete to which it is anchored; replacing any damaged concrete or concrete removed to perform repairs; and conducting touch-up painting of any connections after final tightening of nuts and connections (hot-dip galvanizing is preferred to painting when practical).

Aluminum Railings: Aluminum railings are susceptible to collision damage, anchor bolt and connection damage, oxidation (aluminum corrosion), and original design inadequacies. Because aluminum is a more chemically reactive material than most other structural materials, oxidation protection is required at the contact surface with other materials. Steel-to-aluminum contact surfaces should be caulked with an elastic, nonstaining blend of water-repellent oil, asbestos fiber, and flakes of aluminum (metal or other suitable materials). The contact surface of each aluminum railing post attached to concrete should be separated from the concrete with a nonreactive bedding material such as 14-kg (30-pound) nonperforated, asphalt-saturated felt; galvanized or painted steel plate; or an elastomeric caulking compound. General repair procedures include repairing or replacing any collision damage to return the railing to its original design strength (new railing rather than repair of damaged sections is usually better because of the skill level needed to properly straighten or weld aluminum). Anchor bolts and connections should be inspected and repaired in the same manner as is done for steel railings. When the damaged area is extensive, the possibility of replacing an aluminum railing with a "New Jersey–type" concrete railing should be considered for maintenance and safety considerations.

3.1.4 Superstructure
3.1.4.1 Jacking and Supporting the Superstructure

Bridge rehabilitation and superstructure repair often require jacking to provide load transfer and bridge support while repairs are made. If jacking is required, both the safety of any vehicular traffic continuing to use the bridge while maintenance and repair are conducted and the safety of all working personnel must be considered.

- Jacks and jacking supports must be straight, plumb, and of sufficient capacity to support the portion of the bridge being lifted.
- Place jacks at points that will not damage the structure; reinforce jacking points on the structure if necessary.
- Before jacking, check deck joints for offsets that might be damaged from differential movement between spans during jacking; to prevent damage, check railings and disconnect them if needed.
- Uniformly raise and lower jacks to distribute the jacking load evenly and prevent overstressing or twisting the bridge.
- Position blocks adjacent to the jacks to increase their height as the structure is raised, minimizing any loss of support if a jack fails during the jacking operation.
- Do not permit traffic on the bridge while it is supported by jacks. If traffic is permitted on the bridge while it is supported by blocks, a vertical transition slope should be provided to avoid abrupt changes in the road surface, to provide a safe riding surface, to prevent damage to the bridge, and to minimize any vertical acceleration loading from the traffic.

Jacking the Superstructure: The procedure should be designed and reviewed by a qualified bridge structural engineer for each jacking setup, ensuring that it is adequate for the job to be undertaken. This design and review must account for the following factors [1]:

- Dead-load reaction to bear on the jacks.
- If the bridge cannot be closed to traffic during jacking (preferable), then the expected live load on the jacks must be included.
- The size and number of jacks required.
- The location of the jacks.
- Any temporary bents or cribbing required to support the jacks.
- Any modifications to bridge structural members required at the jacking points so that the bridge members can sustain the jacking pressure.
- Sufficient space at deck joints to permit differential movement between spans.
- Defining the height to which the structure needs to be jacked, jacking only as high as is absolutely necessary to conduct the required maintenance.

Generally the steps and precautions included within a jacking procedure to ensure successful, safe operation are as follows:

(1) Construct the necessary bents and cribbing to support the jacks when it is not possible to locate supports on the existing substructure. An adequate foundation to prevent differential settlement is very important.
(2) As necessary, reinforce bridge members to withstand the force of the jacks.
(3) If necessary, disconnect railing and utilities.
(4) Place jacks snugly in position.
(5) Restrict vehicular traffic on the span while it is supported by jacks.
(6) Raise the span by jacking. Pressure gauges should be used to ensure that all of the jacks are lifting the span evenly.
(7) Use observers placed at strategic points to watch for signs of structural distress because of jacking.
(8) Jack, block, and rejack until the required position is achieved.
(9) Protect joints and provide a transition to the span with steel plates if traffic is maintained while the span is on blocks.
(10) Check periodically to ensure that there is no differential settlement.
(11) After the repairs are completed, remove the blocks using the jacks.
(12) After the span has been lowered into place, ensure that the deck joints are functioning properly, that the alignment has not been changed, that there is adequate space for expansion without debris or restriction in the joints, and that the joint seal is watertight.

Using a Carrier Beam: If jacking the superstructure presents great difficulties because of the superstructure's height, the traffic below the superstructure, or perhaps the depth of the water under the bridge, an alternative to jacking the superstructure is to use a carrier beam to support the portion of the superstructure to be repaired from the bridge deck. An appropriately sized beam that has been salvaged from other bridge operations can be used. The safety of traffic over the bridge is a major concern if the bridge is left open to traffic flow while a carrier beam is in use. A maintenance planning evaluation of the suitability of using a carrier beam should include the following activities [1]:

- Conduct a traffic flow assessment of the options for controlling traffic during the rehabilitation, including the potential to detour traffic completely away from the bridge. Using a carrier beam reduces the clear roadway on the deck and requires special provisions for safe vehicle travel.
- Conduct a design analysis to determine the length and size of the carrier beam required,

the size of the tension rods required to support the dead load plus any live load if traffic is allowed on the bridge while the repairs are made, and the connections of the tension rods to the carrier beam. A professionally qualified structural engineer should conduct this design analysis.
- The support brackets should be designed by a professionally qualified structural engineer to be the appropriate size and placed at the appropriate location, preventing damage to the bridge structural member to be supported.

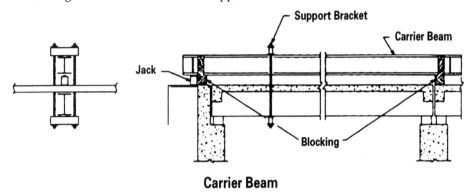

Carrier Beam

Generally, the steps and cautions included in the process of successfully using a carrier beam are as follows:
(1) Restrict traffic on the bridge deck.
(2) Place the carrier beam on blocks above the area to be supported.
(3) Drill holes through the deck to accommodate the rods for the support bracket, avoiding damage to the existing reinforcing steel. (A pachometer can be used to locate the reinforcing steel bars.)
(4) Install the support bracket.
(5) To obtain the required clearance, place a jack under the end of the carrier beam and raise the superstructure.
(6) Shim blocks to support the carrier beam.
(7) Conduct the needed maintenance repair and rehabilitation activity.
(8) Lower the superstructure onto its permanent support and remove the carrier beam after the repairs are completed, including any curing period required for any new concrete work to gain required structural strength.
(9) Repair the holes drilled in the deck through which the support rods passed.

3.1.4.2 Bearing Maintenance and Repairs

Bearings transmit the dead load and the live load on the superstructure to the substructure (abutments and piers) while also allowing the superstructure to move without exceeding its design stress limits. For example, movement can result from temperature changes, wind pressures, substructure movement, and live-load deflections. A bearing assembly that is frozen (corroded or fouled, not moving as intended), out of position, damaged, or for any other reason not operating properly can cause the stress limits to be exceeded in a bridge seat, in beam ends, in supporting columns, or in other bridge members, which in turn will result in structural damage requiring repair or replacement. The cause of most bearing problems are open or leaking deck joints, substructure movement, or bridge approach pressure on the superstructure. Good bridge inspection practice will include recording the position of all bearings and temperature at the time of the inspection so data from a sequential series of inspections will facilitate determination of the degree to which the bearings are functioning properly. Two general types of bridge bearings are "fixed" bearing devices and "expansion" bearing devices.

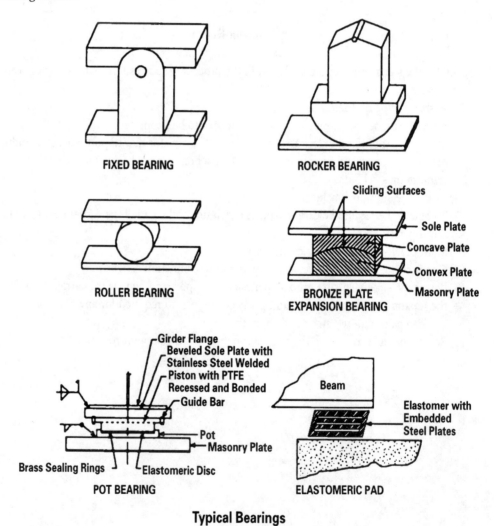

Typical Bearings

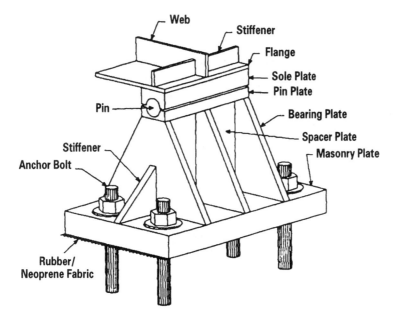

Fixed Bearing Device

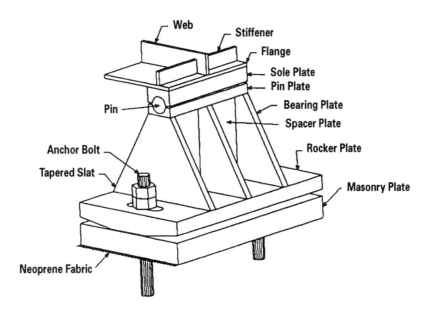

Expansion Bearing Device

Sliding Plate Bearing Maintenance: Sliding plate bearings are plates of similar or dissimilar metal that slide on each other. Early versions were steel on steel, steel on bronze, or bronze on bronze. Refinements that developed later include slightly rounding one bearing surface to reduce frictional binding, inserting a thin sheet of lead or asbestos between the plates, using a graphite-impregnated bronze plate, and incorporating a pad of elastic material (e.g., polytetrafluoroethylene) with a stainless-steel surface. Corrosion that increases friction is the most common problem with sliding plate bearings. If movement in the bridge exceeds that anticipated in the design, sliding plate bearing problems can also arise. Maintenance of a sliding plate bearing requires jacking the structure, except for cleaning the bearing area, removing debris from the bearing area, and lubricating a bearing that has a grease fitting. If a

Chapter 3 Bridge Maintenance and Management

sliding plate bearing surface is corroded or will not allow movement, the bridge must be jacked to remove the bearing surface for cleaning, lubrication, and repositioning (if needed). Lubricants on the bearing plate surface can include a good waterproof grease, oil, or graphite, or lead sheets can be placed between steel plates, or a bronze plate can slide against a steel plate. If a sliding plate bearing assembly was not designed and constructed to include a grease fitting, consider installing one while the assembly is being maintained so that the plate surfaces can be lubricated without jacking the structure.

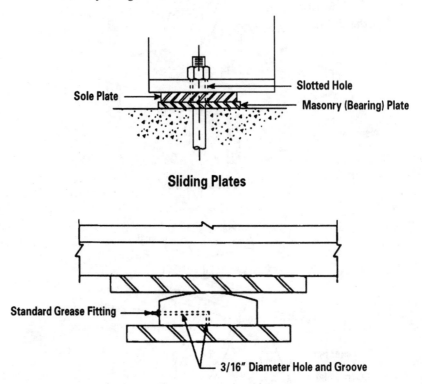

Grease Fittings on Sliding Bearings

Roller Bearing Maintenance: Roller bearings are devices that incorporate a horizontal roller to permit longitudinal movement of the bridge structure. They range from a simple cylindrical roller to variations of segmented rollers and pinned rockers. In some designs, steel balls have been used with lateral restraints to prevent lateral movement. Many roller bearing devices are enclosed units and therefore difficult to gain access to for maintenance. Some units are housed in a sealed lubricant enclosure. Routine maintenance consists of keeping the bearing area clean and painted. Lubrication is generally limited to keeper links and nesting mechanisms that require disassembly. When roller bearings stop functioning as designed and intended, they must be removed and refurbished, which in turn requires jacking the end of the span connected to the bearing. Since the nests and mechanisms normally require off-site shop maintenance to be rebuilt, many agencies fabricate spare units so that a complete change out of the roller bearing assembly can be performed in the field to avoid closing a bridge for an extended period of time.

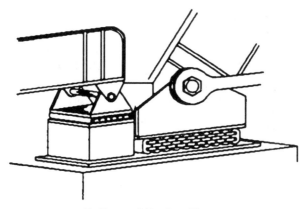

Roller and Rocker Nests

Rocker Bearing Maintenance: Rocker bearings are pedestals with circular bottoms that support a pin. Since the weight of the bridge is transmitted to the rocker through the pin, it is a critical part of the assembly. The pin may become excessively worn or corroded (and then freeze up). The surface of the rocker and bearing plate can be limited in its range of movement by debris and dirt buildup. If the bridge moves beyond the intended design range of the rocker bearing, the rocker can become unstable and fail. Maintenance generally includes keeping the assembly clean, lubricated, and painted. Debris and dirt should be removed from under the rocker, and if such a problem persists, a cover for the rocker assembly should be designed and installed to keep debris and dirt away from the rocker. When pins become worn or corroded, making lack of movement a problem, they should be removed and either replaced or cleaned and lubricated, which requires jacking the bridge. Special pins can be fabricated with grease fittings and installed to permit lubrication without disassembly.

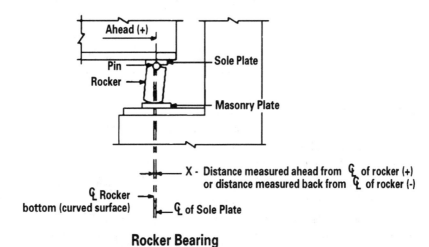

Rocker Bearing

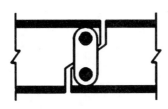

Detail of Typical Link

Chapter 3 Bridge Maintenance and Management

Pin-and-Hanger Bearing Maintenance: Pin-and-hanger bearing devices are used where the load is transmitted from the end of one span (i.e, a cantilever span) to another (i.e., a suspended span). Thus, there is not a substructure element directly under the bearing. Most modern designs do not incorporate pin-and-hanger bearings, but many existing steel beam and truss spans have these type of bearings. Because these bearings are located under a deck joint, the pin-and-hanger assembly is particularly susceptible to corrosion. Corrosion tends to cause the assembly to freeze up, preventing movement, which in turn transmits very large forces to the assembly that can result in pin or hanger failure. The pin frequently breaks while resisting torsional forces. Hangers most often fail at either end adjacent to the hole through which the pin is fitted. A serious maintenance difficulty with pin-and-hanger bearings is that any corrosion of the pin and its bearing surfaces is not easily detected without disassembling the bearing, which is also generally not feasible because these bearings are not located over substructure elements, so are difficult to gain access to. Many pin-and-hanger installations were not equipped with lubrication fittings, which further complicates maintenance and often contributes to the freeze-up of the bearing. Failure of a pin-and-hanger bearing can lead to catastrophic failure of the bridge; therefore, it is very important to inspect such bearings and watch for any cracking or other stress failure indications. In many cases, redundancy has been added to pin-and-hanger bearings that were not initially designed and constructed to be redundant to prevent a sudden failure of the bridge. However, even with redundancy improvements, bridge failures can still happen if pin-and-hanger bearings are not regularly and carefully inspected in an attempt to identify frozen or damaged assemblies.

Elastomeric Bearing Maintenance: Elastomeric bearings incorporate an elastic material, such as neoprene, in either single or multiple pads, with or without steel plates embedded into the laminations, that compress and deform under both longitudinal and rotational movement. This material typically has a long life and is able to withstand repeated deformation cycles. Consequently, maintenance is rarely needed unless the bearing completely fails or gradually works out of position. The bridge must be jacked to replace or reposition the bearing. If the bearing slips out of position on a recurring basis, an abrasive material can be added to the contact surface or a keeper plate can be attached to the bearing seat.

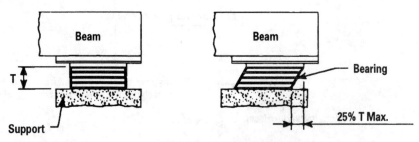

Elastomeric Bearing Pad

These bearings have been designed to accommodate movements up to 75 mm (about 3 inches) and several degrees of rotation. The bearings may fail because of deterioration of the material, excessive crushing, separation of composite pad laminations, or excessive shear forces. Excessive shear is normally considered to result from longitudinal movement exceeding 25% of the bearing height. Uneven compression and twisting of an elastomeric bearing can also contribute to problems.

A recent study and survey of state DOTs indicates that high-load elastomeric bearings have not yet been placed in many service applications even though they appear to offer greatly improved bearing response over other types [17].

Pot Bearing Maintenance: A pot bearing is a special adaptation of the elastomeric bearing that incorporates a steel ring to limit deformation of the elastomeric material. The elastomer rests on a stainless-steel plate to reduce friction and to permit sliding. Pot bearings can be damaged by movement or loading that results in uneven compression across the bearing. A hydraulic pot bearing can lose its operational effectiveness if a piston seal leaks or a piston or pot is cracked. Elastomeric pot bearings experience bond failures between the Teflon and the substrate, bond failures between the stainless-steel plate and the sole plate, deterioration of the stainless-steel plate, and cut or deteriorated Teflon. Routine maintenance is normally limited to keeping the bearing area clean and free of debris. When a pot bearing fails or malfunctions, the bridge must be jacked so that the bearing can be removed for repair or replacement.

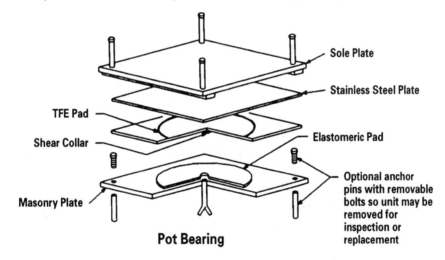

Pot Bearing

Resetting or Rehabilitating Bearings: Frozen or misaligned bearings must be repaired or replaced. Good maintenance planning for this process generally includes the following items:
- Determining the root cause of the bearing problem. Good bridge maintenance needs to address the cause of the bearing failure if it extends beyond the bearing itself. Otherwise, bearing maintenance is treating a symptom and not a cause.
- Evaluating the existing bearings to determine if repair, replacement in kind, or upgrading to a different type or newer design bearing is needed.
- Positioning a bearing, when setting or resetting it, so that it is centered at the median temperature for the geographical area in which the bridge is located.

To ensure a successful maintenance operation, bearing rehabilitation generally involves the following steps:
(1) Constructing temporary supports for the superstructure.
(2) Closing the bridge to traffic.
(3) Jacking the structure up to remove the load from the bearings.
(4) Removing or repositioning the existing bearing assembly.
(5) Making bearing set modifications as needed.
(6) Positioning and securing the new or rehabilitated bearings.
(7) Ensuring that the bearing assembly is properly positioned for the ambient temperature with respect to the median temperature for the geographical region.
(8) Removing temporary supports.

Elevating Bearings to Increase Vertical Clearance: As roadway pavements deteriorate and overlays are added, the clearance under bridges commonly decreases. Replacement or repair of overpass bridge bearings is a good opportunity to consider modifying the bridge bearings to raise the vertical clearance under a bridge, if it would improve safety and commodity movement efficiency. The usual steps are as follows:
(1) Jack and support the bridge slightly above the new elevation while seat and bearing modifications are made. Jack the spans together to avoid damaging the joints or if the superstructure is continuous.
(2) Modify the bearing seats to raise the bearings the required distance. For adequate superstructure support, the seat modification should be designed by a professionally qualified structural engineer. A steel pedestal or a reinforced-concrete addition are typical methods of modifying the bearing seat.
(3) Modify bridge approaches to match the new elevation of the bridge deck.

3.1.4.3 Beam Repair

Beams under deck joints, under drains, and along any bridge facia are potentially exposed to corrosive elements. Bridge beams on overpass bridges are subject to damage from collisions caused by vehicles and loads exceeding the underpass clearance height. When bridge beams have been damaged by such collisions, it is especially important to inspect for collateral damage to bearings, piers, and other elements of the structure so a complete repair can be made. If your agency has a policy of recovering damages for over-height vehicle collisions, all associated damage should be documented at the initial inspection.

Repairing Damaged Reinforced-Concrete Beams: Reinforced-concrete beams should be kept clean and protected from saltwater, saltwater spray, or water from the deck containing de-icing chemicals. Concrete sealers and coatings protect concrete exposed to saltwater if they are applied to a clean surface before the concrete surface is contaminated. If the concrete surface has already been contaminated, the surface must be blast-cleaned before the sealers are applied; even then, the effectiveness may be reduced because of chlorides penetrating below the surface. Before agency personnel attempt to repair any reinforced-concrete beam that has been damaged, a professionally qualified structural engineer should evaluate the damage, determine the type of repair needed, and evaluate if the agency can perform the repair or if it should be contracted to a firm having capacities not present in the agency. Such repairs include repairing impact damage from over-height vehicles and spalling from corrosion of reinforcing steel. While many factors must be considered when determining the best repair method, two key factors are the location and the severity of the damage. For instance, a small area of damage is not structurally significant in a tension area of a beam if the reinforcing steel is not damaged; thus, it may be sufficient to protect the steel from exposure to elements that cause corrosion. This protection can be provided by placing a surface patch with hand methods or even by coating the exposed reinforcing bars (if appearance is not an issue) or applying shotcrete to the area (if it is cost-effective).

If a beam has a large damaged area that will permit a formed repair to be made, make the formed repair because it will likely be a more permanent patch. Generally, placing the aggregate inside the formed area and injecting a grout into the formed aggregate area gives the best results.

Structural repairs to beams may require supporting the superstructure during the repair process. Temporary bents or carrier beams designed by a professionally qualified structural engineer will provide the necessary support. If the reinforcing steel is damaged or if the damage extends into a compression area of the beam, however, it may be necessary to jack the dead-load stresses out of the damaged area before repairs are made.

Concrete beam ends can be repaired by observing the following generally accepted steps:
(1) Direct traffic to the far side of the bridge until repairs on the beam end are completed.
(2) If the superstructure cannot be jacked and supported from the existing substructure, construct a temporary bent for this purpose.
(3) Place jacks and raise the entire end of the bridge a fraction of an inch. The lift should be only that required to insert a piece of sheet metal as a bond breaker for the new concrete.
(4) Place the sheet metal bond breaker on the beam seat.
(5) After cutting with a saw to avoid feathered edges, remove the damaged concrete in zigzag steps to provide horizontal bearing surfaces.
(6) Place new reinforcing steel as needed, making certain it is properly lapped, anchored, or mechanically attached to the existing reinforcing steel.
(7) After forming, clean the surface of the existing concrete and apply a bonding material.
(8) Place new concrete with an additive to reduce concrete shrinkage during the curing process.
(9) After the concrete has reached sufficient strength and the repaired area has been visually inspected and sounded, simultaneously jack all beams to sufficient height, allowing placement of the elastomeric bearing pads.
(10) Uniformly lower the entire end of the bridge.

Post-reinforcement of Concrete Beams: The Kansas DOT [1] has developed a repair system for post-reinforcement of concrete beams. Several bridges that had shear cracks near the ends of the girder beams were repaired with the following process:
(1) Locate and seal all of the girder cracks with silicone rubber.
(2) Mark the girder centerline on the deck.
(3) Locate the transverse deck reinforcing steel.
(4) Vacuum drill a series of 45° holes that avoid the reinforcing steel bars. The drilling is directed perpendicular to the shear cracks.
(5) Pump the holes and cracks full of epoxy material.
(6) Insert steel reinforcing bars into the epoxy-filled holes.

Repairing Damaged, Prestressed-Concrete Beams: Prestressed-concrete beams should be protected and maintained just like other reinforced-concrete beams. However, the consequences of poor maintenance can be more serious in prestressed-concrete because the prestressing strands are designed to keep sufficient pressure on the beam so the concrete never experiences tension forces even under the expected live load.

Prestressed members have thinner sections than conventional reinforced concrete and are more susceptible to deterioration if cracks in the concrete allow any contaminants to generate corrosion of the steel strands. Since the tension is high in the tendons and the concrete is squeezed together, the loss of a significant amount of concrete can cause the remaining cross section of the concrete to be overstressed and crushed to failure. A single tendon breaking because of an impact loading on the beam or because of corrosion reducing the tendon strength can create a minor explosion within the beam, and if several steel tendons snap, a bridge can fail suddenly and catastrophically.

Special procedures are required to repair a prestressed-concrete beam and restore the original dynamic state of stress that gives the beam its high load-carrying capacity. If a significant amount of concrete is to be replaced, the compressive forces must be removed from the concrete beam near the damaged area by applying a calibrated load on the bridge while the new concrete is placed and cured to reach its design strength. A professionally qualified structural engineer should design and review this process, determining the load needed and the proper placement of the load to create the desired stress effect in the beam.

Tension must be put back into any broken tendons as part of the repair process. The Florida DOT uses cable splices to reconnect tendons and to develop the required tension stress. The splices are anchored to the ends of the damaged cable, and a threaded coupler between the two anchors is screwed down, applying the torque necessary to produce the required level of tensile stress in the cable.

The Pennsylvania DOT has used a post-tensioning process external to the beam being repaired. Post-tensioning strands are placed in ducts outside the beam just above the lower flange of the beam. These strands are anchored to the existing concrete. New concrete is formed and placed around the ducts. After the new concrete has been cured and has attained the design strength, the strands in the ducts are tensioned to the specified load to replace the tensile forces lost in the damaged tendons. Any cracks associated with the damage are epoxy injected as part of the prestressed-concrete beam repair.

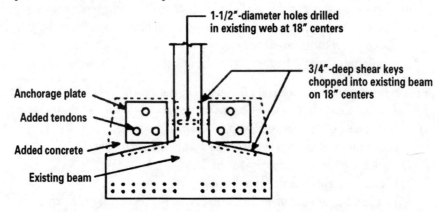

External Post-tensioning of Prestressed Girder

Repairing Steel Beams: The three general types of damage experienced in steel beams are corrosion damage, impact damage, and fatigue damage. Good maintenance practices can prevent or drastically limit corrosion damage, but if corrosion is allowed to develop it can reduce a bridge's load-carrying capacity as much as any other type of damage. Repairs to steel beams frequently require adding structural material, such as plates or angles, to provide structural strength that compensates for the damage. Because of poor experience with field-welding added material to a steel beam, some agencies prefer to bolt additional steel material to increase the strength of a damaged beam. If a repair to a steel beam is at a location where a fatigue failure is unlikely and if the weld quality can be highly controlled, good performance can be obtained from field-welding.

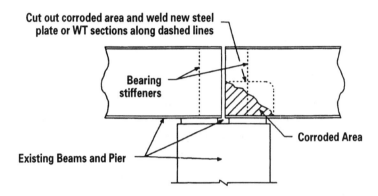

Corroded Beam End Rehabilitation

Repair of corrosion damage to steel beams generally observes the following process:
(1) Relieve the load at the bearing by jacking under the sound portion of the beams.
(2) Cut out the corroded area, rounding the corners to a minimum radius of about 75 mm (about 3 inches) to avoid abrupt changes (re-entrant corners that create stress concentrations). Remove bearing stiffeners if they are present in the existing installation.
(3) Weld the new section into place, using full-penetration welds. The new section may be either a suitable rolled beam section or it may be shop fabricated from other suitable shapes. Replace bearing stiffeners where they are required.
(4) Lower the span to bear and check for any immediate distress.
(5) Remove jacking equipment and other temporary supports.

The same procedure can be adjusted to replace damaged portions of steel beams at other locations on the bridge. A bolted splice that has been properly designed is generally preferred to a field-welded splice, and all of the load (live load and dead load) needs to be removed from the beam before the new section is added (not doing so will result in the old portion of the beam cross section being subjected to a much higher stress level than the new material after the repair is completed). When plates or angles are added to a steel beam for strengthening rather than repair purposes, the dead load is not commonly removed.

Heat Straightening a Damaged Steel Beam: Impact damage to steel beams from over-height vehicles is a common problem. Many agencies use a heat process to straighten beams if the amount of beam deformation is not excessive. If the steel properties are well-known, it is possible to apply heat to the beam with oxyacetylene torches in such a pattern that the repetitive heating and cooling of the steel will cause an expansion and contraction of the metal, "working" the deformation out of the beam. The original properties of the steel have to be such that treating the beam in the field to remove the deformation does not also reduce the strength required in the original bridge design. Thus, a professionally qualified structural engineer should design and monitor the procedure. The damaged area needs to be inspected very carefully before, during, and after the repair process to identify any flaws, tears, or cracks. The heat and the resulting forces in the beam need to be applied in such a way that the warping stresses to reshape the beam result in gradual beam movement rather than any abrupt new distortion that may tear the steel's molecular bond. The total quantity of heat applied to the beam needs to be controlled so that the steel is not heated above about 1200°F to avoid damaging its material properties. Heat sticks can be used to indicate when the steel reaches the maximum allowable temperature. Anytime a steel member is bent and deformed, the steel has been stressed beyond its yield point and its mechanical properties have been altered even if no cracks or tears appear. When a bent steel beam is straightened, the increase in the "working" of the molecular structure reduces the steel's ductility, making it more susceptible to fatigue failure. Thus, even if a qualified structural engineering assessment indicates that a beam was designed with excess load-carrying capacity that negates the need for strengthening modifications, the inspection process following beam straightening should emphasize checking for cracks or other signs of subsequent weaknesses.

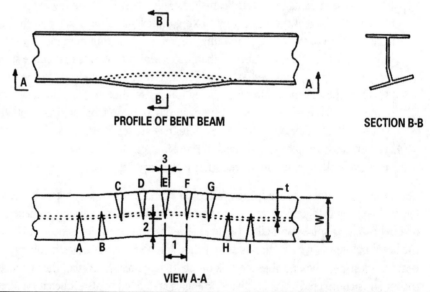

DIMENSIONS	SIZE	GENERAL PATTERN SEQUENCE
1	Varies as necessary - 2/3 W to W generally	E D F B H or
2	W/4 to W/2	E C G B H D F A I
3	About 6", or W/3 to W/2	Note: Other sequence options may be used also.

Locations, Size, and Sequence of Heat Application for Flame Straightening

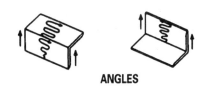

ANGLES

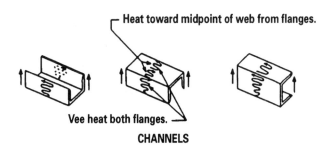

Heat toward midpoint of web from flanges.

Vee heat both flanges.

CHANNELS

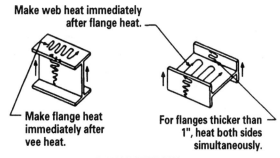

Make web heat immediately after flange heat.

Make flange heat immediately after vee heat.

For flanges thicker than 1", heat both sides simultaneously.

FLANGE SECTIONS
Arrows show direction of bending due to flame straightening.

Torch Paths for Various Shapes

Repairing Timber Beams: Any crack in a timber bridge beam (stringer) reduces its capacity to carry loads and therefore may require immediate repair or replacement before traffic is allowed to use the bridge again. Timber beams with a longitudinal crack can be repaired by placing steel plates on the top and bottom of the beam and clamping the plates together with bolts that extend the full depth of the beam.

Generally, replacing a decayed or damaged timber beam is the most cost-effective maintenance practice. In some cases, a temporary repair can be made by turning stringers with small damaged areas (e.g., around the planking spikes), but in many cases this repair will be as costly in time and effort as replacing the stringer. Furthermore, not replacing the stringer leaves a weakened structural member with a short service life until it must be replaced.

In many cases, it is easier to replace a damaged timber beam if the deck is removed, but if the deck is heavily spiked to all of the beams and only one beam is damaged, this may not be practical. To deal with the situation where it is not practical to remove the deck, a new beam is installed next to the existing damaged beam (or perhaps a new beam on each side of the old beam). The new beam can be trimmed at an angle on one end, providing space to insert enough of the trimmed end between the pier cap and the underside of the bridge deck so that the new beam can be rotated up against the deck and slid into place over the other pier cap. (If a new beam is placed on each side of the existing beam, it may be better to trim the opposite ends of each new beam.) After the new beam is in position over both pier caps, it should be jacked tightly up against the deck and shimmed with hardwood or metal shims between the beam and pier cap. (The material trimmed off the beam end can be used to make shims.) The deck is then spiked to the new beam (or beams).

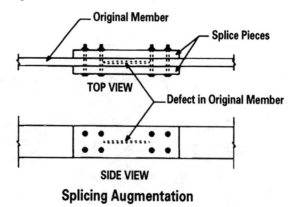

Splicing Augmentation

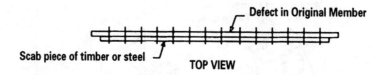

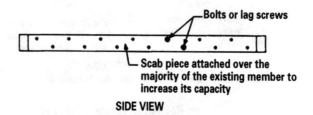

Scabbing Augmentation

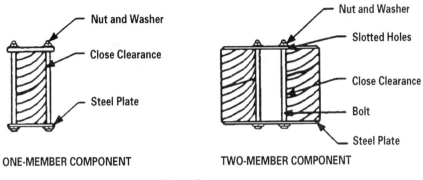

Clamping

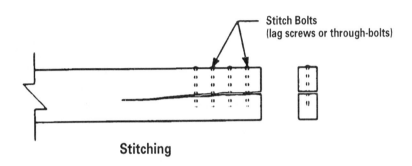

Stitching

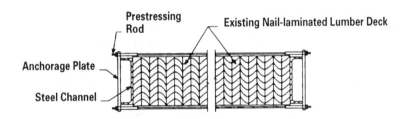

External Stress-Laminated Deck

3.1.4.4 Truss Repair

Truss bridges are usually made of steel, although some old trusses constructed of wrought iron still exist. Generally, repair consists of replacing a damaged member or of strengthening a weakened member by adding steel plates to it. Welding should be avoided if at all possible in repairing older trusses since the steel typically has a high carbon content. A professionally qualified structural engineer should make an assessment of the appropriate repair process to be applied to a truss bridge, especially since the methods used to support the structure may be much different for repair of a member carrying compression loads than for one carrying a tensile force.

Maintenance and Repair of Steel Trusses: All steel trusses are susceptible to damage from rust and corrosion. In most truss designs, the main load-carrying members cannot suffer a substantial loss of cross-sectional area without beginning to lose capacity to withstand design stress levels. Through trusses and pony trusses are generally narrow and especially susceptible to damage from vehicle collisions because the exposed truss members are close to the traveled roadway. Portal and sway bracing of through trusses is also susceptible to collision damage from over-height loads. The condition of connecting pins is critical in maintaining the structural integrity of a truss with pin-connected eye-bars. Since all loads pass through the pin, failure of the pin can cause a complete structural failure of the entire span. A regular, thorough cleaning and spot-painting program is necessary to prevent rust and corrosion damage in all types of truss bridges. Cleaning of pin-connected joints is a critical maintenance activity in pin-connected trusses to ensure that the joints are free to move as the loads are transferred across the bridge. Repair procedures commonly applied to truss bridges to correct deficiencies are as follows:

- If truss members have been damaged by rust or corrosion and a structural engineering analysis indicates the member is overstressed for its reduced cross section, plates are typically added to the member to compensate for the lost section area.
- Truss members that have impact damage from collisions may be temporarily repaired using a strut to reinforce damaged members carrying compression loads and using cables with turnbuckles to reinforce damaged members carrying tension loads. If a collision impact has fractured a member, the member usually must be partially replaced and perhaps the entire truss member will have to be replaced.
- A fractured eye-bar of a pin-connected truss can also be temporarily repaired with a cable to accept part of the stress. However, permanent repair will require that the fractured member be replaced.

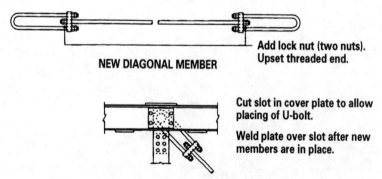

Truss Diagonal Replacement for Pinned End Connections

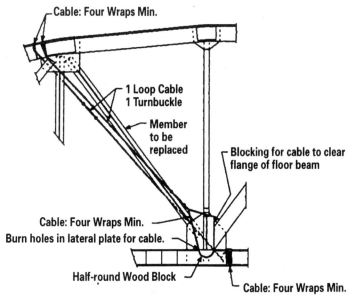

Truss Diagonal Replacement for Riveted or Bolted Connections

Damage from collision impacts to truss bridges can often be reduced by upgrading the bridge railing to provide structural protection to the truss. Sometimes it is possible for bridge design engineers to develop a revised network of overhead lateral bracing to increase overhead clearance if a truss is frequently being hit by over-height loads. Some agencies have also found it helpful to install overhead devices that warn truck drivers that their load will not clear the truss overhead bracing.

Repairing Tension Members: Diagonal tension members may be damaged by corrosion, overload, or impacts from oversized vehicles. A general procedure to guide the repair of tension members should contain the following activities:
- Design and acquire a replacement member that meets the load capacity requirements, including any additional load created by the repair process.
- If the bridge is open to traffic, restrict traffic to one lane on the opposite side of the bridge.
- Cut and install wood blocking.
- Install a cable having the capacity to carry the full dead load in the diagonal plus any live load distributed from the restricted traffic flow.
- Tighten the cable system.
- Remove and replace the damaged member. If the old and the new member consist of more than one section, the sections should be removed and replaced individually, using high-strength bolts and keeping the load as symmetrical as possible.
- Install batten plates or lacing bars at the required intervals along the diagonal and tighten all high-strength bolts in the new member.
- Remove the cable support and other temporary components, and restore the bridge to normal traffic flow conditions.

A vertical tension member can be repaired using the same general process with cable supports. Damaged diagonal tension members consisting of two eye-bars can be replaced by using rods with U-bolt end connections. With no traffic on the structure, one of the eye-bars can usually be replaced at a time.

Chapter 3 Bridge Maintenance and Management

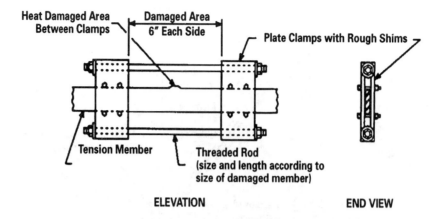

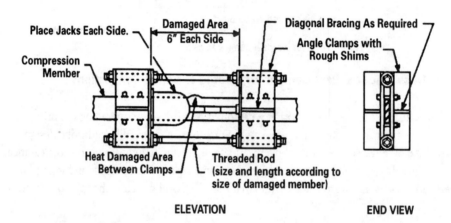

Yoke for Restoring Tension and Compression Members by Heat-Strengthening Method

Replacing Compression Members: The difference between replacing a compression member and replacing a tension member is that a heavier, more expensive temporary support system is needed for the compression member.
- The temporary support system must be capable of supporting the compression load while the member is removed and replaced. Therefore, each member that makes up the temporary support system will consist of a large, "column-like" section.
- To provide room for removing the old compression member, the temporary support must be framed around the old member. It is not mandatory that the temporary support be more than one column; however, since the temporary support is offset, it must be able to resist the unbalanced load that is created when the old member is removed.
- As the temporary support is installed, the load on the damaged member must be relieved by jacking. The temporary support is designed to support each (and all) jack(s).

Repairing Damaged Truss Members: If the damage is localized, such as a transverse crack in one of two channels that make up the bottom chord of a truss, a professionally qualified structural engineer can design a splice as a repair solution. The basic repair steps are as follows:

(1) Bolt a side splice plate to the damaged member. Remove any rivet heads that would interfere with the splice plate if the repair is near a connection. The plate should be centered over the damage and sized by a professionally qualified structural engineer.

(2) If necessary, remove the old tie plates (or lacing bars) from the two channels in the area of the crack, facilitating placement of the bottom tie plate. Some temporary lateral bracing may be required before completing the repair.

(3) Bolt the bottom tie plate to the member.

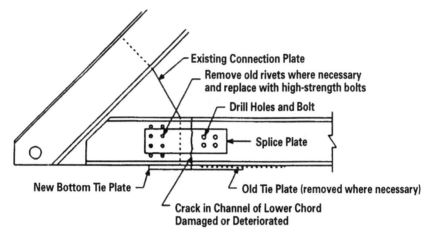

Typical Truss Repair Where Complete Removal Is Not Necessary

Increasing Vertical Clearance by Modifying the Portal Bracing: On many older through trusses, vertical clearance is restricted as allowed by modern transportation regulations. The clearance is controlled by the portal brace connected to the top chord, or the end post at the end of the span, or both. The primary purpose of the portal brace is to support the truss against lateral wind loads; however, it may have been designed also to provide support against buckling in the end post. The portal bracing can be modified to increase vertical clearance by replacing it with a shorter depth truss pattern bracing that has an equivalent or greater strength to resist expected wind loads. Thus, it is important that changes to the portal bracing be designed by a professionally qualified structural engineer, and the load-bearing capacity of the end post needs to be checked since the unsupported length may be changed.

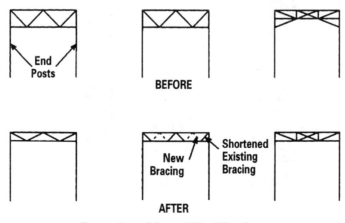

Examples of Portal Modifications

Chapter 3 Bridge Maintenance and Management

Heat Straightening Damaged Truss Members: The Minnesota DOT has successfully repaired damaged truss members using heat-straightening methods. The load is removed from the member to prevent stretching or buckling of the heated steel. Yokes are installed to accomplish this. The yokes are designed to act as either compression or tension members. Generally, the steps conducted to repair a bridge are as follows:

(1) Exclude vehicle traffic from the damaged side of the bridge.
(2) Identify whether the damaged member carries tension or compression and obtain or fabricate the necessary yoke.
(3) Tighten the clamps of the yoke around the damaged portion of the member.
(4) On tension members, progressively tighten the threaded rods to remove tension stress in the damaged member. On compression members, simultaneously apply jacking forces parallel to the member until the compression stress is relieved in the damaged member.
(5) Using an oxyacetylene torch, apply heat to the member until a dull red color is reached. Do not exceed 650°C (1200°F). First apply the heat at the bottom point of a V pattern; then slowly work back and forth and progressively outwards in the V.
(6) Hammer-peen the elongated side to the bent member, eliminating high residual stresses.
(7) As the heated area cools, shrinking occurs that tends to return the bent portion to its original shape.
(8) If the first attempt at straightening is inadequate, this process can be repeated after the steel cools.
(9) Remove yoke and paint exposed steel.

3.1.5 Substructure

Substructure caps of both the abutments and the piers provide seats or bases upon which bearing systems rest that directly support the superstructure. Many bridges have open joints located over the abutments and piers. Some bridges have drainage troughs located below the open joints to intercept the runoff and debris that falls through the deck expansion joints. These troughs discharge the debris beyond the substructure units into the watercourse or onto the ground below.

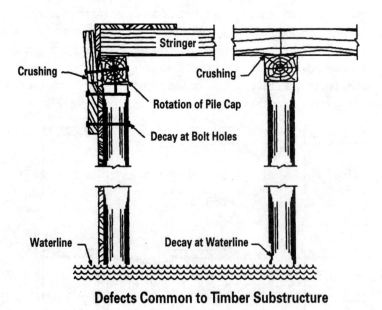

Defects Common to Timber Substructure

3.1.5.1 Problems Associated With Substructure Caps

The presence of debris on the caps results in corrosion of the bearing systems since the debris tends to hold water (and associated de-icing chemicals in freeze-thaw climates) for extended periods of time. The penetration of de-icing chemicals into concrete can corrode reinforcing steel, which in turn expands and breaks out (spalls) the covering concrete.

When the bearing system is frozen by corrosion, additional stress is introduced into the substructure cap that can cause spalling or damage to the bearing system. In addition, if the superstructure or deck joints are not kept clear of debris, the joint may not close as the deck expands, and the deck can be prevented from expanding as it was designed to do. Deck expansion is generally in the range of 50 mm per 100 meters of length (about 5/8 inch per 100 feet). If no other expansion joint can compensate for the jammed joint, the expansion force will be transmitted to the bearing system. Overloading the bearing system can result in fracture of the substructure cap and loss of bearing for the beam ends.

3.1.5.2 Maintenance

Maintenance of the substructure should include routine cleaning of the substructure cap. Using a high-pressure pump with sufficient hose length to flush out the substructure cap from every angle is an effective method of cleaning. The superstructure expansion joint troughs and drainage system downspouts should be cleaned at the same time. This procedure should be scheduled in the spring as soon as the winter de-icing program is over in freeze-thaw climates and at regular intervals in warm climates, depending upon local conditions. Flushing the bridge deck should also be included as a part of this maintenance. The flushing and cleaning process proceeds from the top down. If water supply is limited, the substructure caps can be cleaned first and then the downspouts can be flushed. Rainstorms will flush the deck and the drains, but do not contribute to cleaning of the substructure caps.

3.1.5.3 Preventive Maintenance

A program of effective preventive maintenance generally includes the following items:
- If they can be accommodated in the structure, drain troughs can be installed in deck expansion joints to protect caps for bridges that do not already have such troughs.
- Surface protection of the concrete in the substructure caps should be applied, using one of the commonly available surface treatments.

  Sealants: The concrete surface can be impregnated to reduce surface porosity and stabilize the surface layer of concrete. Typical sealants include silane, silicate, siloxane, and high-molecular-weight methacrylate (HMWM).

  Coatings: Coatings that will adhere to the concrete surface, seal it, bridge small cracks, and provide some resistance to de-icing chemicals include epoxy resins, hard urethanes, and methacrylate.

  Membranes: Membranes that can bridge cracks and provide good protection of the concrete surface from de-icing chemicals include elastomeric urethane, vinylester, and polyester.

Some agencies apply the protective coating to the beam-bearing systems as well as to the substructure caps. This procedure can only be effective if the protective coating is compatible with the paint system used on the bearings. When coating concrete caps, extend the protective film on the abutments a minimum of about 0.3 meter (1 foot) below the bearing seat. On piers, the extent of the coating may depend on the appearance that needs to be maintained for aesthetic considerations.

3.1.5.4 Repair Process

Problems often found in concrete bridge seats include deterioration of concrete and corrosion of the reinforcing steel. Such problems are caused by moisture and contaminants falling through leaking deck joints. A horizontal crack along the face of the pier cap, 75 to 100 mm (about 3 to 4 inches) from the top, normally indicates that the top mat of reinforcing steel has expanded because of corrosion and has forced up the concrete (i.e., delaminated it).

When a superstructure moves beyond the space that is provided for it in the bearing assemblies, pressure is created on the anchor bolts. This can be caused by an inadequate design, improper placement of the bearing assemblies, or corrosion of the sliding surfaces, which produces friction. Occasionally, lateral forces from large chunks of debris hitting a bridge during flood flows or high water levels, or an over-height vehicle hitting a beam can also create large forces on the anchor bolts. The pressure from the anchor bolts is then transmitted to the substructure cap, which can damage the bridge seats or cause cracks in other parts of the substructure (such as the columns).

No bearing device was provided on some older concrete bridges except for a thin fabric or paper bond breaker. Friction created by the beam or bearing device sliding directly on the bridge seat can cause the edge of the seat to shear if insufficient reinforcing steel exists in this area.

Planning for maintenance repair of the substructure should consider the following:
- Identify the extent of the damage by sounding the concrete and marking the areas of unsound concrete.
- Make provisions to correct the cause of the damage.
- Plan to remove vehicular traffic from the bridge during jacking operations.
- Determine the size, number, and location of jacks that will be required.
- Ensure that jacking will not damage joints, bearing assemblies, or the area supporting jacks.
- Define the needed resources, which generally include jacking equipment, form carpentry, concrete sawing or chipping equipment, and any necessary staging.

The substructure cap can be repaired or a new cap cast to offset any settlement that may have occurred at the substructure. Reconstruction of a bridge cap requires raising the superstructure to provide work space as well as to take the load off the bridge cap.

(1) Construct a temporary bent for supporting the jacks and blocking if jacking from abutment of pier elements cannot be done.
(2) Remove vehicular traffic from the bridge while jacking the superstructure.
(3) Lift the jacks in unison to prevent a concentration of stress in one area and possible damage to the superstructure.
(4) If the bridge will carry vehicular traffic during repairs, restrict the traffic away from the repair area as much as possible.
(5) Saw cut around the concrete to be removed and avoid cutting any reinforcing steel.
(6) Remove deteriorated concrete to the horizontal and vertical planes, using pneumatic breakers.
(7) Add new reinforcing steel where it is required.
(8) Apply bonding material to the prepared concrete surface that must interface with the new concrete to be placed.
(9) Build forms for the new concrete as required and place the new concrete.
(10) Service, repair, or replace the bearings as needed.
(11) After the new concrete has cured and reached its required strength, remove forms, blocking, jacks, and temporary supports.

3.1.5.5 Repairing Substructures Above Water

Repairs to the substructure are usually done with basic materials and processes. Repairs underwater require special considerations, as do pile and pile bent repairs. Substructure problems include deterioration (especially at the water line), cracking (usually related to settlement), impact damage (associated with traffic under the bridge), and shear damage (associated with movement or approach pavement pressure). Since most substructure units are concrete, repairs are often concrete-related processes. If the concrete substructure is exposed to saltwater, either from the deck or from below, problems such as those found in concrete bridge decks, including corrosion of reinforcing steel and concrete spalling, are likely to have occurred. Timber substructures can be damaged by decay and vermin attack. Substructure repairs are generally very costly because of the extensive temporary supports needed to carry the superstructure. Thus, preventive maintenance is often a very cost-effective approach to limiting these expensive repairs, especially a program that removes debris and pressure-washes seats, caps, and other substructure surfaces exposed to salt.

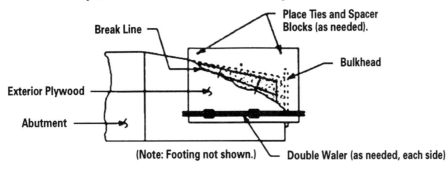

Reconstruction of Wing Wall

Repairing Broken or Deteriorated Wing Walls: Portions of an otherwise sound wing wall may be broken off by frost heave, ice that forms in voids created by fill settlement adjacent to the wall, ice in cracks, voids in the concrete, or insufficient air entrainment voids. Deterioration may result from de-icing and from salt-rich snow and ice being plowed onto the wing walls where it piles up. Weak aggregate in the original concrete mix can also contribute to wing wall deterioration. Losing portions of a wing wall can result in erosion of the fill and further damage to the bridge approach. The cause of the wing wall failure must be determined so it can be corrected as part of the wing wall repair process. Concrete forming should be preplanned and the forming materials cut to size in advance, if possible. Any excavation required to gain sufficient working access and to facilitate removal of defective concrete can be accomplished in advance of the wing wall repair. Materials and equipment typically needed to make this type of repair vary widely but often include excavating equipment such as a backhoe, an air drill, tie screws or equivalent bolts, wood spacers (walers, etc.), reinforcing bars, granular backfill material, hand tools, concrete removal equipment, anchor bolts and anchors, plywood forming, portland cement concrete, epoxy bonding agent, nonshrink grout, and miscellaneous hardware. Everything needs to be readily available to limit the exposure of maintenance personnel to traffic and to expedite the repair operation.

(1) Excavate as required to be able to set the dowels and the concrete forms.
(2) Remove all fractured or deteriorated concrete to sound concrete by chipping and then blast-clean to remove all loosened surface material.
(3) Drill and set form anchor bolts and dowels. Typically, dowels 13 mm in diameter (no. 4 bars) are placed a minimum of 225 mm (9 inches) into sound concrete and set with nonshrink grout, 150 mm (18 inches) center-to-center, both front and back.

(4) Crosslace the 13-mm (no. 4) reinforcing steel bars and set the concrete forms.//
(5) Just before placing the concrete, apply an epoxy bonding agent to all existing concrete that will contact the new concrete.//
(6) Before backfilling with granular material, cure the new concrete a minimum of 7 days or until the concrete has developed sufficient strength to resist the lateral pressures of the backfill.

Repairing Abutment Faces: Concrete in abutments may deteriorate from the effects of water, de-icing chemicals, freeze cracking, or debris impact, any of which can result in portions of the edge or face of the abutment breaking off. Repair is necessary to prevent continued deterioration, especially increased spalling due to moisture reaching the reinforcing steel and causing corrosion. Equipment and materials typically needed include an air compressor, concrete drilling equipment, tie screws (and lag studs), reinforcing steel-wire mesh, forming material, reinforcing steel, portland cement concrete, epoxy bonding agent, and gravel (stone or riprap also).

(1) Establish vehicular traffic control if needed.//
(2) Remove deteriorated concrete and loosened face surface material by chipping and blast-cleaning.//
(3) To support the form work, drill and set the tie screws and lag studs.//
(4) Set reinforcing steel and forms.//
(5) Apply epoxy bonding agent to the concrete surface just before placing the concrete.//
(6) Place the cement concrete, allow it to cure, and remove the forms.//
(7) Install any needed erosion control materials.

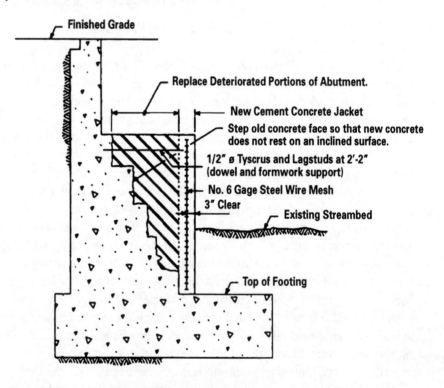

Repair of Abutment Face

Repairing Spread Footing: Deterioration of a concrete spread footing can include breaking off of the footing projections or spalling of the sides. Severe deterioration may be caused by collision of ice and/or debris against the upstream side of the footing, water penetration resulting in corrosion of reinforcing steel, or initial construction using poor materials. Because the load of the bridge was originally designed to be supported by a uniform distribution of pressure on the material under the footing, the area of the footing must not be reduced. Installing a cofferdam, pumping, and dewatering, as needed, will allow the repair to proceed in a "dry" working environment.

- Keep the work area clear of water with diversion channels, cofferdams, sandbags, or sheet piling, as required.
- Chip away the deteriorated concrete until sound concrete is reached. Clean away all loose concrete with air-blasting or other methods.
- Install reinforcing steel, anchors, and rods, as needed.
- Construct forms that adequately restore the footing dimensions of the original design size. Commonly, the footing is extended to cover a large area and the sides of the footing are extended downward if any undermining has occurred.
- Apply any bonding compounds or a neat cement paste for bonding just before placing the new concrete into the forms.
- Mix and place the new concrete, using a strong mix with a low slump. Vibrate the concrete thoroughly to ensure dense placement and a good bond to the existing concrete surface.
- After the new concrete has cured for at least 3 days, remove any cofferdam and restore the stream channel to its proper course. Where shotcrete is used extensively on other parts of the structure, the repairs may be made using shotcrete.

Substructure Cracks: A footing may crack transversely because of uneven settlement of the pier or abutment. This crack is often accompanied by a crack continuing up through the pier or abutment. It is advisable to seal the crack, preventing further intrusion of silt, debris, and water that will attack reinforcing steel. If the crack is moving, it should be filled with a flexible material; otherwise, it will open again. If the crack is not moving, it can be bonded back together. Cracks in substructures are generally vertical. Typically, the most effective method of repair is to inject epoxy into the cracks. To get maximum penetration of the epoxy filler, the first injection is made at the bottom of the crack. Starting at the bottom and working up in gradual increments toward the top increases the pressure needed to apply the epoxy and should result in greater crack-filling penetration.

Another repair method that prevents moisture from entering the crack is to chisel a V into the opening and fill it with grout.

(1) Cut a V-shaped groove at the surface along the crack approximately 50 to 75 mm (2 to 3 inches) in width, using a small pneumatic chisel.
(2) Thoroughly flush and blow out the crack, using water and high-pressure air blasts.
(3) Secure a retaining form on the face of the footing over the vertical portion of the crack.
(4) Wet the surfaces of the crack thoroughly by pouring liberal quantities of water into it. Fill the crack with cement (or epoxy) and fine sand grout in a 1 to 2 mix that runs freely.
(5) Clean out the V portion of the surface after the grout has partially set and apply bonding compound or a neat cement base to the surface of the V; then fill the V with a stiff grout mixture.

Surface Deterioration of Abutments and Piers: To repair any deteriorated concrete, completely remove all unsound concrete. Clean, sound concrete must be exposed to bond with the new concrete. Air tools are typically the most efficient means of removing the deteriorated concrete. The edge of any area cut out should be undercut enough to allow for deep patches that aid retainment of the new material to the existing concrete.

Effective bonding of the new concrete to the old concrete is usually accomplished with a bonding material and is particularly important when deep cracks need to be filled with a large volume of concrete. A grout of neat cement base can be used as an effective bonding agent. Grout can also be used when the form of the concrete is so inaccessible that an epoxy material cannot be effectively applied to the existing concrete surface. The exposed area can be sloshed liberally with grout just before placing the new concrete. Shotcrete may be used to fill the crack after it has been properly prepared.

Shotcrete is a concrete mortar pneumatically projected at high velocity onto a surface area. Shotcrete repair is effective for the repair of bridge beams, caps, piers, abutments, wing walls, and decks. Since forms are not generally used for shotcrete, it is particularly effective for an overhead patch on the underside of a deck where a form cannot possibly be used. Shotcrete gives a superior surface bond, provides great strength because of its high density, exhibits low shrinkage, and does not require any formwork; however, it requires lots of space to be applied, demands a high skill level of the application operators, leaves a shoddy appearance, and is costly (especially in small quantities). When shotcrete is used, no bonding agent is necessary if the patch does not exceed a depth of 75 mm (3 inches). For deeper patches, hook anchors are installed in the existing concrete on 300-mm (12-inch) centers, and 50-mm × 50-mm (2-inch × 2-inch) wire mesh is hooked and wired to the anchors. This anchoring system may be repeated for every 75 mm (3 inches) of shotcrete depth applied. Whenever possible, use form and placing rather than shotcrete. Shotcrete tends to waste cement and requires a much higher worker skill level to obtain satisfactory results.

Surface deterioration in reinforced concrete will frequently reach the first layer of reinforcing steel. When this has occurred, the concrete should be removed to a depth of 40 mm (1.5 inches) to 60 mm (2.5 inches) below this layer of reinforcing steel to provide an excellent anchor for the new concrete or shotcrete. If concrete removal is stopped at the surface of the reinforcing steel, a cleavage plane may develop at the interface of the new concrete and the old concrete, reducing the strength of the structure. All rust and other harmful materials should be removed from the exposed reinforcing steel. Where reinforcing steel is exposed, clean both the concrete and the steel by sandblasting the cracks for a good repair.

Water Line Deterioration: Deterioration at the water line is unique to abutments or piers in streams or marine environments. A depression or cavity forms in the concrete, extending some distance above and below the average water level. Deterioration at the water line usually occurs on the upstream face or along the sides of the pier. Repair is very similar to any surface deterioration repair, except that it is necessary to control the water flow so the work can be kept dry.
  (1) Dewater the abutment or pier.
  (2) Chip away all loose concrete in poor condition.
  (3) Clean the reinforcing steel of scale and loose rust.
  (4) Clean the surface in all areas where new concrete will be placed.
  (5) Chip or roughen the surface of the existing concrete, providing a better bond between the old and the new concrete.

(6) Treat the entire area with epoxy or grout before placing the new concrete.
(7) Construct a form of adequate strength and place concrete in the form.

Since repair of deterioration at the water line can be time consuming and expensive, maintenance supervisors should have the damage situation evaluated by a professionally qualified structural engineer or materials engineer to determine if this work should be done by maintenance personnel or contracted to a firm with special skill and experience in such repairs.

3.1.5.6 Underwater Repair of Substructures

Less than 20 years ago, only a few transportation agencies routinely inspected bridge substructures below the water line. Now, it is recognized that failure of structural elements under the water line can lead to bridge failures. Agencies now must identify all public highway bridges for which underwater inspections are necessary, define an appropriate inspection procedure for each bridge, and determine the necessary frequency of underwater inspection, which must not be less than every 5 years. Increased underwater inspection of bridges has made agencies more aware of the extent of deterioration below the water line, and consequently, has increased the need to respond with maintenance and repair of such damage.

Engineering the Repair: Solutions to underwater problems must be based on sound engineering. When underwater repairs have been necessary, bridge maintenance supervisors have tended to think in terms of dewatering the site. Cofferdams are installed and the damaged area is dewatered so that workers can perform the repair work using the same methods that would be used above the water line. The use of conventional above-water methods provides confidence that quality of construction can be controlled. However, dewatering is not always feasible. Additionally, when underwater repairs are undertaken, there is sometimes a tendency to use one of a few commonly available techniques marketed for underwater applications, without considering alternatives that may be superior.

A major concern of underwater repair is that it may only hide a fundamental structural problem while this problem continues to grow worse. Repairing a condition that hides an uncorrected problem may be worse than doing nothing to the existing damage. For example, when contaminated concrete and corroded reinforcing steel are left in place to interface with new concrete, the corrosion process accelerates. Consequently, covering a reinforced-concrete member with a form or jacket that stays in place will not stop or prevent corrosion of the reinforcing steel; the member may look satisfactory while it is deteriorating to an unsafe condition. The repair should address the deterioration process in the structure's environment. If structural members are involved, the repair should be designed to provide the appropriate safety factor and structural redundancy needed. Unless special monitoring can be guaranteed, the repair should provide dependable service within the normal inspection cycle and the normal maintenance that can be reasonably expected from the responsible maintenance agency.

Since most bridge engineers are not divers, it is important that they understand the problems and limitations of performing repairs underwater. To improve the estimates of long-term maintainability in suggested repairs, some state DOTs have developed underwater inspection teams that include professionally qualified structural engineers [3, 4]. All repair schemes used above water are not cost effective when performed by divers underwater. The time and cost of labor is more of a consideration for underwater work. For example, it may be less expensive to accept that deterioration will continue and modify the load path by designing the repair to support the total load or by designing a supplemental supporting system than to remove and replace the damage.

Control of Work in Waterways: The restrictions and permits required by agencies such as the U.S. Environmental Protection Agency (EPA), the U.S. Army Corps of Engineers (CoE), and the Occupational Safety and Health Administration (OSHA) must be considered when planning underwater repair. A permit is required from the CoE if the work will involve a navigable waterway. EPA restricts introducing pollutants and contaminants into the water or air. OSHA regulates safe working conditions. Regulatory restrictions may significantly affect the cost of repair alternatives. For example, fewer environmental problems may result if an old element is left in place and supplemental support is added rather than the damaged material removed. It may also cause fewer environmental problems to prefabricate the new element before it is placed in the water. Most maintenance activities are allowed under CoE regulations in "nationwide permits." However, the local CoE district office rulings and regulations should also be consulted.

Protecting Underwater Bridge Elements: Cementitious and epoxy coatings have been applied to underwater surfaces to protect concrete against abrasion and to cover cracks and make small repairs. The material is usually a thick mortar that is applied by hand. The cementitious products often include anti-washout admixtures and cure-set accelerators. Cementitious materials are mixed above the water line and delivered to divers in plastic bags. Epoxy resins applied underwater must perform satisfactorily and cure under both wet conditions and low surface temperature conditions.

Underwater cathodic protection (CP) has achieved favorable results in preventing and halting corrosion of reinforcing steel. Cathodic protection can be provided for the reinforcing steel by either a sacrificial anode or an impressed current flow from a rectifier power source. In a marine environment and an underwater application, the sacrificial anode CP is often recommended because of the low-resistivity environment. This system uses a metal (sacrificial) anode higher in galvanic series (metal activity scale) than the reinforcing steel to be protected. Zinc is often used as a sacrificial anode, but magnesium and aluminum are also sufficiently higher in galvanic activity than steel to be effective as sacrificial anodes.

Pressure Injection of Underwater Cracks: When cracks expose the reinforcing steel to moisture, the corrosion process may begin. In saltwater environments, corrosion can occur quite rapidly. With the proper selection of a water-compatible adhesive (normally an epoxy resin), dormant cracks (cracks that are not moving) that are saturated with water can be repaired. The procedure can also repair other small voids such as delaminations or honeycombed areas near the concrete's surface. Within limits, pressure injection can be used against the hydraulic head, provided the injection pressure is adjusted upward to counteract the pressure of the hydraulic head. The material must displace the water as it is injected into the crack to ensure that the crack is properly sealed, resulting in a watertight, monolithic structural bond.

Epoxies must have certain characteristics to cure and bond the cracked concrete. Many adverse elements are present inside the concrete crack (e.g., water, contaminants carried by water, dissolved mineral salts, debris from the rusting reinforcing steel, etc.). The typical low surface temperature of concrete underwater makes many repair products unsuitable candidates because of their inability to cure properly. The epoxy injection resins for cracks are formulated for low viscosity, and they do not shrink appreciably. The surface wetability of epoxy resin is a major concern because the resin needs to displace all of the water in the crack, adhere to a wet concrete surface, and then cure in the wet environment.

The procedure involves cleaning the crack with a high-pressure water jet system and shaping the surface of the concrete directly above the crack so that it can be sealed with a grout. Using a hydraulic or pneumatic drill, holes are drilled to intersect the crack and then injection ports are installed in the holes. Subsequently, the surface of the crack is sealed with a grout material suitable for underwater use, such as cementitious or epoxy mortar. The purpose of the grout is to retain the adhesive as it is pumped into the crack. The adhesive is pressure-injected into the crack through the ports that are embedded in the grout at regular intervals. The injection sequence begins at the bottom and advances upward. The injection moves up when the adhesive reaches a port and begins to flow out of it. Epoxy resin is mixed either before or after pumping begins. Cracks varying in width from 0.005 cm (0.002 inch) to 0.6 cm (0.25 inch) have been successfully injected.

Epoxy pressure injection has become regarded by the maintenance communities of a number of state DOTs as a cost-effective method to bond and seal cracked structural members under the water line. However, some precautions must be considered before using this repair method.
- Contaminants growing inside the crack, especially those found underwater, can reduce the ability to weld cracks.
- Corrosion debris can reduce the effectiveness of pressure injection.
- Time and patience is required for a successful injection repair.
- An injection repair is a labor-intensive process. (As the temperature drops below 10°C (50°F), it becomes more difficult to pump epoxy resins into fine cracks.)
- A diver experienced in the injection process and in the formulation of an epoxy resin is very important for high-quality work.

Concrete Repair Underwater: While the methods and materials for underwater concrete repair are nearly identical to those used above the water line, the adverse working environment creates special considerations.

Concrete Removal: Unless CP is used, the salt-contaminated concrete and rust must be removed from contact with the existing reinforcing steel to ensure that the corrosion damage will not continue. Concrete removal underwater is labor-intensive, difficult, and expensive work to perform. Consequently, structural jackets or auxiliary members should be evaluated as alternatives to removing concrete. Concrete can be removed with high-pressure water jets or with chipping hammers. Construction joints between old and new concrete should be saw cut prior to removing concrete, preventing feather edges at the joint. Hydraulic or pneumatic-powered concrete saws and chipping hammers may be adapted for use underwater. Mechanical grinders are also available for cleaning concrete surfaces. Surface preparation is required after concrete is removed and before making the repair. High-pressure water jets, abrasive blasting, or mechanical scrubbers can remove all loose and fractured concrete, marine organisms, and silt. Where the water is heavily laden with contaminants, repairs should be made on the day of final surface preparation. This will reduce accumulation of new surface deposits.

Forms: Forming materials and forming techniques have been developed specifically for application underwater. Forms are used to encase damaged concrete areas or masonry substructure units. Some pile-jacketing forms are proprietary and marketed as a repair package. The shape, size, and location of the damaged element will often dictate the forming system to be used for the repairs. In work underwater, the cost of the forms is less of an economic factor than the ease of installation and the suitability to the repair. Commonly available polyethylene drainage pipe is also used as a form. In some repairs, it may be more economical and quicker to encase all piling with a jacketing wall in one concrete placement process rather than trying to jacket individual piles. Fabric forms are relatively inexpensive and can be easily handled by a diver. With a fabric form, the final repair may appear irregular in thickness, but shape and texture are not concerns for underwater repairs. Fabric forms are available with zippers for ease of installation, with spouts for pumping the repair material into the form, and with pressure seals to hold the material inside the form.

The Mix: Anti-washout admixtures are used to: (1) minimize washing the fine aggregate and the cement out of the concrete while it is in contact with flowing water, (2) prevent segregation of the concrete, (3) reduce bleeding, (4) decrease migration of moisture within the concrete mix, and (5) inhibit water entrainment as the concrete is placed. These admixtures tend to make the concrete sticky. A water-reducing agent or a high-range water reducer may be necessary to maintain proper concrete slump. Slump values up to 20 cm (8 inches) are possible. Experience has indicated that concrete mixes containing anti-washout admixtures with either silica fume or fly ash can produce higher quality repairs at equal or lower cost than similar concretes containing higher silica fume content or high cement content but not incorporating anti-washout admixtures.

Underwater Placement: When a limited amount of water is needed for cement hydration and for concrete workability, additional water will damage the mix. As the ratio of water to cement increases, the permeability of the concrete increases and the strength decreases. If conventional concrete is dumped into water with no confinement as it falls through the water, it will lose fine particles, become segregated, or be completely dispersed, depending upon the distance of the fall and the velocity of the water current. Saltwater mixed with the concrete will corrode the reinforcing steel. Special techniques are needed to protect the concrete as it placed underwater.

Bagged Concrete: Concrete placed in bags can be used to repair deteriorated or damaged portions of concrete or masonry substructure elements underwater. Conventional bagged concrete repairs are made with small fabric bags prefilled with dry concrete mix (often only sand and cement is used) and anchored together to form the exterior of the repair. The bags are small enough to be placed in position by hand. The interior portion of the repair is then filled with tremie concrete or dewatered by small crews. This requires minimal skill and equipment. This process is often used when the water is so shallow that special underwater diving equipment is not required. Bagged concrete application was expanded when it became possible to take advantage of the durability and high strength of synthetic fibers to produce forms for casting concrete underwater. These bags possess sufficient durability to be used in marine environments exposed to cyclic changes (tidal flows) and provide abrasion resistance (to floating debris and to particulates carried in the water flow). The properties of fabric-formed concrete are essentially the same as those expected from concrete cast in conventional rigid forms, with one exception: The water-cement ratio of the concrete can be quite low at the surface since the permeable fabric allows water to bleed through the bags.

Prepackaged Aggregate Concrete: After the repair area is properly prepared and the forms are in place, graded aggregate is placed in the form. A cement-sand grout is then injected into the area containing the aggregate, displacing the water and filling the voids between the aggregate particles. This method is particularly effective for underwater repairs where it would be difficult to place premixed concrete because of forming restrictions. Generally, an expansive grout with a fairly high water-cement ratio is used to provide fluidity. When anti-washout admixtures are added to the grout, the forms do not have to be as watertight as is otherwise needed.

Tremie Concrete: Tremie concrete is placed underwater by gravity flow through a pipe called a tremie. The underwater portion of the pipe is kept full of plastic concrete at all times during placement of the total quantity of concrete. Concrete placement starts at the lowest point and displaces the water as it fills the area. A mound of concrete is built up at the beginning of the placement. To seal the tremie, the bottom of the tremie must stay embedded in this mound throughout the placement. The concrete is forced into the occupied area by the force of gravity from the weight of concrete in the tremie. The thickness of the placement is limited to the depth of the mound of concrete. Tremie concrete is best suited for larger volume repairs where the tremie will not need to be relocated frequently or for deep placements where it would be impractical to pump the concrete. The method of placing concrete with a tremie is simple and requires few pieces of equipment, minimizing potential malfunctions. Thus, it is one of the most common methods used to place concrete underwater.

Pumped Concrete: Pumped concrete is placed underwater using the same equipment that is used to place concrete above water. The placement process is similar to the process of using a tremie except that the end of the pump line does not need to be in the concrete as with a tremie. A direct transfer of the concrete is provided, and the pump forces the concrete through the supply line. The placement of concrete must start at the bottom of the area and the hose or pipe must stay submerged in the fresh concrete during placement. However, the pipe does not need to be lifted as much as with a tremie. A handle on the end of the pipe or hose will help the diver position it.

Free-Dump Concrete: Anti-washout concrete admixtures have been developed to minimize loss of fine aggregate and reduce segregation when the product is placed underwater. This admixture makes the concrete more cohesive yet sufficiently flowable for placement. However, the concrete mix loses some of its normal self-leveling properties and tends to stick to equipment. It is not clear that free-dumping concrete underwater with admixtures to improve placement quality is as effective as other more controlled methods of placement. Therefore, this method should be used with caution. It may be appropriate when high-quality concrete is not a primary consideration, where there is low water current velocity, and where the free-drop distance is limited to about 1 meter (about 3 feet).

Hand-Placed Concrete: Hand-placed concrete is mortar or concrete that the diver places by hand and then packs or rams for consolidation. This method is best suited for isolated repair sites. Use of accelerators, anti-washout admixtures, and a low water-cement ratio is recommended. The method is best suited for deep, narrow cavities. The concrete can be delivered to the diver by a bucket on a rope conveyor assembly. It can also be dropped to him in baseball-sized quantities through a pipe with holes cut in the sides to allow displaced water to escape, easing descent of the concrete. Small quantities needed for patching can also be delivered to the diver in plastic bags.

3.1.5.7 Pile and Bent Repair

Most piles require little maintenance because the material into which they are driven protects them or deterioration is not common. Where piles are exposed (whether by design or by scour), there are potential problems. These problems include scaling and spalling of concrete piles, corrosion of metal piles, or decay in timber piles, and buckling in all types if the unsupported length of the pile becomes excessive.

Preventive Maintenance: Preventive maintenance of exposed piles is the same as for other substructure elements constructed of the same material. For example, steel should be painted and concrete coated to protect against deterioration and corrosion. Timber piles are often used in bents to support small bridges. New cuts and bolt holes in treated timber should be thoroughly coated with preservative materials to prevent moisture from entering the wood and causing decay.

The area around timber pile bents and abutments must be scalped, and all weeds and brush removed from the vicinity to limit the fire hazard. Where timber pilings are subject to frequent damage from ice, some sort of armor plating (e.g., discarded motor grader blades) should be used to protect them. Scour of the streambed may expose piling below the flow line to a degree that additional cross bracing may be necessary to maintain structural stability.

Jackets for Pile Protection and Repair: Jackets are a common type of pile protection and repair. They are used for protection of all types of piles: concrete, steel, and timber. The jacket can protect against abrasion damage, repair lost cross section, or accomplish both purposes. If the jacket is for protection only, it typically consists of a liner placed around the area to be protected with a cementitious grout or epoxy resin pumped into the annular opening between the existing concrete and the liner. If the jacket is intended to repair structural damage, the liner will provide space for reinforcing steel, and the space between the liner and the old pile will be filled with concrete. The liner (form) is often premolded fiberglass. However, it could be steel or fabric. Old drain pipes have been used as jacket liners.

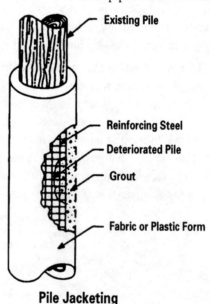

Pile Jacketing

Deteriorated reinforced-concrete and prestressed-concrete piles can be encased with a concrete jacket after all unsound concrete has been removed and the surface prepared as for other repairs. Encasement will compensate for the cross-sectional loss and strengthen the pile. Reinforcing steel cages or reinforcing wire are placed around the pile before the forms are placed. The reinforcing steel is usually epoxy coated for protection against corrosion. Standoffs are placed on the reinforcing steel before they are drawn tight to the pile. Forms, either rigid or flexible, are then installed and sealed. Concrete is placed in the form either through a tremie or by pumping underwater. After placing the concrete, the forms are either left in place permanently for further protection of the pile or removed when the concrete is cured.

The North Carolina DOT uses fiberglass forms to construct pile jackets. This method is useful when damage extends above and below the water line. Deteriorated concrete is removed using high-pressure water jets. The jacket extends approximately 600 mm (about 2 feet) beyond the damaged area at each end of the pile to account for any concrete segregation near the bottom or loose materials at the top of the new concrete. Welded wire fabric or a reinforcing steel cage is wrapped around the repair area. The figerglass form has a vertical seam so it can be fitted around the pile. Subsequently, top and bottom centering devices and a bottom seal are placed. The form is secured in place with bolted bands and tightened to ensure full enclosure. If the length of the repair exceeds the length of the form, the piles may be repaired in two lifts. If the damage extends below the mud line, trenches are dug at the bottom to extend the repair into the mud zone.

Jacketing of steel piles is basically the same as that described above for concrete piles. Both flexible and rigid forms can be used. Often fiberglass and plastic forms are used because of their ease of installation for underwater applications. Prior to pile jacketing, remove marine growth and corrosion to clean all steel. Next, standoffs are placed on the pile flanges before the forms are installed and the concrete is placed. Welded wire fabric is typically used to reinforce the concrete against cracking. Concrete jackets can cause accelerated corrosion on a steel pile when both concrete and water are in contact with the steel. A corrosion cell will develop either below the bottom or above the top of the jacket. Thus, concrete jackets should be extended well into the mud zone and also well above the high water line.

Special considerations for installing concrete-filled pile jackets include (1) using qualified divers for underwater survey and repair work and (2) having concrete pumps available for underwater placement. The general process to jacket a pile includes the following steps:

(1) Scrape the surface of the pile clean, removing deteriorated concrete or wood.
(2) Clean the exposed reinforcing steel in concrete piles above the water line. Splice the existing reinforcing steel with new reinforcing steel, if needed. Install a steel mesh reinforcing cage around a timber pile or concrete pile. Use spacers to keep the forming in its proper place.
(3) Place the forming jacket around the pile and seal the bottom of the form against the pile surface.
(4) Pump suitable concrete into the form through the opening at the top. Sulfate-resistant concrete should be used in saltwater environments.
(5) Finish the top portion of the repaired area.

Steel piles must be protected with coatings that prevent dissolved oxygen from contacting the steel. Epoxy coating systems and polyvinyl chloride barriers have been used.

Pile Replacement: If the necessary equipment is available, either in the agency equipment fleet or through equipment rental shops, it may be easier to replace a damaged pile rather than repair it. Replacement is accomplished by cutting a hole in the deck and driving the pile through the hole. Since the pile is driven from the deck, the deck must be capable of supporting the necessary pile-driving equipment. Maintenance operation planning includes the following steps:

- Determine if the bridge will support pile-driving equipment. If it will not, an alternative is to drive from a barge or from dry land if conditions permit.
- Make provisions to restrict traffic from the work area.
- Typical equipment needs include pile-driving, wood-cutting, and deck-patching equipment; piles; come-alongs; jacks; flashing; fasteners; cutting torches; and pavement-breaking, concrete-sawing, and welding equipment.

The general process includes the following steps:
(1) If the deck is overlaid with asphalt or concrete, locate the centerline of the stringers closest to the pile replacement.
(2) Cut through the overlay and the deck along the centerline of the stringer. Remove sufficient deck to permit the pile to go through the hole adjacent to the cap.
(3) If cross bracing is present, remove it from between the piles and on the side of the bent where the new pile will be driven.
(4) Set the pile at a slight batter so it will be plumb when it is driven and pulled under the cap. Drive the pile to the specified bearing.
(5) Install U-clamps and blocking around the pile to be replaced. Place a jack on blocking and jack the cap up approximately 13 mm (about 0.5 inch).
(6) Cut the pile 6 mm (about 0.25 inch) below the cap and pull the pile into position under the cap.
(7) Positioning the pile can be accomplished by using come-alongs to pull against adjacent bents. Place copper sheeting on a timber pile head. On a steel bent, the pile is welded to the cap.
(8) Lower the jack and strap in place.
(9) Reconstruct cross bracing on bent piles.
(10) Close deck holes and restore normal traffic flow.

This process may also be used for strengthening an existing bent.

Piling Repairs: Pile repair depends upon the type of pile: steel, concrete, or timber.
  Steel Piles: Steel piles may be damaged, particularly if they are located in waterways where they may be struck by heavy barges or if they are placed near roadway work zones where they may be struck by heavy equipment. The latter scenario is less common. Damage in the form of bent, torn, or cut flanges may reduce the cross section, and hence the load-bearing capacity, of the pile, requiring repair. More commonly, steel H-piles may become severely corroded in a relatively short section near the main water line in a waterway or as the result of unusual conditions such as broken drains. An otherwise sound steel pile or one that cannot be easily replaced or supplemented because of access or scheduling may be strengthened by repair with bolted channels as a temporary measure. Maintenance operation planning should include the following actions: (1) select appropriate channel size to meet strength and dimensional requirements; (2) determine length of damaged area and secure steel channels of selected size that have been fabricated in appropriate lengths with necessary hardware; and (3) assemble equipment and tools needed (such as equipment to drill bolt holes), protective coating materials, and necessary staging.

The repair process includes the following steps:
(1) Clean the damaged pile.
(2) Locate the extreme limits of the deteriorated section. The repair channel section should be about 0.5 meter (about 18 inches) longer than the distance between these limits.
(3) Thoroughly clean the area to which the channel is to be bolted.
(4) Clamp the channel section in place against the pile.
(5) Locate and drill holes for high-strength bolts through the channel and the pile section.
(6) Place bolts and secure the channel.
(7) Remove the clamps.
(8) If the pile repair is above the water, coat the entire area with a protective coating material.
(9) For long-term rehabilitation, steel piles should be encased with a concrete jacket.

This procedure can also be used to add a new section of pile above a damaged area.

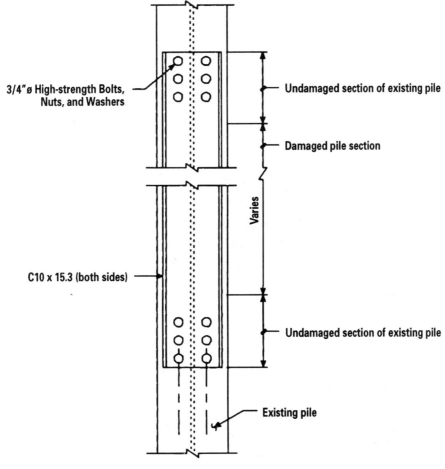

Steel Pile Strengthening

Patch Concrete Pile: Deteriorated concrete in a concrete pile should be removed and replaced with sound concrete. After the deteriorated concrete is removed, the reinforcing steel should then be cleaned of all rust and scale, and the new concrete placed. Sufficient concrete should be removed so that the thickness of the new concrete placed is a minimum depth of about 50 mm (about 2 inches). After forms are placed, all of the old concrete surfaces that adjoin new concrete should be covered with a bonding material. The new concrete is then placed, or grout-injected dry-packed aggregate is used.

Chapter 3 Bridge Maintenance and Management **3-73**

Casting Subfooting to Cap Piles: Badly deteriorated piles that are exposed under a footing must be repaired before the void under the footing is filled. If accessible, timber or steel piles may be spliced. Where there is no room to implement a repair by adding sections, a possible alternate repair process may be used. An alternate process is outlined in the following steps:

    (1) Cut out the deteriorated portion of the pile from the footing bottom to the depth of the sound piling material.

    (2) Form and place new concrete from the footing bottom to about 150 mm (about 6 inches) below the new top of the pile, using Class I concrete.

    (3) When topping out the new concrete, maintain a hydrostatic head at the interface between the fresh concrete and the old footing. Attempt to eliminate voids at the interface. Pumping or pressure grouting may be required after the concrete cures.

    (4) The repair must be phased so that there is always sufficient support to the structure to carry the required load.

    (5) Fill all of the voids.

When the void that exposed the piles is caused by erosion, some measures must be taken to prevent its recurrence.

Splicing Steel H-Piles Under Footings: Deteriorated steel can be repaired by adding sections, if there is sufficient working space to weld or bolt sections in place. To proceed with splicing, loss of cross section should be less than 50%. The web and flanges of the pile should be strengthened by welding steel plates extending far enough above and below the deteriorated area to carry the full load on the pile. Where extreme section loss is present in the pile at the interface with the footing, repairs can be made by welding plates to form an angle with one leg against the footing and the other leg against the pile. Stiffeners should be placed across the angle as needed. These angle plates should be placed on both flanges. When the welding is completed, all of the exposed piles should be given a heavy, protective coating. Fill should then be placed around the piles up to the bottom of the footing.

Pile Shell Repair: Filled shells are cast-in-place concrete piles. A metal shell is driven, the mandrel is withdrawn, and the shell is filled with concrete. Sometimes a problem develops when the shell is corroded as the concrete deteriorates. Temporary support should be erected to carry the load when necessary. Rust and scale must be removed, and the steel primed and painted. When both the shell and the concrete are damaged, the deteriorated portion of the shell must be removed and a collar of sufficient strength and diameter placed around the pile. The collar is extended well above and below the affected area, and a high-quality, low-shrinkage concrete; epoxy mortar; or cementitious grout is pressure-injected inside the collar to fill the voids. The material should be well compacted so that the voids are completely eliminated.

Repairing Intermediate Bents: The maintenance repair process varies somewhat depending upon the material used to construct the original bent.

Steel Pile Bents: Corrosion or deterioration in steel piling usually occurs at the water line or ground line where wet and dry conditions alternate. Collision damage may also be a problem where piles are near roadways or navigable waterways. Rust and corrosion should be removed by sandblasting or with pneumatic needle scalers. When possible, damaged areas should be straightened and the deteriorated areas should be strengthened by welding steel plates extending far enough above and below the deteriorated area to restore the full load-carrying capacity of the pile. If deterioration is minor, metal plates may be added by welding. All repaired areas should be cleaned and painted as a preventive maintenance practice.

Adding Sections of Timber Pile: Treated timber piles that have decayed or been damaged by fire or impact can be repaired without having to drive into the old pile, if the portion of the pile below the ground is still sound. These steps must be followed:

(1) After the required auxiliary support is in place, the old pile is cut off below the decayed or damaged area.

(2) A new section of pile is cut about 150 mm (about 6 inches) shorter than the section removed. Plates are put in place on top of the existing pile and on the bottom of the new section of piling.

(3) A 19-mm (3/4-inch) bolt with a nut is welded to the bottom plate, extending through a hole in the top plate. By adjusting the nut on the bolt, the new section of pile can be raised until it is securely seated against the bent cap. Care must be taken to raise the new section of piling far enough to cause the bent cap to lift from the adjacent piling.

(4) After the new section is in place, 6-mm (1/4-inch) thick angles are welded between the plates at each corner.

(5) Used girder plates or flat stock are then bolted to the timber pile and extended down on the original pile about 300 mm (about 12 inches). Using straps, the top of the pile is secured to the bent cap.

(6) A 1 m × 1 m × 0.6 m (3 feet × 3 feet × 2 feet) block of dense concrete is placed around the pile. Any falsework erected may be removed after the concrete has been placed and has cured.

Splicing Timber Piles With Steel Columns: Timber pilings that have decayed, have been weakened by insects or marine organisms, or have been structurally damaged by collision or overloading may be replaced with steel columns. Maintenance planning for such repair should (1) evaluate the condition of existing caps and existing piles below the surface; (2) determine any need for a cofferdam to dewater the work area; (3) determine a method of temporary support for the superstructure during repairs; (4) make provisions to restrict traffic from the work area during repairs; and (5) procure necessary equipment and tools such as wood-cutting tools, welding and steel-cutting equipment, light lifting equipment, wrenches, and other small hand tools. The following steps are generally necessary to ensure a satisfactory maintenance process:

(1) Determine all cutoff points on the existing piles and the column length needed for the repair.

(2) Construct temporary support for the superstructure before beginning the repair.

(3) Construct a cofferdam (if needed) and dewater the work area.

(4) Excavate so that the top of the footing will be a minimum of about 225 mm (about 9 inches) below the ground line.

(5) Cut existing piles off so that they will project at least 300 mm (12 inches) into the new footing.

(6) Separate the old sections of piling from the pier cap.

(7) Form and place concrete footings over the existing pile stubs.

(8) Place the anchor bolts in the footing concrete before the initial set of the concrete.

(9) Cut the steel columns to the proper length and weld on the base plates.

(10) After the concrete has reached the required strength, attach the new steel columns to the footings with nuts and washers.

(11) Attach the top of the new columns to the existing pier caps with lag screws.

(12) Remove all temporary supports, backfill where necessary, and remove the cofferdam (if one was used).

Repairing End Bents: Wing walls and abutments that are exposed along the bottom of their backing can be repaired temporarily by driving pieces of plank on end, vertically, along the bottom of the wing abutment. When a plank decays, the best method of repair is to replace the plank.

Pile Splice: The timber splice method should not be used when replacing piling in the abutments or end bents because it will not provide sufficient resistance to the overturning moment produced by the force of the fill against the back wall. Steps generally included in the maintenance process to repair the top of an abutment pile are as follows:
  (1) Place the required falsework.
  (2) Cut the pile off below the decayed or damaged area.
  (3) Split lengthwise and place around the pile a cylindrical steel pile shell, long enough to extend from the bottom of the cap down about 600 mm (about 2 feet) on the remaining pile.
  (4) After the steel pile shell is pulled tight around the pile and welded back together, fill it with concrete.
  (5) Remove the falsework after the concrete has cured.

Where it is not practical to repair the piling with this process, a new section of pile can be spliced to the old section, using steel pipe or a band to hold the two butt ends together.

Installing Helper Bents: An existing substructure unit that is not capable of supporting its required load may be supplemented with a timber helper bent. One example is a concrete bridge pier in which seat damage is so acute that the bearings are affected and the beams may dislodge. In this case, a timber helper bent adjacent to the pier would support the load and preclude bridge failure if bearing failure did occur. The timber helper may also be used to reduce the span length and increase the bridge load-carrying capacity in a situation where the beams are weakened or were not designed for current legal loads. A professionally qualified structural engineer should determine the size and location of the helper bent. The engineer must also determine if the bridge can support a pile-driving rig or if the equipment can be located off the deck. A professionally qualified hydraulic engineer should determine if the additional restriction to stream flow that the helper bent will create is acceptable for hydraulic efficiency. Provisions should also be made to maintain traffic safely away from the work area. Equipment systems that may be needed include pile-driving equipment, lifting equipment, deck-cutting equipment, and perhaps scaffolding. Steps generally included in a maintenance process are as follows:
  (1) If the deck is timber with an asphalt or concrete overlay, locate the centerline of the stringers closest to each pile location for the helper bent. If the deck is reinforced concrete, locate the piles so that the deck beams will not interfere with pile driving.
  (2) Cut holes in only one traffic lane at a time.
  (3) If a timber deck does not have an overlay, cut the timber decking at the centerline of the stringer near the centerline of the bridge and remove the decking across the traveled lane.
  (4) Cut through any overlay and the timber deck along the centerline of the stringer. Remove sufficient amount of the timber decking to permit a pile to go through the hole. For a reinforced-concrete deck, remove a sufficient amount of concrete in a square pattern to permit a pile to go through the hole. Cut any reinforcing steel at the center of the hole and bend it out of the way of the pile to be inserted through the hole.
  (5) Set the piling and drive it to the required bearing capacity.
  (6) Cut off the piling approximately 6 mm (about 0.25 inch) above the bottom of the existing cap. If the existing cap has settled, allowance must be made for grade differential.

(7) Place cover plates over deck holes, open the lane to traffic, move to the adjacent lane, and repeat the process as needed.

(8) After all piles have been driven and cut off, jack up the superstructure approximately 13 mm (about 0.5 inch), using the existing pile bent. This may be accomplished with U-clamps and blocking supporting jacks against the beam bottoms.

(9) Place timber caps over both rows of pilings. For end bents, only one row of piles and caps is required.

(10) Lower the superstructure onto the new caps and strap each cap to its piling. Shimming may be required to obtain bearing between the superstructure and the timber caps.

(11) Remove the deck plates and reconstruct the deck. If the deck is reinforced concrete, splice all reinforcing bars that were cut. Replace the deck one lane at a time. Sections of the deck under repair may be reopened to traffic if protected with steel plates.

(12) Erect cross bracing on the new pile bent. For intermediate bents, cross bracing between the two new bents is also required.

An alternative temporary bent may be constructed with the following procedure:

(1) Set piles and drive to the specified bearing capacity.

(2) Cut off piles to allow for beams, neoprene pads, and wedge plates.

(3) Connect 300-mm × 300-mm (12-inch × 12-inch) timber caps to plates with 22-mm (7/8-inch) × 530-mm (1-foot 9-inch) headless drive spikes.

(4) Connect timber brace to piles.

(5) Set steel beams (such as W21 × 55 or W21 × 62) into position.

(6) Jack beam from cap to obtain temporary bearing against superstructure.

(7) Set neoprene pads into position. If necessary, cut them to obtain 225-mm × 225-mm (9-inch × 9-inch) pads.

(8) Remove jacks and drive lag bolts.

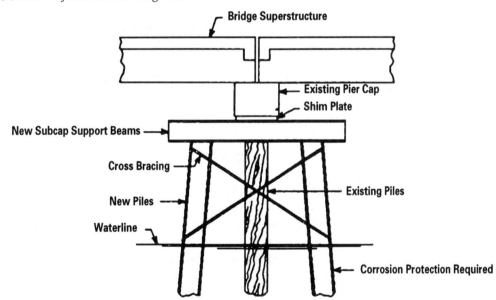

Strengthening an Existing Bent

Chapter 3 Bridge Maintenance and Management

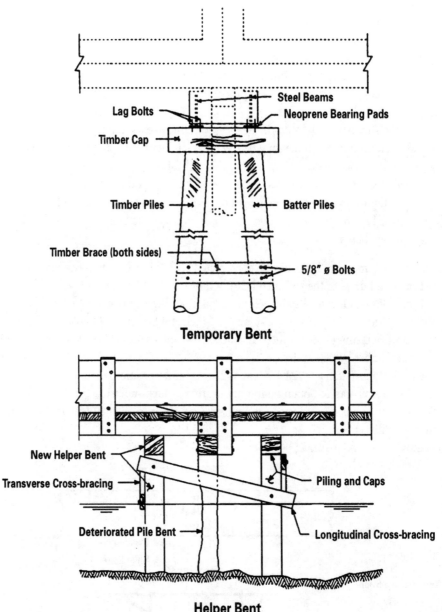

Temporary Bent

Helper Bent

Repairing Timber Caps: Repair of timber caps can involve repairing and rotating the cap, replacing the cap, or strengthening the cap.

Rotating Caps: When timber pile abutments are pushed forward by the earthen fill that the abutment is retaining, causing the pile cap to rotate, either the pile stays are broken off or the abutment was constructed without stays. Abutments that are too high can also cause this problem. Generally, the maintenance process to correct this problem includes the following steps:

(1) Remove all earth from behind the abutment. (Excavation behind an abutment may present a high potential for a cave-in. Shoring may be needed [5].)
(2) Pull the piles and caps back into position.
(3) Repair or install the pile stays.
(4) Bury or drive deadmen behind the abutment. Then fasten and tighten the cables to the piles with eyebolts, using the size of cables specified by a professionally qualified engineer.

Replacing Timber Caps: A common maintenance problem with pile caps is decay followed by longitudinal cracks and crushing from the load on the cap. When these problems arise, the pile cap must be replaced. The superstructure is jacked either from the existing columns or from a temporary bent. The decayed cap is removed and a new cap is secured in position. To avoid future decay, timber pile caps that have deteriorated may be replaced with 300-mm (12-inch) steel beam caps. Stiffeners may be welded between the flanges directly over the piling. The steel cap is secured to the piling with a piece of 75-mm (3-inch) flat stock bent to encircle the top of the pile and welded to the bottom of the cap.

Strengthening Timber Caps: The need to increase the load capacity of a bridge may arise from improper sizes or defects in a particular member. An example is wood pier caps that have developed large lengthwise shrinkage cracks or a large number of splits near bolt fasteners. A structural analysis may indicate such caps rate significantly lower than other members because of these defects. If the original cap is still in good structural condition and no decay is evident, strengthening the cap is often easier and cheaper than replacing it. In some cases, access limitations or other factors may make it very difficult to replace a cap. Thus, strengthening the cap is the most logical maintenance process. Good maintenance planning for cap strengthening should ensure that the existing cap and columns are in good condition, that the new section (as strengthened) will meet the design analysis load conditions, and that the appropriate resources are available (heavy-duty drilling equipment, light lifting equipment, access to the pile cap, wrenches, and small hand tools). The general process usually conducted to ensure a successful maintenance operation is as follows:

(1) Construct scaffolding, as needed, around existing bent.
(2) If the pile diameter is wider than the existing cap, notch the existing piles or columns to accommodate the new timber cap members.
(3) Snugly place the new members against the existing cap and stringers, and temporarily clamp them in place.
(4) Drill 21-mm (13/16-inch) holes for 19-mm (3/4-inch) bolts.
(5) Insert the bolts, tighten them down, and remove the clamps.
(6) Remove scaffolding, if any was used.

Installing Deadman Anchorages: The force of earth and stone in the bridge approach behind the bridge abutment tends to push forward and rotate (tip over) the abutment, especially if the fill behind the abutment is unstable or the abutment is not adequately anchored. A deadman is a heavy mass (weight)—usually concrete blocks—attached to the abutment with a long steel rod and located in stable earth far behind the abutment. This provides an anchor that prevents overturning of the abutment. Good maintenance planning for installation of a deadman anchorage includes (1) using a professionally qualified engineer to calculate the magnitude of the forces to be resisted by the deadman, to determine the required size of the deadman and the restraining rod, and to determine if piles are required; and (2) providing the necessary resources including excavation equipment, light lifting equipment, concrete, drills, miscellaneous hand tools, and any shoring as needed. The general procedure to install a deadman anchorage includes the following steps:

(1) Excavate the area where the deadmen are to be placed and provide a trench for the restraining rods. (Excavation behind an abutment may present a high potential for a cave-in. Shoring may be needed [5].)
(2) Drive piles for the deadmen, if required.
(3) Place formwork and concrete for the deadmen. The side of the deadmen facing the abutment should be cast without forms. No formwork is required if the soil conditions

are stable enough for the walls of the excavation to act as earthen forms.
(4) Drill through the wing walls and place the restraining rods. Wrap and coat them with tar (or provide other means to protect the rods from corrosion).
(5) Bolt the restraining rods at the deadmen.
(6) Place the waler beams and tighten the rods.
(7) Grout holes in the wing wall.
(8) Backfill the excavated area.

3.1.6 Watercourse and Embankments
3.1.6.1 Removing Debris From Channels

The terms "drift materials" and "debris" can be used interchangeably to describe any floating or submerged materials that are transported by flowing water. Vegetation, sediment, trash, man-made materials, rocks, and ice are all considered debris or drift. Vegetation debris and sediment are the primary debris materials that are a maintenance problem. Ice is a secondary, specialized problem.

Bridge Damage Caused by Debris: Many bridges have been damaged or have failed as a result of problems caused by debris. The types of damage caused are usually categorized as scour, impact, drag, and miscellaneous.

Scour: Floating debris can catch on bridge members, and these obstructions in the stream channel will tend to catch and accumulate additional debris. These accumulations divert and constrict the flow of water, which increases the velocity of the flow and creates turbulence in the flow, both of which increase the potential for bank and bottom erosion (scour) to occur.

Impact: During peak flows, a waterway can carry massive debris such as car bodies, entire trees, superstructure elements of other bridges, etc. Debris in fast-moving streams may strike a bridge and damage its timber piles. If the water level reaches the superstructure, the beams and bottom chords of trusses may be damaged. Impact damage to solid piers and other structural members is rare but not unknown, especially if barges, floats, or heavy watercraft are swept into them by fast water.

Drag Forces: Debris caught on the substructure or superstructure members will trap additional debris. Depending upon the size of the debris accumulation and the water velocity, extremely large forces can be applied to the bridge. For instance, the massive floods that hurricanes can generate have frequently moved bridge superstructures off of their centered position on a pier and have sometimes even dropped bridge spans.

Miscellaneous Damage: Less common types of damage include fire damage during dry seasons caused by vegetation caught on or accumulated beneath the bridge; abrasion damage from sand, gravel, or other suspended particles, especially in streams with high-velocity water flows; and scars on the bridge surface caused by debris scratching and marking the substructure and superstructure units.

Debris Removal: After heavy storms, structures should be routinely checked to determine the need to remove debris. All structures should be checked at least once per year. If maintenance records indicate that debris accumulation is a common problem, bridges should be checked more frequently. Factors that affect the likelihood of debris problems include the following:
- *Type of structure:* Open web piers and pile bents are more likely to catch debris than piers with solid webs.
- *Length of structure:* Short spans are more likely to be damaged by debris than long spans.

- *Height of Structure:* Bridges with low clearance will usually collect debris more readily than high bridges.
- *Unstable Banks:* Bridges on streams with easily eroded banks are more likely to accumulate debris than bridges on streams with stable banks.
- *Land Use:* Heavily forested areas will usually generate more floating debris than nonforested areas. Logging and mining operations will increase the chance of debris problems as will some land development and farming activities.

Emergency Debris Removal: During peak flows, it may be necessary to have maintenance personnel available for emergency debris removal. This work involves removing floating debris that has lodged against the piers and abutments. This type of work can be hazardous, and when appropriate, workers should be required to wear safety belts, buoyant work vests, hard hats, and other safety gear. A flashlight that will operate when submerged is an excellent safety item for nighttime work. Having the necessary resources available to close a bridge when it is in danger of flow-induced damage during peak flow can prevent traffic hazards.

Routine Debris Removal: Routine debris removal should include removal of all floating debris such as logs, brush, tree limbs, and fallen trees that have lodged against the bridge or have been deposited in the immediate vicinity of the bridge. It may be necessary to obtain EPA and Coast Guard (if applicable) permits or approvals before performing this work. Sand or gravel bars and other deposits of debris should be removed on a high-priority basis when it is likely that the deposited material will cause turbulence near the piers or abutments. Otherwise, the material may direct stream flow into the piers or abutments or into the embankments behind the abutments. Trees that have been undercut along a bank should be removed or protected with riprap, preventing them from falling into the stream. Obstructions in the streambed that can trap or catch debris—such as trees, sand bars, or boulders—should be removed. All debris that is removed should be hauled away from the bridge and disposed of in a manner that will not allow it to contribute to debris accumulations at downstream bridges.

Preparation: The methods used for debris removal may depend on environmental regulations. Before performing routine removal tasks, maintenance supervisors should contact appropriate environmental and water resources agencies to ensure clearance for the work required. The agency's maintenance supervision policies should provide guidance to supervisors, assisting them in determining what types of work can be performed in a stream channel without a permit, with a blanket permit that applies to many sites, or with a permit for a specific site.

Environmental restrictions generally apply to those activities that involve major disturbances of a natural streambed. Although maintenance personnel have a different objective because they want to protect the bridge rather than the stream, alternative courses should be evaluated when extensive work must be performed to remove sand bars or to reshape or regrade a stream channel. Removing debris with light cranes, subdividing debris by sawing, or other similar type of equipment processes may be necessary. Schedule variations (e.g., scheduling for fall work rather than summer work) may also be needed. Locations requiring extensive work should be referred to both hydraulic engineering staff as well as conservation and environmental analysts to evaluate the maintenance situation and provide advice on the preferred option.

Maintenance workers should obtain permission from private property owners when it is necessary to remove debris from private property. In some agencies, maintenance personnel handle this task through formal or informal agreements with property owners, while other agencies require that a right-of-way office make arrangements.

Methods: A winch or crane is often the most effective type of equipment to remove vegetative debris such as logs and fallen trees that have accumulated near a bridge. A crane with a clamshell bucket can remove trees, stumps, and other debris. At bridges that span a flood plain, a bulldozer or front-end loader can be used during dry periods to push debris on the flood plain into piles that can be reached by a crane.

When work in the stream channel is required, a crane with a drag line can be used for reshaping and regrading. During low flow, it may be possible to reshape and regrade the channel with a bulldozer. The need for regrading and reshaping or other extensive work in the stream channel must conform with environmental regulations and should be evaluated by a professionally qualified hydraulic engineer as well as environmental conservation analysts. An appropriate solution to the problem may require a great deal of coordination, planning, and review before approval to proceed is received. However, if all needed changes are not accomplished, the cause of the problem cannot be corrected.

Debris Countermeasures: The effectiveness of debris countermeasures has not been proven. However, some state DOTs believe that debris problems can be reduced by implementing or installing one or more of a variety of countermeasures to bridges that are critical transportation network links.

Structure Modifications: Piers that have rounded edges and solid webs reduce debris accumulation. Solid webs can be constructed in multiple-column piers to prevent debris from lodging between the columns. Timber or metal cribs have been installed by some agencies on bridges with open pile bents. When selecting the cribbing, ensure that it does not increase the problem. Cribbing with large mesh may actually snag and hold debris rather than reduce debris accumulation.

Debris Deflectors: Debris deflectors may consist of piles immediately upstream from a pier or fins attached to the upstream end of a pier. The piles and fins are intended to divert and guide debris through the waterway opening of the bridge. The effectiveness of deflectors depends upon the direction of water flow. In situations where the flow is not parallel to the pier, deflectors may create hydraulic problems. In such cases, flow control devices, such as spur dikes, may be necessary to allow the deflectors to function properly.

Flood Relief: When the approach roadway geometry permits, flood relief sections can be designed and constructed. These sections allow water to flow over the approach roadway when or before the bridge waterway opening is blocked by debris. Although relief sections often require extensive repair after flooding, their maintenance is usually less costly than replacement of the bridge.

Debris and Sediment Traps: Debris and sediment traps have been effective in forest management projects but are not generally used on bridge approaches. Sediment traps and other erosion control measures are commonly used in highway construction projects to reduce the amount of sediment washed into streams when bare earth surfaces are exposed during the construction process. They are not generally considered a permanent installation of the facility. All measures that reduce the sediment load in a stream are helpful to bridge maintenance efforts, even if the effect is a secondary one.

Land Use Regulations: Sediment control and land development regulations can be effective means of reducing the amount of vegetative debris and sediment that is washed into streams. In many areas, strict sediment control regulations have been enacted to apply to various types of land use, land development, and construction activities. For example, some agencies limit the square footage of disturbed land area and the amount of time that can elapse before re-establishing ground covers. In some areas, logging operations are required to leave undisturbed buffer strips along streams to catch debris.

3.1.6.2 Scour Protection and Repair

A wide variety of methods have been developed for scour control and slope protection. Commonly used methods range from routine maintenance activities (e.g., filling and stabilizing washed-out areas) to major construction of extensive slope protection and flow control systems. In extreme cases, modifications such as constructing additional spans to bridges may be necessary to provide a larger flow channel.

Role of Maintenance Forces: Maintenance crews are primarily involved with bank erosion problems that can be corrected by minor or routine types of maintenance. The selection, design, and construction of scour control and slope protection measures for major or chronic problems generally should be conducted under the design and supervision of professionally qualified hydraulic engineers. However, once any major protective measures have been designed and constructed, they become the responsibility of the maintenance division. Thus, maintenance supervisors need to have a basic understanding of the types of devices available, how these devices function, the most likely types of failure, and how the devices should be maintained. Maintenance activity generally falls into the following four areas:

(1) Response to emergency situations.
(2) Maintenance of the protective devices.
(3) Assistance in identifying the need for protective devices.
(4) Provision of input on the selection of alternatives.

Area maintenance personnel should know the bridges that are susceptible to problems from a major storm. They are generally the first on the scene and the first to know when bank erosion is occurring or when slope protection and flow control measures have been damaged. They should also look for damage to the bridge or the bridge foundation. Often during major storms, bridges are damaged and later fail under traffic. The need for repairs must be recognized as early as possible. Minor repairs made in a timely manner are likely to reduce the need for major repairs that may result if the problem is left untreated. Many slope protection and erosion control devices are expected to be damaged by hydraulic forces under severe conditions. This prevents more costly damage to the bridge or highway facility that these devices were designed to protect. Maintenance personnel must be aware of these situations so that needed repairs can be identified and scheduled on a routine basis.

Identifying the Need for Scour Countermeasures: Slope protection and scour control devices are usually provided at the time the highway or bridge facility is constructed. Natural or man-made changes in the stream channel or drainage basin can change runoff characteristics, the type and amount of sediment carried by the stream, stream alignment, and other stream characteristics. Early identification of potentially severe problems can provide the lead time necessary for analysis, design, and construction of appropriate erosion control and slope protection measures. Sites where problems recur frequently, worsen in spite of frequent maintenance, or where the problems shift to another unprotected location when eroded areas are repaired, should be reported to the agency engineering division for a hydraulic and geotechnical evaluation.

Selection of Alternatives: The selection and design of appropriate slope erosion control and bank protection measures usually involves analysis of complex stream characteristics and can best be accomplished by professionally qualified hydraulic engineers. Several practical matters require input from maintenance personnel.

Cause: Maintenance personnel are usually the best source of historical data that might pinpoint the cause of the problem, such as when shifts in channel alignment have occurred. Repairs are more likely to be successful when the cause of the damage can be determined.

Past Practices: An important consideration is how successful or unsuccessful different treatments have been under similar conditions in the past.

Materials: The use of locally available materials will usually reduce construction and maintenance costs.

Maintenance Requirements: Ease of maintenance should be an important consideration. Access and frequency of maintenance expected are key elements to a good design.

The many different methods to control erosion and protect slopes from stream flow, tidal flow, and wave action are generally classified as revetments (slope protection), flow control (stream training), and structure modification or channel modification.

Revetments: A revetment is a facing, such as of stone or concrete, used to add support to an embankment. This engineering application includes a wide variety of longitudinal structures, coverings, or linings used to protect the banks of stream channels. Revetments can also be used to protect the channel bottom from scour. Revetments are also referred to as slope protection, bank protection, bed armor, and bank armor. Revetments are generally classified as either flexible or rigid. Flexible revetments conform to minor changes in the underlying bank surface with very little damage to the covering material, while rigid revetments do not. Flexible revetments include rock riprap, wire-enclosed rock riprap, planted vegetation, precast concrete shapes, and grout-filled bags.

Dumped Riprap: Dumped riprap revetments consist of graded stone (or broken concrete) placed on a prepared surface, usually with some type of filter layer between the stone and the supporting bank. Primarily the stone is placed by dumping, although hand and mechanical placement is also used to achieve even distribution of the various sizes of stone. Dumped riprap is flexible, easy to repair, and the most widely used type of revetment in the United States. It is used primarily to protect banks that run parallel to the direction of flow. It is also a common protection for ditch and drainage channels within the highway right-of-way. When properly sized and properly installed, dumped rock riprap has been highly effective. Failures in dumped rock riprap include the following:

Inadequate Cut-off Walls and Toe Walls: Failures caused by inadequate protection of revetment ends is common. Sandy and silty soils require that the toe walls and cut-off walls extend 5 feet below the streambed. Riprap walls should be deeper on the outside of curves and in other locations that are prone to scour than on straight waterway sections.

Inadequate Coverage of the Bank: Failure to extend the revetment far enough causes problems similar to inadequate toe walls. Revetments must be extended far enough upstream and downstream from curves and flow constrictions so that they do not end in an area that is likely to scour. They should also extend far enough up the bank to protect against the expected high-water level.

Inadequate Size of Stone: Failures from stone that is too small are less common than failures from inadequate coverage or inadequate toe walls. Larger stone and thicker layers of stone should be used in areas where the flow is not parallel to the revetment. When broken concrete is used, flat-shaped pieces usually are not satisfactory.

Erosion of the Supporting Soil: A filter layer must be used in sandy, silty soils to prevent loss of the material supporting the riprap. Aggregate layers or filter cloth may be used. Filter cloth is usually selected since it is easier to install.

Slope Failure: Before installing riprap, the slope must be shaped to a stable grade. The limiting stable slope is a function of the internal frictional characteristics of the soil, but generally no slope should be steeper than 1.5 to 1 (horizontal to vertical).

Maintenance of dumped riprap consists primarily of replacing riprap that is displaced or damaged by storms.

Hand-Placed Riprap: Hand-placed, ungrouted riprap is similar to dumped riprap except that stones are carefully placed by hand to produce a tightly fitted, fairly smooth surface. Hand-placed riprap lacks the flexibility of dumped riprap and does not have the strength of a rigid revetment. Experience has generally shown that hand-placed riprap is much more likely to fail than an equal thickness of dumped riprap.

Wire-Enclosed Rock Riprap: Wire-enclosed rock riprap is installed in one of two ways: in wire mattresses usually less than about 0.3 meter (1 foot) thick or in rectangular wire baskets called gabions. This type of revetment is used in similar ways to dumped riprap. Since it has the advantage of not being easily displaced, it is appropriate for use on steep or unstable slopes, or when large riprap stone is not available. The primary disadvantage of this type of revetment is potential failure of the wire mesh from corrosion or abrasion. Wire mesh coated with polyvinylchloride (PVC) should improve corrosion and abrasion resistance.

Vegetation: Planted vegetation is generally not suitable for bank protection at locations with high water flow velocities. Plantings of sod, grass seed, or willow stakes have been used successfully for smaller streams and in combination with other types of bank protection. A number of products exist to aid in establishing seedlings and plantings, including biodegradable materials such as paper and mats made of jute that last only long enough to permit the vegetation to become established and synthetic materials such as nylon fabric that reinforce the plant root system on a longer term basis.

Concrete Block Matting: Precast concrete blocks can be obtained from several manufacturers to form flexible bank protection mats. Most concrete block mats (CBMs) use interlocking blocks with spaces for vegetation to grow between and through the blocks. The vegetation helps anchor the blocks in place. Some CBMs have cables or synthetic fiber ropes that interconnect the blocks.

Grout-Filled Bags: Grout bags are individual nylon or acrylic bags fabricated from panels of material to create a rectangular block form. Every bag is large enough to resist movement from high-velocity water flows. The bags are positioned like tiles around a bridge pier, forming an erosion-resistant channel floor. After they have been located on the channel bed, the bags are pumped full of grout. Because the bags are formed in position, they fit snugly against each other.

Tetrapods: Tetrapods are precast concrete shapes with extending legs that are randomly placed like riprap for channel protection. The details of tetrapod casting and placement can be found in the *U.S. Army Corps of Engineers Shore Protection Manual*. The general principle in their use is that they interlock (like a handful of jacks) to form a channel surface barrier that absorbs the friction energy of the water flow and creates a stilling basin effect between the legs of the tetrapods at the bank soil surface.

Concrete Pavement: Plain and reinforced-concrete pavement (channel lining or slope protection) is a rigid revetment form used by many state DOTs. The concrete is cast in place on a prepared subgrade. This kind of revetment is also used in areas where suitable riprap stone is not readily available. Concrete is typically applied at the outlet end of a culvert or where a diversion ditch drops overland flow into the roadside ditch leading to a culvert or bridge. The sides and bed of the stream or flow channel are paved for up to about 20 meters (about 60 feet) to dissipate the energy of the stream turbulence as the water exits the culvert or diversion ditch. Typically, the paving thickness may range up to 300 mm (12 inches).

The primary disadvantage of concrete slope protection is that it is very susceptible to failure by undercutting (scour and undermining at the toe and ends of the pavement slabs). Once a failure begins, it is generally progressive in nature and leads to complete failure of the revetment. It is extremely important to protect the toe and ends of a concrete revetment. Extending the riprap beyond points that are likely to scour helps protect the ends. The toe of the slope protection must be extended vertically below the level of any possible scour. Concrete slope protection should be inspected regularly, particularly after major storm flows. Sealing cracks and joints in the concrete will help prevent infiltration of water underneath the slope protection and subsequent erosion of the subgrade.

In cases where scour has occurred around the concrete slope protection but has not yet caused undermining, dumped riprap can be used to prevent further erosion. When undermining has occurred, pressure grouting or underpinning with cast-in-place concrete may be used. Such remedial measures are often not practical unless accompanied by other work since the voids may be difficult to plug or they may be very large before they are detected. Before filling voids, it is usually necessary to backfill the exposed scour areas and then extend the concrete slope protection to solid foundation material. The voids beneath the slope protection may then be filled through holes drilled or punched through the existing concrete slab. Some agencies install anchors at the tops of slabs in areas likely to be undermined, preventing slippage when undermining does occur.

Concrete-lined channels may also fail because of hydrostatic pressure from water behind the lining or from water-saturated soil behind the lining. The installation of weep holes during the construction process is commonly used to provide a relief mechanism for hydrostatic pressure. In areas where hydrostatic pressure is a problem, maintenance crews should periodically check weep holes to make certain they are not plugged and are functioning. Holes drilled through the slabs may reduce hydrostatic pressures if weep holes were not installed at the time of construction. Underdrains can also be installed to intercept water seeping through the soil behind the slab.

Grouted Rock Riprap: Grouted rock riprap is another rigid form of revetment. Grouting prevents the rock from being displaced and provides an impervious surface. Like concrete slope protection, it is subject to undercutting and should be inspected after major storm flows.

Sacked Concrete: This rigid-type revetment is constructed by stacking burlap (or other porous fabric) sacks filled with concrete. The concrete oozes through the burlap and cements the sacks together, creating the rigid revetment. In some applications, the sacks are connected by dowels driven through them. Sacked concrete slope protection can be placed on fairly steep slopes when the soil is stable. However, the slope grade should not exceed 1 to 1. Ease of construction and maintenance is the advantage of sacked concrete. Because sacked concrete develops into a rigid revetment, it should be inspected for undercutting after major storms. Large cracks should be repaired to prevent loss of embankment material.

Concrete-Filled Fabric Mats: This type of revetment uses synthetic fabric mats that are pumped with grout in the field. Concrete mats are similar to concrete slope paving and are considered a rigid revetment. Advantages include reduced construction time for forming and simple slope preparation.

Bulkheads: Bulkheads are vertical walls constructed from either plain or reinforced-concrete sheet piles, timber piles, or cribbing and are rigid forms of slope protection. They act as retaining walls and as slope protection. Bulkheads are primarily used to protect the embankments at abutments and approach roadways where fill slopes are highly erodible or have failed by slumping. They may be used when there is no room to grade the banks to a stable slope. Failure of bulkheads is generally because of undermining. Erosion of the fill material behind the bulkhead and lack of foundation support are other causes of failure. Maintenance involves repairing erosion at the base of the bulkhead, keeping weep holes open, and repairing cracks and other damage.

Flow Control Measures: Flow control or stream-training measures control one or more of the following characteristics: direction of flow, velocity of flow, and depth of flow. Commonly used flow control structures include spurs, retards, dikes, spur dikes, jack fields, and check dams.

Spurs: A spur is a linear structure that projects into a channel from the stream bank or embankment. Spurs change the direction of flow and reduce flow velocity along the bank or embankment, protecting the bank from erosion. Spurs are also known as jetties, groins, deflectors, and wing dams. Common materials used to construct spurs are rock and earth embankments, timber piles, steel piles, and sheet piles. Spurs can be divided into two groups: those that allow water to pass through the spur (permeable) and those that do not (impermeable). Permeable spurs are generally fence-type structures and are best for water flows that carry a lot of silt or sediment. Maintenance of permeable spurs consists mainly of replacing pile bracing and fencing material. Drift debris that could develop into a fire hazard should also be removed.

Retards: A retard is a linear flow control structure that may use the same types of structures as spurs, revetments, and other flow control devices. The main difference between retards and spurs is that retards run parallel to the bank rather than project from it into the stream. They are also not supported by the bank, are usually placed at the toe of the bank, and may be either permeable or impermeable. Construction methods, materials, and maintenance needs are similar to spurs. Retards are used to correct alignment problems and to reduce flow velocity along the stream bank. Retards may be used at curves to keep flow from directly striking a bank.

Dikes: Dikes are usually used on flood plains to control the flow of water that overflows the banks. For example, dikes may be used on the upstream side of a bridge to prevent high water from bypassing the waterway opening of the bridge. Maintenance is similar to that required for corresponding revetment types.

Spur Dikes: A spur dike is a projecting dike usually located on the upstream side of a bridge. It projects out from the approach roadway embankment, reducing erosion caused by water flowing along the upstream side of the embankment. Spur dikes also direct the flow of water in the channel into the bridge waterway opening. They are generally used when the bridge waterway opening is narrower than the channel, as when approach roadways are on embankments, and when scour is occurring at the upstream ends of the abutments. Scour may still occur with spur dikes, but it will be moved upstream and away from the abutments. Spur dikes are usually earth or rock embankments protected with some type of revetment. Maintenance is similar to the maintenance for revetments.

Jack Fields: Jacks, or tetrahedrons, are devices used to reduce flow velocity along stream banks. Each jack consists of several struts that are bolted or fastened together to form six perpendicular arms. Jacks may be made from concrete or steel struts. They are usually arranged in fields consisting of several rows of jacks fastened together with cables. Jacks may also be arranged to serve as retards and spurs. Jack fields work best in streams that carry a lot of floating debris. If debris does not collect on the jacks, they will have little effect on bank erosion.

Check Dams: A check dam is a low dam or weir built across a channel to control water flow velocity and prevent erosion of the streambed. Check dams are usually built downstream from the structure that they are intended to protect. Check dams may be built from rock riprap, gabions, concrete, sheet piles, or concrete-filled mats. Maintenance includes replacement or repair of displaced or damaged portions of the check dam and removal of debris (such as trees) that may affect the flow, as needed.

Structure or Channel Modifications: Changes to the stream channel or the bridge structure may sometimes be necessary to control erosion. Structure modifications include increasing the height of the structure above the expected high water level and adding open spans to increase the flood waterway opening. Channel changes include modifications to horizontal alignment, vertical alignment, cross section, or roughness. Structure and channel modification projects are considered beyond the scope of responsibility for maintenance division crews. However, maintenance personnel should be aware of these options since poor channel alignment or a restricted bridge waterway opening may be the cause of continuing maintenance problems.

The practice of relocating streams for better alignment is no longer permitted in many areas. If permitted, it requires involving other agencies, such as the U.S. Army CoE. Most changes to horizontal alignment, such as straightening a channel, will also involve a change in slope. Change in slope can result in increased erosion when the slope is steeper or an increased accumulation of sediment when the slope is flatter.

Local channel modifications in the immediate vicinity of the structure do not always have the intended effect. For example, if the water level is controlled by downstream conditions, modifying the channel cross section at the bridge will not lower the depth of flow. Minor channel modifications in the immediate vicinity of a bridge may be effective but should not be made without a detailed engineering analysis of the situation.

Repair of Scour and Substructure Undermining by Replacement of Material: Most repairs for scour around bridge piers and abutments will involve replacement of the eroded material. In many cases, both riprap and concrete will be used. Several special considerations apply to the placement of these materials around bridge substructure units.

Riprap: Factors that should be considered when placing riprap around piers or abutments are as follows:
- Dumping riprap around existing structural units must be done carefully. The large stones can easily chip or break the concrete elements. Placement of riprap around piers must often be done from a barge.
- Placement should be made in even lifts to avoid unbalanced loads on footings.
- A structural analysis should be conducted to ensure that the footings or pile supports will not be overloaded.
- The stone used must be of suitable size for the anticipated flow conditions.
- Riprap should not extend above the original streambed. Riprap that does extend above the original streambed will act as an obstruction. The turbulence resulting from improperly placed riprap around a pier can cause localized scour at other piers or at the abutments.

Concrete: Placement of concrete to repair scour or undermining of substructure units usually requires either dewatering of the area or underwater placement of the concrete. In either case, proper placement, good forms, and skilled personnel are essential. Dewatering may be accomplished by constructing cofferdams or, if environmental regulations permit, by diverting water flow away from the repair area. The primary disadvantage of dewatering the area with cofferdams is the clearance required to drive the sheet piling. General precautions for concrete placement under wet or dry conditions are as follows:
- A structural analysis should be performed to ensure that the pile support will not be overloaded.
- Forms should be prevented from moving and should not be removed prematurely. Prebagged concrete or concrete pumped into fabric forms is often used in place of conventional forms.
- Concrete should not be placed in running water or allowed to drop through water because this could wash the cement out of the concrete.

Underwater placement can be accomplished by using tremie or pumping methods, by using bagged concrete, and by using prepackaged concrete.

A tremie is a tube with a hopper at one end and a discharge gate at the other end. It is used to place concrete underwater by gravity flow. The discharge gate end is closed until the tremie is filled with concrete and lowered to the point where the concrete is to be placed. Once the concrete placement begins, the discharge end must be kept filled with concrete. If this is not done, the pipe will fill with water. Multiple tremies are needed to reduce lateral travel and to compensate for conditions where reinforcing steel or pilings under a footing restrict movement of the tube.

Placing concrete by pumping is similar to tremie placement. However, since the concrete is pumped rather than gravity fed, it is somewhat easier to control the discharge. A manifold connects multiple hoses when needed.

Concrete can be placed underwater in bags made of porous material such as burlap. The bags are partially filled with concrete, then placed by workers in shallow water or by divers in deep water. Bags may be filled with dry concrete and wetted during placement. Concrete may also be pumped into fabric forms specially sized to fill a void under the footing.

The prepackaged concrete method involves filling forms with coarse aggregate and pumping in a grout. The grout displaces the water and fills the voids in the aggregate. Special equipment and special techniques are required to use this method.

Rebuilding Scoured Streambeds: The effectiveness of rebuilding a streambed depends on whether the scour problem is isolated in an area around the bridge or whether the streambed has been generally lowered. When the streambed has been generally lowered, it is best to extend the concrete footings below the scour line, taking advantage of the deeper channel to reduce water flow velocities. This procedure also aids in minimizing reconstruction of the channel by placement of riprap. Maintaining the maximum possible depth and width of a channel aids in reducing water flow velocities, which in turn reduces the scour action and the probability that any riprap used will be undermined or dislodged.

Maintenance Planning: Determine if the streambed has been generally lowered. Determine if the extension of concrete footings below the scour line is feasible. Determine the velocity of the stream flow and the size of riprap (and crushed stone) necessary to resist scour.

Resource Requirements: Large quantities of riprap, stone, or crushed stone will likely be required. Front-end loaders or cranes will be needed to place the materials.

Construction Procedures: The general process typically includes the following steps:
(1) Place crushed stone in the channel to an elevation approximately 600 mm (about 2 feet) below the proposed top of the streambed, taking care to fill voids below the footings.
(2) Add a 600-mm (2-foot) layer of large stone riprap in the 70- to 225-kg (150- to 500-pound) size range over the crushed stone. The riprap should extend above the high water line at both sides of the channel.

Repair of Settled and Tilted Piers: Erosion that undermines a footing, particularly if no supporting and anchoring piling has been used, may permit the pier to settle and tilt, damaging the superstructure. The causes of this problem are basically the same as those for all other undermining problems. Stabilizing the footing is also similar to other undermining situations, but the repair is complicated by the need to repair the damaged superstructure.

Maintenance Planning: Evaluate the condition of the structure and the foundation to determine if it is possible to make the repair and if the pier can be stabilized in its present location. If traffic must continue to use the bridge, evaluate the need for temporary supports until repairs are complete. Design the stem enlargement and fabricated reinforcement.

Resource Requirements: Equipment will be needed for light lifting, dewatering, jacking, tremie concreting, and for installing a cofferdam if it is required. Concrete will be required for riprap, tremie seal, pier repair, and superstructure repair.

Construction Procedures: The generally executed process includes the following steps:
(1) Proceed as outlined earlier under the section on repair of eroded areas at pier foundations.
(2) Construct the first stage of the pier enlargement to above the water line.
(3) After the new concrete has attained its design strength, it can be used as a seat upon which to jack the superstructure back to its original grade.
(4) A sufficient amount of the second stage portion of the pier is placed to permit removal of the temporary supports.
(5) After the concrete has attained sufficient strength to support the superstructure, the remaining concrete is placed in the pier, and the deck and the superstructure are repaired.

3.1.7 Protective Systems
3.1.7.1 Protecting the Substructure

Deterioration at the Water Line: Several deterioration mechanisms act at the water line. Wet/dry and freeze/thaw cycles accelerate substructure deterioration. Material carried by the water can cause an abrasive action that contributes to surface damage at the water line. To protect against these kinds of deterioration, protective coatings and fiberglass liners can be applied at the water line.

Settlement: Where serious settlement is present, a professionally qualified soils engineer, structural engineer, or both should be consulted before repairing damage to any abutment or pier. Movement of an abutment or pier should be stabilized before making any repairs, such as by filling cracks or releveling bridge seats and bearings. Leveling the substructure cap and bearings is required so that when the superstructure load is placed on the bearings there will be no unexpected stress in the structural members. A substructure unit that is resting upon piles may require additional piles to stabilize it. Additional width of a spread footing may have to be constructed to stabilize the footing. In a spread footing, additional support can usually be gained by underpinning. That is, sectional piles are jacked to bearing under the footing, then a short steel column is inserted and wedged tight after the jacks have been removed. When an abutment has been repaired as a result of settlement of the back wall, it should be checked to ensure that it is not binding on the substructure member. Deck expansion joints and seals should always be checked and adjusted after major repair of a settled substructure unit. When serious settlement occurs in large structures, a professionally qualified soils engineer should be consulted before any attempt is made to correct the situation.

Impact Damage: Guardrails and energy dissipaters can be installed to protect against damage from highway traffic under a bridge. On navigable waterways, where ships and barges may come in contact with piers and abutments, adequate fenders are necessary to protect against mechanical and impact damage. Steel plates can be installed on the upstream edge of an abutment or pier to reduce the damage caused by ice or flood debris. Any plates installed should be of sufficient thickness to withstand the expected forces and should be well anchored.

Salt Damage: Protection against saltwater damage to the substructure is the same as for a bridge deck. Coatings can be effective for steel or concrete if applied early in the life of the structure and reapplied as needed to maintain the integrity of the coating. Dense concrete, such as that attained with a pozzolanic additive, can also aid in resisting the damaging effects of saltwater exposure.

Pressure From Approach Pavement: Abutment damage caused by pressure from approach pavements can be minimized or even eliminated by installing relief joints in advance of the bridge abutment.

3.1.7.2 Spot Painting to Protect the Superstructure

Spot painting normally involves cleaning the surface, removing corrosion, and replacing the paint system on selected areas of the bridge. The replacement paint is selected for color match (as nearly as possible) and for chemical compatibility with existing paint. Environmental concerns continue to grow with respect to bridge painting. It has been estimated that about 80% of the steel bridges in the United States are coated with lead-based paint [6]. Paint removal presents several problems: (1) collecting the by-product of the paint removal process including blasting materials; (2) disposing of any hazardous waste present in the by-product; (3) protecting workers exposed to the cleaning process, especially from lead and noise; and (4) monitoring lead levels in workers' blood. Because of these problems, most transportation agencies now conduct bridge painting as a contract maintenance activity. Contract maintenance allows the agency to use organizations with specially trained personnel and specialized equipment systems to deal with the environmental problems. Because of the increasing costs of proper environmental controls for painting a bridge, the decision to repaint must include a discussion of whether to defer painting until the bridge is scheduled to be replaced, even if deferring the painting moves up the expected replacement date. The best preventive maintenance approach may be to schedule bridge repainting before the condition of the existing paint system deteriorates to the point that paint removal is necessary. If the existing paint contains lead, it can be encapsulated with an environmentally friendly paint such as a zinc-rich or vinyl-based product.

Paint Systems: Paint consists of two basic parts: a pigment made of fine particles to provide coloring, and a vehicle that is the liquid portion carrying the pigment. The paint vehicle generally consists of a binder and thinners. The binders and the embedded pigment remain as the paint coating after the paint has dried (i.e., after the thinners have evaporated). Chemical reactions, in addition to the drying process, cause the binder in some paints to harden when the paint is exposed to air. Rapidly evaporating thinners are sometimes called dryers.

Red lead, titanium oxide, zinc chromate, and silicates are typical pigments. Some pigments, such as zinc chromate, also increase steel resistance to corrosion. Most agencies no longer use lead-based paints because of concerns and restrictions related to worker safety and environmental damage. Binders typically include linseed oil, alkyds, latex, polyurethane, epoxy, or other chemicals. Thinners typically include turpentine, mineral spirits, acetone, water, or other substances. New pigments, binders, and thinners are continually being developed to improve ease of application, quality and durability of the paint surface, and protective qualities of the paint.

The maintenance purpose of painting is to protect the bridge from corrosion. Spot painting consists of painting localized areas where the paint has been damaged, has failed, or corrosion has begun. Performing spot painting as soon as defects are noted stops or reduces corrosion before it progresses and before significantly more time and money are required for a larger painting effort.

Preparation of the Surface: Preparation of the surface before painting is the most important element of the painting process. Surface preparation involves removal of all corrosion, paint, or deposits that may interfere with the adhesion and covering ability of the paint to be applied. Paint should not be applied over loose, scaly, or flaking paint. Mill scale, rust, dirt, oil, and other foreign substances that prevent paint from adhering or covering must be removed. Water and dirt may be removed from the surface by air-blasting and wiping. Grease-like contaminants are most successfully removed by scraping, if the accumulation is large, then wiping or scrubbing with a petroleum-based solvent. Make certain that oily substances are not simply diluted and then spread over a larger area. Vigorous wiping with a clean rag may be desirable to ensure that residues will not interfere with paint adhesion.

Paint can be removed by a variety of methods. Each method differs in its cost of implementation, degree of containment required, quality of surface prepared, and amount of debris generated.

Cleaning With Solvents: Heavy oil or grease accumulations on a surface must be removed before abrasive blasting or other surface preparation gets underway. Oil and grease may interfere with surface preparation or be spread further by it. In some areas, removal of lighter accumulations may be the only surface preparation required. In other areas, heavy oil or grease deposits and incorporated dirt may have solidified over time so that the material is hard and thick enough to require scraping to remove. Preliminary scraping of the heaviest deposits usually saves time and solvent. The solvents that may be used range from water in combination with special soaps to kerosene to some complex, hazardous chemicals such as di-isobutyl ketone. Properties of solvents that are relevant to their use in surface preparation because of health and fire considerations are as follows:

Relative Evaporation Time: This is the time required for the solvent to completely evaporate, based on a scale value of 1.0 for ethyl ether. The higher the number, the longer the time required for evaporation.

Flash Point: This is the temperature at which the solvent releases sufficient vapor to ignite in the presence of an open flame. The higher the flash point, the safer the solvent.

Explosive Limits (Flammable Limits): This is expressed as a percentage of solvent vapor in a total volume of vapor plus air. Minimum concentrations below this percentage will not ignite. Maximum concentrations above this percentage will not ignite. Concentrations between the minimum and maximum will ignite or explode.

Maximum Allowable Concentration: This is the concentration of solvent vapor in the air that can be tolerated by workers throughout an 8-hour day, expressed in parts per million. The higher the value, the safer the solvent. Common solvents such as carbon tetrachloride and benzol are very toxic.

Paint should not be applied over solvents, so evaporation times should be considered when selecting a solvent. Safety is the first consideration in using solvents because of health and fire hazards. Some general precautions in the use of any solvent include the following:
- Wear goggles, protective clothing, rubber gloves, and barrier cream (petroleum jelly).
- Do not breathe the fumes. (Proper ventilation is always required.)
- Do not use benzene and carbon tetrachloride since they are poisonous.
- Do not use gasoline or solvents with low flash points since they might catch fire or explode.
- Do not smoke or use solvents near fire, flame, or electrical connections.
- In case of skin contact, clean thoroughly with soap and water.
- In case of eye contact, rinse with water immediately and contact a physician.

Hand Cleaning: Hand cleaning is laborious and used to remove old paint or corrosion only from small areas. Hand cleaning greatly reduces the amount of material for disposal, and it does not require large or expensive equipment. Hand cleaning is used to prepare areas where the paint is in fairly good condition with only a few bad spots around rivets, welds, and joints; in corners and "blind spots" that other methods do not reach; or on larger areas where traffic does not permit the use of other methods. Hand cleaning can lead to eye injuries from flying debris particles, to cuts from sharp edges, and to falls from slipping. Goggles should be worn at all times. Heavy-duty clothing and leather gloves are also needed. Work must proceed carefully to avoid slips and falls. A minimum of dust is generated by hand cleaning, so hanging a few tarps around the work area can generally satisfy containment requirements. Typical tool requirements include wire brushes, scrapers, chipping hammers, sandpaper, slag hammers, chisels, painters, putty knives, dust brushes, brooms, and sanding blocks.

Power Tool Cleaning: Power tool cleaning removes rust, loose paint, and mill scale. As with hand cleaning, productivity is low and the compatibility of the paint system must be assessed. The dust generated by power tools is generally greater than that created by hand tool cleaning, but it is still considerably less than that generated by grit-blast cleaning. Containment consists simply of tarps placed around the work area. Typical tool requirements include impact tools, rushing tools, grinding tools, needle guns, and rotary scarifiers.

Power Tool Cleaning With Vacuum Attachments: Power tools can be equipped with vacuum attachments to collect the dust and debris. The degree of dust generated is minimal, but some dust will escape in areas of difficult configuration or where complete seals are difficult to attain. The shrouding can also restrict access in hard-to-reach areas. Debris consists of only the products removed from the steel surface. Productivity is generally lower than for power tools without the vacuum attachments.

Chemical Stripping: Chemicals can be applied to the steel surface to soften the paint before scraping or water washing. Chemicals such as sodium hydroxide or methylene chloride are applied to the surface and are allowed to remain in contact with the surface for a few hours or overnight. The stripper and the wash water must be collected for proper disposal. Dust-tight containment is not necessary, but the containment must be capable of capturing the stripper debris and the wash water. The used stripper will be hazardous because of the lead particles in the paint removed, and it may also have a hazardous pH value. Some strippers are classified as hazardous chemicals because of other characteristics. The volume of waste may be increased if the rinse water tests hazardous and it cannot be filtered from the debris. Strippers will not remove rust or mill scale. To properly prepare the surface, blast cleaning may be required after the strippers have been used. If abrasive blast cleaning is needed, some containment will be required for the nuisance dust, even though the stripper has removed the lead paint. Productivity with strippers can be slow, especially if repeated applications are necessary and if additional mechanical surface preparation is required.

Overcoating: An alternative to cleaning lead-based paints and repainting steel structures is to overcoat the bridge. This process involves applying a surface-tolerant coating over an existing coating containing lead, after minimally preparing the surface. The surface is typically prepared using power water washing to remove dirt, paint chalk, and chlorides. In isolated areas, a combination of hand cleaning and power cleaning may be used. Overcoating eliminates grit blasting in the open air. Prior to overcoating, steel surfaces are spot-painted with a one-component, moisture-curing polyurethane aluminum primer. A polyurethane intermediate coat that meets environmental and safety standards for volatile organic compounds (VOCs) can then be applied to the entire bridge surface. The repainting is completed with a light-stable polyurethane topcoat that meets VOC environmental and safety standards.

Painting costs may be reduced 30% to 75% using the overcoat method. Overcoating is generally applicable to bridges with a maximum of 25% to 30% surface corrosion. It may also be used if the bridge paint has broken down. Important to the success of overcoating is the special surface-wetting, edge-sealing, and curing capabilities of the moisture-curing polyurethane spot primer. The low-viscosity primer can penetrate and wet out the old paint, tightly adhering rust. In addition, the primer can penetrate under old paint and can be used for spot-cleaning areas. To cure, the primer scavenges moisture from the rust, the atmosphere, and the existing paint. When determining if a bridge is a suitable candidate for overcoating, the following factors should be considered:

- The percentage of the bridge surface that is rusted.
- The degree of rusting.
- Structural steel condition.
- Adhesion of the coating.
- Adhesion between the layers of the coating.
- Paint type of the undercoating. (It may be difficult to develop proper adhesion between leafing, pigmented paints and the new coating.)
- Reparability of the coating.
- Compatibility of the existing coating system. (Patch areas may need to be tested.)

Spot-Painting Guidelines: Spot painting involves painting damaged, repaired, or corroded members of a bridge where less than 35% of the paint on the bridge has deteriorated. Generally, if more than 35% of the bridge needs to be painted, the whole bridge should be painted since it will all need to be painted soon. The first consideration is to select a paint type compatible with the existing paint. Paint formulas are constantly changing, and many newer paints will not adhere, cover, or endure if applied over an older formulation. Generally, it is best to spot-paint with the same type of paint already on the bridge. When this is not possible, consult the paint manufacturer's technical data to find a compatible paint. Whenever possible, spot painting should be done with a matching color to enhance the appearance of the bridge. Only the part of the structural member that has corroded is cleaned to bare metal, and only that part is given a prime coat, followed by a final coat.

- *Weather Conditions:* Weather is an important consideration in producing a high-quality paint job. Painting is best done in warm, dry weather with little or no wind. Avoid painting when the wind velocity exceeds 7 meters per second (15 miles per hour) or when the temperature is below 4°C (40°F) or above 50°C (125°F) unless the paint is specially formulated for more extreme temperatures. Avoid applying paint when the relative humidity exceeds 85%.

- *Thinners:* Thinners should be used only when specified or necessary. Too much thinner results in a coating that is not thick enough to protect the steel properly. The paint manufacturer's recommendations should always be followed. If cold weather conditions require increased use of a thinner, extra coats of paint may have to be applied to obtain the necessary coating thickness for proper steel protection. Do not thin lead-based paints. If necessary, they may be heated with hot water or steam radiators.

Paint Care and Storage: Proper care and storage of paint is essential for maintaining the quality of the paint and for safety reasons, since many paints are toxic or are a fire hazard. Keeping a reasonable inventory of paints promotes efficiency.

- Store paint at temperatures between 18°C (65°F) to 30°C (85°F) in a dry, well-ventilated area where it will not be exposed to excessive heat or cold, explosive fumes, sparks, flames, or direct sunlight.
- Because of fire hazards, paint and solvents should be stored in a location apart from other combustible materials. If possible, store them in separate buildings.
- Store paint neatly. Keep aisles and walkways clear for safety.
- Ensure labels are intact and legible. Relabel containers accurately.
- Containers should remain unopened until required for use.
- Previously opened containers should be used first.
- The oldest paint should be used first.
- If a skin of dried paint has formed on the surface of the paint in a previously opened container, cut the skin out, dispose of it properly, and thoroughly mix the remaining paint.
- Pour partially used containers of paint of the same type and color into one container to reduce the amount of air space in a container. Air space causes a paint skin to form.
- Ensure that partially used containers are sealed tightly to prevent contamination and drying of the paint.
- Recheck container lids periodically to ensure that the lids are tightly sealed.
- Invert containers in storage each month or two to prevent pigments from settling and caking on the bottom of the container.
- Do not try to salvage improperly stored paint. Return it to the stockroom for disposal and notify the supervisor of the condition and identification of the paint.

Inspection of Painting: Painting is not completed until the coating has been inspected and any deficiencies have been corrected. The inspection techniques are designed to reveal defects (e.g., porous areas, pinholes, blisters, unpainted areas, thinly coated areas). A flashlight-equipped magnifying glass can be used to detect and examine surface irregularities. An electric current measuring device can be used to locate thin paint areas and pinholes. A wet-mil gauge can be used to test the thickness of paint before it has set up. A dry-mil gauge can be used to measure the paint thickness after it has set up and before the next coat is applied. Nondestructive dry-mil gauges generally use magnets to test for thickness, while destructive gauges require a scratch through the paint. Before applying successive coats of paint, it is necessary to touch up damaged areas, repaint areas with insufficient thickness, and repair and repaint all unsatisfactory areas.

Painting Defects: Some common painting defects and possible repair options are as follows:

Alligatoring: This is a mesh of paint cracks that resembles alligator hide, with the coating pulling away from the surface and causing a rough finish. It is usually caused by not allowing sufficient drying of the paint before recoating, by extreme temperature changes, and by incompatibility between coats of paints (e.g., when a vinyl paint is applied over an alkyd paint). Remove the finish down through the damaged paint film and refinish the area. Use a solvent recommended by the paint manufacturer. Paint should be mixed thoroughly before applying. Sufficient time for drying should be allowed between coats of paint. Compatible paints should be used.

Blistering: There are many causes of blistering: The topcoat did not adhere to the primer, paint was applied over oil or moisture, too much paint was applied at one time, steam cleaning caused disbonding, fingerprints were on the metal, or air was trapped under a very thick coating of paint. Correction requires removing the blisters by sanding with #400 paper or a ball of screen wire, then refinishing. Products should be properly thinned, and sufficient drying time should be allowed between coats. There should be no water in the air lines when spray-painting.

Lifting: Incompatible coatings may not wrinkle or alligator. Instead, the incompatible coating may cause the coating beneath it to lose its adhesion, resulting in both coats peeling from the surface. This is caused by the solvent in the topcoat acting as a paint remover on the coating beneath it. This is likely to occur when paints containing a strong solvent such as xylene are applied over soft, oil-based paints. Lifting may also occur if an undercoat is not allowed to dry properly before the next coat is applied. Painting over dirty, oily, or greasy surfaces may also cause lifting. Removing the finish down through the damaged paint film and refinishing the area should repair lifting. A solvent recommended by the paint manufacturer should be used. The paint should be mixed thoroughly before applying. Sufficient drying time should be allowed between coats of paint. Compatible paints should be used.

Pinholing (Bubbling, Solvent Pops): These defects are quite common in coatings. Pinholing is often the result of water contamination in the air line of the sprayer or a solvent imbalance (a solvent that is drying too quickly). The coating does not have enough time to flow out before it dries, and little holes are left in the coating. Trapped solvents, settling of pigments, and insufficient atomization of the paint may also cause pinholing. If pinholing occurs, it may be necessary to consult a materials laboratory so that the cause can be determined and eliminated. One successful remedy is to use a considerably thinned tie coat or primer or to thin the topcoat 25% to 50%, sealing the porous surface of a zinc-rich primer. When the use of a tie coat is not acceptable, a mist coating of the topcoat paint should be applied over the surface. A full topcoat should follow this. In extremely severe cases of pinholing, it may be necessary to sand to a smooth surface and refinish. Pinholing can be prevented by keeping water out of the sprayer air lines, by not applying paint too heavily, and by allowing proper evaporation of the solvents. Recommended thinners and sufficient air pressure for proper atomization should be used.

Runs: Runs are rivulets of wet paint film. This defect is caused by overthinned paint, by slowly evaporating thinners, by improperly cleaned surfaces, or by surfaces being too cold. Holding the spray gun too close to the surface and depositing too much paint on the surface also causes runs in the paint. Repair runs by sanding or washing off the surface and refinishing. The surface should be thoroughly cleaned. Paint should not be applied over an old surface. The paint should be thinned as recommended, using properly specified solvents.

Sags: Sags consist of heavy thicknesses of paint that have slipped and formed curtains on the surface. This is caused by insufficient thinner, insufficient drying time between coats, low air pressure causing insufficient atomization, holding the paint gun too close to the surface, or by a paint gun out of adjustment. Repair sags by sanding and washing the surface, then refinishing it. The paint viscosity should be reduced as recommended by the manufacturer, using a proper thinning solvent. The air pressure and the gun should be adjusted for correct atomization, and the gun should be kept at the correct distance from the surface being painted.

Improper Repair: Holes and cracks that are not properly filled and repaired will allow moisture to get behind the coating and lead to blistering, flaking, and peeling.

Insufficient Coating Application: If too little paint is applied or the paint has been thinned too much, chalking and erosion will soon deteriorate the paint film. Care must be taken to see that there is sufficient paint over the top of the surface roughness to prevent corrosion from starting at the peaks of the surface variations.

Incompatible Paints and Thinners: The importance of using compatible paints and thinners cannot be stressed too much. Remember that inorganics will not stick over organics.

Weathering Steel: Under certain conditions, weathering steel requires maintenance coatings. The corrosion of weathering steel presents certain unique surface preparation problems. The advantage of weathering steel is that the rust that forms on its surface is stabilized by the effect of alloying elements contained in the steel. Thus, the rust layer thereafter inhibits the corrosion of the metal. However, the rust layer does not properly form in environments containing saltwater spray or salt-laden fog. Weathering steel does not perform properly in the vicinity of bridge deck joints in climates where de-icing salts are used. The uniqueness of weathering steel creates a significant problem when surface preparation is considered.
- Pits that develop in weathering steel are very deep relative to their diameter, making it difficult to clean the bottom of the pits properly.
- Removal of chlorides from the bottom of the pits is difficult.
- The "green mold" phenomenon is a discoloration that appears shortly after blast cleaning.
- Repeated blast cleaning is often required.

3.1.8 Environmental Aspects

Environmental regulation seems to increase every year. Some of the regulatory acts affecting bridge maintenance include the Comprehensive Environmental Response, Compensation, and Liability Act (CERCLA, commonly called Superfund), the Resource Conservation and Recovery Act of 1976 (RCRA), the Hazardous and Solid Waste Amendments of 1984 (amended by RCRA), the Solid Waste Disposal Act (enacted in 1965, amended several times including by RCRA), the Clean Air Act (CAA), the Clean Water Act (CWA), and the Federal Insecticide, Fungicide, and Rodenticide Act (FIFRA). Transportation agencies need to provide maintenance engineers and maintenance managers with training and education on exactly which regulations apply to specific maintenance activities and what is the appropriate response to the applicable regulatory process.

3.1.8.1 Lead-Based Paint Removal

Removing lead-based paint from deteriorating bridges has developed into a major environmental difficulty. The need to provide worker safety and respond to environmental regulation has dramatically increased the cost of repainting bridges that were originally coated with lead-based paints. Containment and disposal of the lead-based paint material removed from the bridge are two additional factors contributing to the increasing cost.

Solid Waste Disposal: EPA rules now require that materials classified as hazardous, such as lead-based paint debris, must be subjected to a test procedure to estimate the toxicity of the leachate (TCLP) and not the solid waste itself. The TCLP (i.e., the test procedure) was developed recognizing that landfills are common final resting places of disposed material. Consequently, the leaching characteristic of a hazardous substance can have an impact on the integrity of surface water and groundwater. If the levels exceed those set by the EPA, then the hazardous waste generator must treat the waste to reduce its leachate concentration before disposing of it in a landfill.

The residue produced as a by-product of the lead-based paint removal process must be collected as it is removed from the structure. If the TCLP test indicates that leachable lead levels exceed 5 ppm (parts per million), the waste is considered hazardous. The waste must then be treated to reduce the leachable lead levels to below 5 ppm before it can be disposed of in a landfill. Disposal of the waste, if it is classified as hazardous, must meet RCRA requirements, which vary depending upon the amount of waste generated per month.

Containment of Residue From Lead-Based Paint Removal: Generally, the working area of the bridge is shrouded with tarps during the process of removing lead-based paint to prohibit contaminants from coming in contact with the soil or water. Removing lead-based paint from a bridge also initiates application of the requirements of RCRA and CERCLA. EPA has established that 4.5 kg (10 pounds) of lead or characteristic hazardous wastes containing lead is a reportable quantity (RQ). If an amount equal to or greater than the RQ of a hazardous substance is released into the environment within a 24-hour period, CERCLA requires that the National Response Center (1-800-424-8802) be contacted. The CWA also requires that discharges of hazardous substances to navigable waters in excess of the RQ be reported to the National Response Center.

Air Quality: The CAA covers emissions of airborne lead and particulate matter. Although it does not address abrasive blasting projects specifically, some states have developed ambient air monitoring requirements for abrasive blasting. These requirements may be a part of a state implementation plan (SIP), in which the CAA requires each state to outline its plans to attain or prevent deterioration of acceptable air quality, as defined by the CAA.

Typically, the SIP covers a broad spectrum of activities, including plans to control industrial pollution and transportation plans to reduce carbon monoxide emissions. State governments have the authority to target certain types of emitters and impose controls. Two pollutants related to the air blasting of lead paint that are regulated by the CAA are particulate matter and lead.

The requirement to monitor air quality for exterior emissions may be based upon the SIP or may be implemented in response to public health concerns. The contractor and owner can be held responsible for polluting the air with fugitive emissions. Airborne emissions can also lead to liability issues for injuries resulting from contact with the contaminants. The reportable quantity (RQ) limit of 4.5 kg (10 pounds) of lead-contaminated material released into the environment, as described above, includes release into the air.

Particulate levels can be monitored with high-volume pumps called PM10 samplers. PM10 refers to particles equal to or less than a nominal size of 10 microns. Air is passed through a filter by a high-volume pump, and the filter residue is analyzed. A total suspended particulate sampler is used to collect suspended particles of 100 microns or less in size, which are then analyzed for lead content. Before an abrasive-blasting project is initiated, air quality agencies responsible for the affected jurisdiction should be contacted. If ambient air monitoring is not conducted before and during the blasting activity, the contractor and the bridge owner may be subject to litigation or fines, even if state regulations do not require monitoring.

The CAA specifies that an average of not more than 1.5 micrograms of lead per cubic meter of air may be released into the atmosphere over a 90-day period. The National Ambient Air Quality Standard specifies that not more than 450 micrograms of particulate matter (particles less than 10 microns in size) per cubic meter of air may be released on average during an 8-hour workday.

3.1.8.2 Other Environmental Concerns

Lead-based paint removal is a serious and significant environmental concern in bridge maintenance. However, a host of other environmental factors can impact bridge maintenance. In addition to federal regulations, state environmental agencies and city or county health ordinances may impose environmental restrictions on work done on or near bridges. Before initiating bridge repair activities, maintenance engineers and maintenance managers should confirm if environmental permits are required and review proposed repair methods to ensure that the method is appropriate and environmentally sound, when this is practicable. The following list provides some of the environmental factors that impact bridge maintenance in some localities. This list is not comprehensive or current because the number of factors considered continues to grow, and the regulations relating to specific factors continue to change. However, this list does provide some insight into the degree to which maintenance is being held to an increasing level of environmental sensitivity.

- State or federal list of threatened or endangered species.
- Species of high interest to state or federal agencies.
- Migratory waterfowl habitat.
- Anadromous fish habitat.

- Trout and other cold water fish habitat.
- Habitat for birds of prey.
- Wetlands and wetland habitat.
- Riparian habitat.
- Migratory corridors.
- Wintering areas and other critical feeding areas of wildlife.
- Important wildlife reproductive habitat.
- Public water supplies, including important aquifers.
- Islands and other coastal barriers.
- Hazardous waste sites.
- Regulatory floodways and other flood plain areas.
- Commercial fish and shellfish production areas.
- Important sport fishing areas.
- Highly erodible soils.
- Listed or proposed wild and scenic rivers.
- Navigable waterways.
- Significant historic resources.
- Natural resource agency holdings or interests (refuges, parks, habitat areas, etc.).

Streambed-Disturbing Activities: Activities that potentially disturb a streambed or wetland are subject to review by the U.S. Army Corps of Engineers (CoE). The CoE is responsible for implementing the U.S. code and associated federal regulations on streams and wetlands. A system of nationwide permits has been developed to define the limits of activities that can be undertaken to make changes and adjustments to streambeds and wetlands without obtaining an individual permit for activity. In most states, internal review processes and coordination schemes have been developed between transportation agencies and environmental quality agencies to ensure that maintenance activity allowed under the nationwide permitting system is clearly understood. Thus, planning and conducting such activity is routine and ongoing. Furthermore, in many states, agreements have been reached between the CoE and the state DOTs that bridge maintenance for all but the most unusual circumstances can proceed without any permit contact to the CoE. Maintenance engineers and maintenance managers should learn what the regulatory process and practice is in the state in which they are working, and they should be kept informed of any changes in the process and practice. It is difficult to properly plan and schedule bridge maintenance without being aware of the regulatory environment in which maintenance has to operate.

Channel work such as installing pilings, placing riprap, stabilizing banks, changing stream alignments, etc., is all covered by U.S. Army CoE 404 permits. Criteria have been developed that define the limits of such activities that are minor in impact and are therefore exceptions to the regulations. These limits may differ from one CoE district to another, depending upon the hydraulic and hydrologic character of the drainage basin involved. The CoE maintains an Internet site (<www.spa.usace.army.mil/reg/>) at which the 404 regulations, nationwide permit process, regional permit process, and individual permit process can be viewed. This Internet site is associated with the Albuquerque District, Regulatory Branch. It includes a delineation of exempted activities and some examples of exemptions allowed in that district to guide any organization in the process. Typically, the form and instructions for an individual project permit are two-page documents. Maintenance supervisors should remember that if public objections arise, even activities allowed under a nationwide or regional permit can be subject to interruption until the CoE conducts an investigation. When in doubt about the degree to which activities in a stream might conflict with the 404 permit process, check with the local CoE district office with jurisdiction over the project area.

If a wetland (not a stream) is involved, the Internet site for the U.S. Army CoE Wetlands Research and Technology Center provides information useful to maintenance engineers and managers (<www.wes.army.mil/el/wrtc.html>). Links are available to the on-line wetland delineation manual, the hydrogeomorphic approach for assessing wetland functions, wetland training information, wetland-regulating assistance programs, and CoE bulletins on wetlands.

Water Quality: The CWA requires that any discharge of a hazardous substance in excess of its RQ into navigable waters be reported to the National Response Center. There may be additional requirements if the water is a source for drinking water or it provides wildlife habitat.

Sedimentation and erosion can occur as a result of bridge maintenance activities such as removing vegetative cover or construction associated with earthwork. The quality of water in the stream, including its turbidity, can be degraded by maintenance activities, such as placing fill at settled bridge approaches. A sediment and erosion control plan is needed if it is anticipated that soil will be disturbed during bridge maintenance activities. There may be a requirement for regulatory authority review and certification of acceptance with respect to regulations intended to minimize any impact on water quality and aquatic environments.

For underwater maintenance repairs, the EPA and the U.S. Army CoE typically issue permits under the nationwide permitting process. Construction activities required to accomplish the needed maintenance repair should be designed to minimize effluents from the repair procedure. Watertight cofferdams can help minimize water pollution. The cofferdams themselves should be designed to be environmentally sound and compatible with the stream environment. All materials should be nontoxic, especially if they are to be applied underwater.

When maintenance is required on movable spans with facilities for operators, bridge improvements must often include new sanitary facilities that do not discharge pollutants directly into the water, if such facilities were not a part of the existing bridge. Incinerator or digester toilets may be an alternative if connecting to an existing sanitary sewer facility is not practicable.

Stream Classifications: A number of environmental classifications may impact bridge maintenance activities. Additional requirements may be placed on structures that fall in areas of unique habitat. For example, maintenance activities disturbing trout streams must be coordinated with the U.S. Fish and Wildlife Service or the appropriate state or local agency. Bridge maintenance activities may be restricted to certain times of the year to avoid interfering with spawning activity in such streams.

Some streams may be classified as "scenic or wild" rivers by a state agency with the authority to make such classifications. Maintenance of bridges over such streams may be subject to additional restrictions to minimize the impact on the character of these streams.

3.1.8.3 Hauling and Disposal Regulations

Local governments typically define hauling regulations. Maintenance supervisors need to be certain that hauling activities are coordinated with or approved by the agency with oversight for any restrictions that affect the bridge maintenance operation. Any such regulations often include restrictions on routes and time of day during which hauling of disposal material is prohibited in an effort to minimize population exposure to any hazard in the event of a spill.

3.1.8.4 Hazardous Wastes

Another issue impacting maintenance activities is the disposal of incidental wastes associated with bridge maintenance and repair. Oil, gasoline, hydraulic fluid, and various cleaners must be disposed of in accordance with federal, state, and local requirements.

Hazardous waste denotes a regulated waste whose treatment, storage, and disposal must meet EPA minimum standards. "Toxic waste" is a phrase used by the media and the general public to mean a substance that has the ability to cause harm.

Bridge repairs that involve excavation may potentially discover hazardous materials from earlier spills or dumps. If this situation develops or is suspected, the discovery must be reported to the appropriate authority, and additional action (i.e., remediation of the site) may be required.

Hazardous waste spills can occur through the actual conduct of the maintenance activity, either from faulty equipment or from accidental spills. An attempt should be made to confine maintenance activities to areas where spills can be easily intercepted and contained before the material penetrates the soil or enters the stream.

3.1.8.5 Historic Structures

While it is not a common problem, sometimes the nature of bridge repair may adversely impact the historic integrity of a bridge listed in the National Register of Historic Resources, such as a covered wooden bridge in the northeastern or the midwestern United States. The EPA maintains a national list, but state-registered bridges are not included on it. The appropriate agency responsible for maintaining a historic register in each state may have its own list of registered bridges. Maintenance activities on a listed or nominated bridge should include notification to the appropriate agency, ensuring that the historical integrity of the bridge will not be undermined.

3.1.8.6 Noise Control

Local ordinances for noise may restrict the hours when construction and maintenance repair may be performed. Some localities may require a permit for heavy construction activity related to noise generation. In urban areas, maintenance supervisors should be sensitive to the concept of being a "good neighbor" (even if no ordinances regulating noise are in effect) by conducting bridge maintenance operations as quietly as possible in areas where a reasonably quiet environment is expected (e.g., near hospitals, schools, elderly care centers, churches, funeral homes, etc.).

3.1.8.7 Creosote-Treated Timber

The disposal of treated timber products has become a concern for states that use (or have used) creosote or other treated timbers in any of their bridges. The wood from these bridges can be disposed of in several different ways.

- The disposal wood can be used as a treated wood product in another application, such as for fence posts or landscape timbers.
- The wood can be used as fuel, with proper permit and approval from local and state authorities. Some power generation facilities have the capability to incinerate treated wood products in an environmentally approved manner and may be considered as an alternative if such facilities are in your area. Generally, such facilities charge a fee to accept treated timber waste.
- The wood can be recycled for use as a wood fiber source. Several methods are available to remove the preservatives. The wood fiber can then be used for other wood products or in the manufacture of certain paper products. A process is available for making wood

crossties in the railroad industry using wood fiber from disposed crossties. In this process, the preservative is not removed since the final product will be placed in a use where a preservative is still needed.
- The wood can be disposed of in a landfill.

As long as treated timbers are being used for their intended purpose, they are exempt from EPA pesticide regulations (e.g., FIFRA). When the timbers are no longer suitable for use in a bridge structure as beams, they can be provided to other users, who can convert them to other purposes. Once the timbers are no longer used for their intended purpose, they fall under RCRA and TCLP regulations. The treated wood itself does not fall into the category of hazardous waste. However, the leaching characteristics of the product do require testing by the toxicity characteristic leaching procedure. The EPA Permit Policy Compendium, Directive No. 9441.1990 [20] includes a memorandum dated July 3, 1990, to Paul Burkholder from David Bussard addressing the issue of creosote-treated timbers. The memo states that these timbers are not likely to fall into the category of hazardous waste.

Most TCLP tests that have been run on treated wood find its leachate does not exceed EPA regulatory limits. Regulated chemical compounds that could be found in wood products include pentachlorophenol, arsenic, and chromium as components of waterborne preservatives; benzene (a possible trace compound in creosote and petroleum oil carriers), and creosote with possible traces of pyridine and creosols. Treated wood products are not a hazardous waste and therefore can be disposed of in a landfill. However, each waste generator must decide how to handle its own waste materials. This may require confirmation of the leaching characteristic by testing a representative sample in accordance with TCLP requirements.

3.1.8.8 Worker Safety

Worker safety is traditionally thought of as a personal concern. However, worker safety results from a safe work site; thus, it is logical to categorize it as an environmental concern. Procedures to prevent equipment accidents, falls, and other traumatic injuries are a major part of worker safety, as are safety programs to educate workers about the materials and chemicals used in maintenance operations. As these operations increase in sophistication, the dangers to workers also increase. Strong safety programs are good financial investments that reduce lost productivity from injuries, increase employee morale, and reduce insurance costs. Safety has to be a priority of maintenance engineers and maintenance managers if it is to be taken seriously by rank-and-file workers. Employees must know the rules, policies, and practices intended to promote a safe work environment if they are to be expected to follow them. Generally, most accidents are preventable if proper safety equipment and safety rules are followed.

Safety at the Work Site: Increase safety at the work site by holding safety meetings; using tools properly, including small hand and small power tools; and storing tools appropriately.

Safety Meeting: A five-minute safety discussion at the beginning of each day can pay dividends in terms of reduced accidents and increased efficiency. A daily orientation provides an opportunity for everyone to remind each other what has changed from the previous day, such as modified rigging procedures or different materials that require special caution in handling.

Tool Use: Proper use of tools and equipment will accomplish the work assignment in a professional, safe manner. Typical program guidelines include the following items:
- Do not use defective tools or equipment.
- Use the proper tools or equipment for each job.
- Safety procedures should be learned and practiced.
- When in doubt, ask for information.
- To prevent accidents, assume a personal feeling of responsibility.
- Caution to a new worker may prevent an injury.
- Negligence causes accidents.

Small Hand Tools: All tools should be in a safe operating condition. Defective tools should be replaced or repaired immediately.
- When a worker is using an ax, pick, scythe, or similar tool, the worker should always use the proper standing position. The worker should have firm footing and be clear of any obstruction or other workers. The worker should be careful to not strike any objects that could produce flying chips. Even a glancing blow from these types of tools can cause a painful and serious injury.
- Only the proper tools should be used for the job. Crowbars or pry bars should not be used as chisels or punches. A file should never be used as a chisel, a punch, or a pry bar, and a file should never be struck with a hammer since the hardened steel may shatter.
- Sharp-edged tools should always be kept sharp. The cutting edge of a sharp tool should always be carried away from the body. Unguarded sharp-edged tools should never be carried in the pocket. The force of a blow on a sharp-edged tool should always be directed away from the body, never toward the body. A draw knife should be held securely and away from the body. Double-bitted axes should not be permitted in the tool inventory. A single-bit ax should never be used as a striking tool or as a maul. Such action spreads the opening, or eye, of the ax head, causing the handle to loosen. When leaving an ax in a tree or stump, it must be firmly embedded so that it cannot fall out and cause injury. Do not fine dress an ax on an emery wheel; instead use a sandstone or oil stone. Chisels and punches should be inspected for tempering and cracking. The heads should be kept well dressed and free of burrs.
- Safety shoes and protective helmets (hard hats) should be worn around overhead work or where there is any danger from dropping of heavy or sharp objects. Face, hand, and arm protectors should be worn when work constitutes a hazard to these parts of the body.

Small Power Tools: Most power tools have significant hazards associated with their operation. Electrical shock is always a possibility if the tool is powered by electricity. Malfunctioning of tools powered by compressed air, gasoline, or explosives can cause serious injuries. Most tools are powered to achieve high velocities and penetrate or scrape away wood, metal, or other hard materials. Shielding, insulation, proper inspection, and careful use of these tools are the basic safety measures that should be applied. High-velocity, hand-operated power tools (chain saws, belt sanders, disk sanders, portable grinders, portable circular saws, explosive-powered fastening tools, portable planes, routers, etc.) should be controlled by a deadman switch that shuts the power off if the tool is not properly held. Protective shatterproof goggles or a face shield should be worn when working with these tools, even when protective shields are provided on the equipment itself. Electric power tools should be either double insulated or grounded by a third wire and three-pole connection plug. The plug should be disconnected when the equipment is not in use or when repairs or adjustments to the equipment are made. Tools should be inspected regularly by a mechanic or electrician to ensure that they are in safe operating condition. Operators should note any defects appearing during operation of the tool and correct them immediately. Old materials should be checked carefully for heavy knots or nails before sawing into them with a circular saw to avoid the saw kicking back or severely damaging the blade. Operators of explosive-powered fastening tools should receive special training, and a licensing procedure is recommended to ensure that only properly qualified persons use such equipment.

Tool Storage: Adequate tool storage is important. Assembly time is saved when tools have assigned spaces and are readily accessible. Good housekeeping is a natural outgrowth of orderly tool storage and pays dividends because the tools are kept in a better state of repair. When not in use, it is important to return tools to their proper storage places.

Shoring and Falsework: The term *falsework* denotes any construction intended for erection use only. In other words, falsework is construction that it is later removed or abandoned. It includes temporary towers, bents, trestles, fixed and floating platforms, staging, runways, ladders, and scaffolding. On major structures, temporary trestles provide quick access to points of construction. Falsework bents provide temporary supports for erection or superstructure spans. Staging is used to provide working platforms. Ladders should be provided for all towers, and safety should be given full consideration in locating and designing falsework. Whenever practicable, falsework should be built of local materials or materials that can later be used in the permanent structure.

Properly designed and installed shoring is critical to work site safety during concrete placement. Shoring must be able to support all of the concrete, forming, and construction loads until the concrete gains sufficient strength to assume the various loads. Formwork must be erected on a proper foundation. General guidelines in the planning and development of shoring and formwork are as follows:
- Instruct workers in proper shoring erection and dismantling, including the importance of strict adherence to the prepared formwork layout.
- Plan the concrete placement sequence to guard against unbalanced loading conditions.
- Do not store reinforcing steel or equipment on erected shoring unless that shoring has been designed for those loads.
- Do not dismantle shoring until the concrete has cured.

3.1.8.9 Work Site Safety Review

All work related to highway maintenance is inherently dangerous because of its proximity to traffic and the very nature of the activities requiring the use of a wide variety of tools, materials, and heavy equipment. This is particularly true for bridge maintenance work since the area where the work is performed is likely to be restricted and a great deal of the work must be done manually. Thus, every precaution possible should be taken to reduce any unnecessary risks. Some questions that help maintenance engineers and maintenance managers think through the safety issues of bridge maintenance work sites are as follows:
- Does the general appearance of the work site demonstrate that tools are used and then replaced, or are tools scattered around the area?
- Are oxygen and acetylene cylinders stored properly in an upright position?
- Are safety glasses, masks, and hearing protection used by workers when appropriate?
- Are proper tools of the correct size available and being used?
- Are tools being used correctly for the purpose for which they were designed?

3.1.8.10 Toxic Materials

The storage of dangerous toxic and corrosive materials should be given special attention. Materials that are especially dangerous should be stored in separate secure areas and only issued to personnel qualified to use them properly. Manuals or charts that deal with dangerous chemicals and materials that maintenance employees work with should be readily available. Identification, qualities, and precautions should be included as a material safety data sheet. Training courses should emphasize the importance of using the manuals or charts for reference. Safety precautions when handling corrosives or irritants require the presence of an adequate supply of water to flush exposed parts of the body. Clothing that may become saturated should be easily removed, and the use of aprons, protective goggles, shields, gloves, and a respirator should be encouraged.

Acids: The strongest corrosive material that maintenance personnel usually handle in their work is sulfuric acid for use in batteries. Some materials can cause severe irritation through prolonged contact with the skin or can cause extreme pain and possible blindness when accidentally entering the eye.

Herbicides and Insecticides: Herbicides and insecticides (including wood preservatives) can be extremely toxic in concentrated forms. Some are so poisonous that a few drops on the skin can be lethal. Handling and mixing these substances should be regulated in accordance with their rated toxicity. Only qualified personnel wearing rubberized clothing and hoods that ensure complete protection should handle very toxic materials. Workers should be aware of the hazards associated with mishandling such materials.

Lead: Workers must not eat, chew gum, or use tobacco products when working with lead paints. They must not wear lead-contaminated clothing home. A worker can be poisoned in three to four weeks when exposed to high levels of airborne lead. All workers involved with lead paint removal, including containment maintenance, are automatically assumed by OSHA to be exposed to high levels of airborne lead. These workers need to become familiar with their rights and responsibilities under OSHA standards. Workers and supervisors of workers exposed to lead through painting or paint removal processes should take particular note of permissible exposure limits, exposure assessment, methods of compliance with standards and regulations, respirator protection, protective work clothing and equipment, housekeeping requirements, personal hygiene requirements, needed medical surveillance, the possibility of removing a person from the work site for medical protection, required employee training, required work site signing, record keeping for exposure monitoring, the right to observe monitoring, and rights to information.

Other Toxins: Epoxy can cause chemical burns on contact with skin, and a respirator must be worn when applying epoxy materials. Guano (bird droppings) can cause lung cancer, so a dust mask should be worn when performing bridge maintenance in areas infested with bird droppings.

3.1.8.11 Confined Spaces

Special procedures are needed when maintenance work must be performed in a confined space. This situation often occurs during the inspection and maintenance of box beam bridges and similar structures.

Training: Before proceeding to the site, personnel should be trained in the hazards of working in confined spaces as well as first-aid procedures, CPR, standard emergency procedures, use of instruments and monitoring equipment, and use and maintenance of protective equipment.

Identify Hazards: When planning for maintenance in a confined space, the following items are generally included to ensure a safe, efficient operation:
- Potential hazards such as oxygen deficiency should be considered.
- Fire safety should be examined when flammable materials are to be used in a confined space.
- Lighting will be required for all confined spaces; hence, an ignition hazard will be present. An oxygen-rich environment can be as dangerous as an oxygen-depleted environment. Use only grounded power equipment.
- Protective equipment and safety procedures for personnel should be outlined.
- Air sampling frequency should be established.
- The type of air testing to be performed should be identified. Usually oxygen, flammable gas, vapor, and toxic gas (carbon dioxide, carbon monoxide, sulfur dioxide, and hydrogen sulfide) tests are performed. Other tests for specific chemicals may be required.
- The type of respirator needed should be selected.
- Sampling equipment should be identified. Calibration will be required as specified by the manufacturer. Before its use, verify that the equipment is functioning properly.
- Test results should be recorded periodically, and alarm conditions should be reported.
- Evaluate the surrounding environment for storage tanks, sewers, chemical spills or bogs that can release poisonous gases or explosive gases; scum, slime, fungus, or decomposing organic matter that can deplete oxygen or release toxic or flammable chemicals; corrosion that can deplete oxygen; and bird droppings that can create a hazard requiring a combination of respirators and ventilation.
- A rescue plan should be developed. The availability of local emergency services must be verified prior to entry. Contact procedures must be included.
- OSHA requires entry permits for confined work spaces. A confined space is characterized as having limited means of entry and egress, having inadequate natural ventilation, or not being safe for continuous human occupation.
- A confined space entry permit is required if the space meets one or more of the following conditions: (1) contains or has the potential to contain a hazardous atmosphere, (2) contains material that has a potential for engulfing an entrant, (3) has an internal configuration that could trap or asphyxiate or a floor that slopes downward and tapers to a smaller cross section, or (4) contains any other recognized serious safety health hazard. If operations are to be performed in a confined space with a possible source of ignition (e.g., riveting, welding, cutting, burning, heating, or any other open flame) a hot-work permit is also required. This permit provides information on fire prevention, protection, and ventilation. The intent of the form is to ensure that all safety precautions are taken.
- Evaluate the work to be done by following these steps: (1) list all tools, equipment, and chemicals required; (2) identify a potentially flammable atmosphere if one can arise in this maintenance activity; (3) locate gasoline- or diesel-powered equipment outside the confined space to be used (not only are exhaust fumes a problem, but noise in a confined space can prevent workers from hearing critical directions or warnings); (4) identify ventilation work that generates dust, fumes, mist, vapor, odor, or smoke; and (5) provide exhaust ventilation for welding or other point sources.
- Identify personnel and ensure that they have received proper training and are medically fit for work in confined spaces. Medical limitations include emphysema (a worker may not be able to breathe adequately against the additional resistance of a respirator), asthma (a worker wearing a respirator would be tempted to remove it during breathing difficulties from an asthma attack), chronic bronchitis, heart disease, anemia, hemophilia, poor eyesight, hernia (may be aggravated by wearing or carrying respirator protective equipment), evidence of reduced pulmonary function, severe or progressive

hypertension, epilepsy (either grand mal or petit mal), diabetes (insipidus or mellitus), punctured eardrum, communication or sinus through upper jaw to oral cavity, claustrophobia or anxiety when wearing a respirator, breathing difficulty when wearing a respirator, lack of use of fingers, scars or hollow temples or prominent cheekbones or deep skin creases or lack of teeth or dentures.

Requirements of Each Entry: Before each entry, personnel shall review the potential hazards, the proper use and maintenance of respirators, blowers, and emergency equipment, the specific entry procedures for that bridge, and the emergency procedures. During each entry, the procedure shall include the following steps:
(1) Atmospheric testing equipment will be tested using test gases to ensure the equipment is functioning properly.
(2) The atmosphere shall be tested at the entrance to evaluate for possible hazards.
(3) Be sure to check all spaces where gases can accumulate. Some gases such as propane and butane are heavier than air and will sink to the bottom of a confined space. Light gases such as methane will rise to the top of a confined space.
(4) After testing for explosive gases, test for toxic gases.

At least one worker should remain outside of the confined space to monitor the test equipment and guard against entrapment. The worker outside should have the entry permit in hand and should have radio communication with the workers inside. The worker outside could also have a lifeline attached to all personnel inside the confined space. The worker outside is responsible for contacting any emergency services required.

Workers shall continuously monitor their test instruments for hazardous atmosphere. If an alarm sounds, workers shall perform the following steps:
(1) Evacuate the space immediately.
(2) Record the event and readings that produced the alarm, including work in progress as well as tools and materials used.
(3) Check the instrument for malfunctions.
(4) Determine what caused the alarm.
(5) Correct the problem. Do not enter the space without a mechanical ventilator.
(6) Reenter the space only when the atmosphere is tested to be within acceptable limits. If the space must be entered and is not within acceptable limits, respirators must be used and escape provisions must be provided. Never reenter a confined space to rescue a worker before contacting emergency services and determining the cause of the emergency. Over 60% of all confined space fatalities occur to rescuers, not the original entrant.

Evacuate the area if odors are present or if workers experience dizziness or shortness of breath.

The entry team will carry a radio so that radio communications can be maintained between the team and the worker outside the confined space.

Requirements for ventilation, respiratory protection, and standby personnel will be specified. Work with chemicals, such as paints, solvents, and epoxies, requires ventilation and standby personnel.

3.1.8.12 Fall Protection, Rigging, Scaffolding, and Hoisting

Occasionally, maintenance work must be performed on parts of the bridge that are high above the ground or water. This activity can be safe when it is performed properly. However, carelessness, inattention, or horseplay can lead to serious injury or fatalities. It is important that workers are properly trained in how to erect safe work platforms and how to protect against serious injury from falls. Maintenance planning and preparation for any high bridge maintenance activity should include a review of the OSHA safety section and any appropriate department regulations on rigging, scaffolding, ladders, and wire ropes.

Fall Protection: Falls from heights are a routine hazard facing bridge maintenance workers. Precautions must always be taken when working from an elevated work surface.

Guardrails: OSHA construction industry standards require guardrails, mid-rails, and toe boards on all work platforms more than 1.8 meters (6 feet) above ground level. Guardrails and toe boards are to be installed on all sides and ends of platforms. Railing should have a vertical height of 1,067 mm (42 inches) from the upper surface of the top rail to the platform. An intermediate rail should be spaced midway between the top rail and the platform floor. The toe board should be at least 4 inches (about 100 mm) in vertical height from its top edge to the platform level. The bottom of the toe board should be no more than 6 mm (1/4 inch) away from the floor. When material is piled on top of the platform and obscures the toe board, screening or paneling should be provided from the platform to the intermediate rail. Vertical posts should be placed at 2.4 meters (8 feet) or less.

Safety Belts, Lines, and Lanyards: In situations where guardrails are not practical or feasible, a secondary means of fall protection must be provided. An alternate system consists of either a harness or safety belt connected by a lanyard to a safety line. The American National Standards Institute defines a safety line as a horizontal rope between two fixed anchors independent of the work. Lifelines must have a minimum breaking strength of 24 kilonewtons (5,400 pounds). This equipment should be inspected for frays and other deformities from stress, dry rot, or damage from chemical attack. Clips and bearings should also be inspected for deformities and alterations. Defective equipment should be removed from service and taken to an authorized service center to be repaired or recertified.

Body harnesses should be rigged to minimize free-fall distances to a maximum of 2 meters (6 feet) prior to their use. Equipment must be inspected for mildew, wear, damage, and other deterioration. Defective components are removed from service. At the end of a job shift, equipment should be cleaned and stored in a clean, dry place. Components subjected to impact loading shall be removed from service and taken to an authorized service center to be recertified.

Safety Belts and Harnesses Attached to Restraint Lines: Anchor points used for fall restraint shall be capable of supporting four times the intended load. All safety lines and lanyards shall be protected against being cut or braided. All hardware shall be drop-forged, pressed to form. Steel hardware shall be corrosion resistant. Vertical lifelines should not be used by more than one worker at a time. Anchors should be capable of supporting 22 kn (5,000 pounds) per worker. Vertical lifelines shall have a minimum tensile strength of 22 kn (5,000 pounds). Horizontal lifelines should have tensile strength capable of supporting a fall impact load of at least 22 kn (5,000 pounds). Lanyards should have a minimum tension strength of 22 kn (5,000 pounds) per worker. Snap hooks should not be connected to each other.

Lifelines should be attached to anchor points that are stapled, rigid, and structurally independent of the work platform. Safety belts and harnesses should be worn. Safety belts should not be used when falls of less than 600 mm (2 feet) are anticipated. Otherwise, a safety belt may actually cause injuries by causing shock to the neck, back, and internal organs. The D-ring of the belt should be worn in the back of the body. For falls greater than 600 mm (2 feet), a harness will distribute the shock of the fall over the shoulders, thighs, and seat area, reducing internal injuries. The lanyard is a short piece of flexible line that secures the belt or harness to the lifeline or to a fixed anchor point.

Safety Net: In some situations, guardrails and safety belts are not practical. In long-term situations, safety nets may be necessary. OSHA requires safety nets when conventional fall equipment is not practical and the work site is in excess of 8 meters (25 feet) off the ground or water surface. Safety nets must be manufactured and tested in accordance with ANSI and OSHA standards. Nets should display the manufacturer's name, date of manufacture, proof of load testing, and a serial number so that repairs and load tests can be recorded.

Injuries: Workers with serious injuries are not to be moved unless their current position entails risk of additional injury. Workers at grade level shall be removed by ambulance. Workers at elevated stations accessible by a lift shall be transported via the lift to grade level. Workers at elevated work stations not accessible by lift shall be removed by ladder.

Ladders: Falls from ladders can be prevented by securing a taut cable or rigid rail running centrally or alongside of the ladder. In this system, the lanyard is attached to the cable or rail by means of a safety sleeve that moves freely along the rail or cable but locks the instant a fall is sensed. OSHA requires protection on all fixed ladders when the total length of the climb equals or exceeds 7.3 meters (24 feet). This protection may include devices described thus far or self-retracting lifeline systems, gauges, or wells.

Precautions for Using Ladders

These specific precautions should be taken when ladders are being used.

- Inspect the ladder before use. If it is not safe, do not use it.
- Do not use a ladder with missing or defective rungs or studs.
- Do not use a wobbly or shaky ladder.
- Extension and straight ladders should be chained fast at the top or held by someone at the bottom. The top of the ladder should be placed so that both sides of the rail have a firm bearing.
- Wood ladders should never be painted. Paint hides defects that otherwise would be noted.
- Ladders should always be placed on firm footings. If the ladder must be placed on soft ground, solid footings should be provided.
- The bottom of a ladder should have safety feet or should be adequately protected against sliding.
- Place both hands on the side of the ladder and face it when going up or coming down. Always have a grip on the ladder with at least one hand and one foot.
- When using a stepladder, be sure the cross braces are locked.
- Keep steps and rungs free of grease, oil, paint, or other slippery substances.
- Heavy and bulky materials or equipment should be hoisted by means of a rope, not carried by hand.
- Do not overreach the safety limits of the ladder. Always move the ladder instead.
- Set the ladder at a safe working angle. The recommended angle is 1 unit out for every 4 units up.
- Do not use metal ladders in the vicinity of power lines because metal conducts electricity.
- Do not work from the top rung.
- Do not lay ladders on the ground or a surface where they can come in contact with water or chemicals.
- Do not use ladders in the roadway without red warning flags and a flagger on the traffic side.

Other Precautions: Additional precautions may be necessary for maintenance personnel on bridges and in high places.
- Work over water may require wearing of a type of life preserver approved by the U.S. Coast Guard. Bulky life preservers may be hazardous. A skiff with oars should be readily available on the shore, closest to where the work is in progress. Ring buoys with adequate throw lines should also be available.
- Use of boatswain's chairs, safety harnesses, and safety nets should be required at heights above 8 meters (25 feet).
- Any rope or cable used to support workers such as guide rails, safety lines, or other safety devices should be inspected for deterioration and integrity at regular intervals, not to exceed 60 days. More frequent inspections should be made if the rope fiber is susceptible to mildew or exposed to corrosive materials or fumes.
- Loose lines, swinging weights, and loose boards or materials should be eliminated.
- Handrails, safety walks, and fixed ladders in the work areas should be checked routinely by maintenance personnel before work starts. Cages or other ladder safety devices should be provided on fixed ladders more than 6 meters (20 feet) high.

Rigging: Rigging is a very specialized activity necessary to perform bridge maintenance. Basic terms and vocabulary for rigging are included below to assist persons new to maintenance engineering and management. Rigging with wire rope (cable) and nonwire rope is discussed separately in this section.

Rigging Terms and Vocabulary

Blocks and Sheaves: Blocks are the parts of tackles that hold the sheaves or pulleys over which a rope turns. This system is commonly known as block and tackle and should be properly sized for the rope or wire rope or cable used.

Cable Clamps: Cable clamps are U-clamps used for fastening wire rope together and are also referred to as clips.

Choker: A choker is a device used to form a loop on the end of a rope.

Coil: To coil is to wind a rope into a circular or spiral form.

Fall: The fall is the part of the block and tackle made up of rope.

Heave: To heave is to haul or pull on a line.

Hitch: To hitch is to attach a rope to a post, pole, ring, hook, or other object.

Hook: A hook is a device for catching and holding a line.

Jam: To jam is to wedge tightly.

Lay: The lay of the rope defines the direction in which the strands pass around the axis of the rope (referred to as right lay or left lay).

Pad Eye: A pad eye is an eye located on a pad for fastening riggings.

Plow Steel: Plow steel is a wire steel of high strength and toughness.

Reeve: To reeve is to pass a rope around the sheaves of a block.

Rigging: Rigging is the manila rope or wire rope used to support booms, scaffolding, or other construction elements.

Shackle: A shackle is a U-shaped device with a pin across the ends.

Size of Rope: The size of rope is given by its diameter, although circumference is sometimes used.

Sling: A sling is a looped or hanging band, strap, etc. used in raising and lowering a heavy object.

Splice: To splice is to join the ends of two ropes or the end of a rope with the body of the rope by weaving the strands over and under the strands of the other part.

Tackle: A tackle is a rope and pulley block or a system of ropes and pulleys used to lower, raise, or move a heavy object.

Taut: Taut is tight or snug (opposite of slack).

Thimble: A thimble is a form device around which a loop of rope is placed. A thimble cuts down on wear and decreases the stress placed on the rope.

Working Loads: When figuring the working strength of ropes (manila and wire), the load should never exceed 1/10 of the breaking strength.

Nonwire Rope: Generally, only no. 1 grade manila rope identified by the manufacturer's trademark should be used for rigging. When rope is attached to a hook, a ring, or a pulley block, a thimble should be used in the loop or eye to reduce wear and decrease stress. Ropes judged to be unsafe should be cut into short hand lines so that they cannot be used for lifting loads.

- In tying lines to needle beams or supporting beams, be sure the hitch is secure and safe before getting on the scaffolding.
- Avoid working rope over sharp edges or chafing it on rough surfaces. Dragging ropes on concrete pavement causes wear and cuts on the underside surface. Use padding whenever necessary.
- When using manila rope for lifts, the tackle should be tested just above the ground with three times the anticipated load.
- Do not place manila rope tackle supporting a worker in the same sling or ring with tackle used to hoist work loads.
- Do not use wire blocks with manila rope.
- Always use thimbles when attaching ropes to a ring, hook, or pulley block.
- Keep manila rope clear of acids or other chemicals.
- Always take the load off a taut rope when it is not in service.
- Never slide down a manila line or a wire cable.
- Do not allow ropes to get wet as this causes the rope to kink. If a wet rope develops kinks, get them out before it dries.
- Remove manila line from a new coil by pulling the inside end out. Never start from the outside.
- When removing ropes from the coil, avoid kinks by determining if the rope is right lay, left lay, or cable lay. Right lay should come out of the coil in a counterclockwise direction, and cable and left lay should come out of the eye of the coil in a clockwise direction. If the rope does not come out correctly, draw the end back through the coil and out the other eye.
- When a rope is twisted continually in one direction, compensating turns should be thrown in the opposite direction to avoid damage to the rope structure.

Storing Ropes: Clean all mud, sand, or grit from rope by washing it down with a hose before drying and storing it. Always dry a rope before storing it, but remove it from the sun when it is dry. Do not let a wet rope freeze since freezing will damage the fibers. Store rope above the floor by hanging it in loose coils on large-diameter wood pegs.

Inspecting Rope: Rope should be inspected daily to ensure it is in good condition. Dirt on the surface does not indicate that the rope is in poor condition if the inside is as bright and clean as a new rope. When the inside of the rope is dirty or the rope has lost its elasticity, it should not be used on scaffolding. Broken yarn inside the rope means that the rope has been overloaded and should not be used.

Knots and Hitches Used With Manila Rope: There many hundreds of different rope knots, bins, and hitches. For most applications, personnel who are setting rigging only need to know a few knots and hitches. For a thorough study of the strengths, weaknesses, and most appropriate knots for a particular need, consult a handbook on rigging. A few of the most important knots and their appropriate uses are as follows:

- An overhand and a figure eight are knots that prevent rope from unraveling.
- A backhanded sailor's knot attaches a rope to a ring.
- A bowline knot is used whenever a hitch that will not slip, jam, or fail is required.

- A square knot joins two ropes or lines of the same size. It holds firmly and is easily untied. It is also a dangerous knot because it is easy to tie incorrectly, producing a granny knot, which will slip.
- A cat's paw knot is used to secure the middle of a rope to a hook.
- A round turn and two half hitches attaches a scaffolding line to a supporting beam.
- A scaffold hitch fastens single scaffold planks to the hang level.
- A back hitch secures the hoist rope to a scaffold stirrup.
- A safety hitch secures a boatswain's chair.

Wire Rope and Cable: Wire rope is stronger and has longer usable life than manila rope. However, it is heavier and harder to handle. The best grade of wire rope is called improved plow steel rope, and this is the type that should be used. Wire rope is regular lay (preferred for general use), meaning the wires are twisted in one direction to form the strands, and the strands are twisted in the opposite direction to form the rope. Lang lay wire rope has wires that are laid in the same direction as the lay of the strands. Lang lay ropes are used when more flexibility is needed. Lang lay has greater wearing surface per wire than regular lay. Lang lay ropes are susceptible to damage from bending over small-diameter sheaths, from pinching in undersized sheath grooves, and from crushing when wound on drums. They can fail from excessive rotation.

Wire Rope Maintenance: Wire rope can deteriorate from neglect and abuse. Poorly maintained equipment can reduce rope life. Lubricants should be applied to the surface of ropes to reduce corrosion from acid and alkaline solutions, gases, fumes, brine, salt air, sulfurs, compounds, and combined high humidity with high temperature. Corrosion causes pits that act as stress concentrations in the same way nicks do.

- Breaking in a New Wire Rope: A new wire rope should be run with a light load for a short period of time after it is installed. This breaking-in process gives the parts of the rope opportunity to adjust to the conditions under which the rope is to be used. Time spent breaking in the rope will pay dividends in extra length of useful life.
- Lubrication: Wire ropes are lubricated as they are manufactured. This initial lubrication is generally not sufficient to last throughout the entire useful life of the rope. A good grade of oil or grease should be applied frequently while the rope is in use. Lubricating iron and steel wire ropes prevents corrosion of the wires and deterioration of the fiber core around which the wire strands are wound. When a wire rope is taken out of service, it should be cleaned, lubricated, and stored in a dry place.

Inspection of Wire Rope: A wire rope in service should be inspected regularly. A broken rope left in service can destroy machinery, curtail production, and result in worker injury. Problems that warrant removal of a wire rope from service include (1) crushing damage that has misshapen the rope, (2) breakage of the wires on the exposed surface from wear or abrasion, (3) kinks in the rope, (4) open strands with displacement of the core, (5) corrosion of the wires, (6) separation of the strands, and (7) damage to the strands from electrical arcing.

When inspecting wire rope, the condition of the drum, the sheaves, guards, cable clamps, and other fittings should be noted. The condition of these parts affects wire rope wear. Any defects noted should be repaired. Equipment operators can be helpful by watching wire ropes under their control. OSHA requires periodic inspections for most wire rope applications. By flagging excessive wear in the wire rope, an inspector can determine where the rope is rubbing on equipment parts so that repairs or modifications to the equipment can correct the situation.

Abrasion Versus Fatigue: A trade-off exists between abrasion resistance and bending fatigue resistance in wire rope. Wire ropes with relatively few wires of large diameter have high abrasion resistance but are very stiff. Conversely, wire ropes made up of a large number of small-diameter wires have poor abrasion resistance but are highly flexible. Improper sheaves and drum size can also reduce the fatigue life of a wire rope. When a wire rope is bent around a sheave, it loses effective strength because the tensile load is not evenly distributed among the individual strands and wires; some wires carry more of the tensile load than others. Shock loading and vibrations can cause rapid fatigue of wire ropes. Wire rope failure is cumulative in that each overstressing brings the entire rope closer to sudden failure. A wire rope may become fatigued to a point close to failure under a heavy load and then actually fail under a much lighter load. While lifting loads, make accelerations and decelerations gradual to avoid overstressing the wire rope. Ropes should not be left on machines for long periods of time. The wire ropes should be removed, cleaned, and thoroughly lubricated. A wire rope takes a "set" to a particular operating condition. Out-of-service wire rope should not be used for a different job because operating it with different bends and stresses can reduce its life. Wire rope should be removed from service when in any length equal to eight diameters of the rope, the number of broken wires that are visible exceeds 10% of the total number of wires in the rope.

Proper Handling of Wire Rope: Careful handling of wire rope will prevent kinks that are usually impossible to remove and that result in permanent damage. There is always the danger of kinks if a wire rope is not unreeled or uncoiled properly. A reel should be mounted on jacks or a turntable so it will revolve as the rope is pulled off. Sudden stresses and jerks can break the rope or cause weak spots. Loads should always be applied gradually to a wire rope.

- Spooling and Coiling: When winding wire rope on a smooth drum, the first layer should be started with the side that causes the coils on the drum to hug together. This produces a uniform, closely wound first layer. The next layer should have the same smooth appearance as the first layer. If it does not, the coils of the first layer may have spread, allowing the second layer to wedge itself between them. This may cause crushing and abrading of the rope. All layers on a drum should wind smoothly. When coiling wire rope by hand on the floor or on a bench, coil it in the direction that will take the twist out of the rope. Improper coiling can ruin wire ropes.
- Changing Stress Points: Wear and fatigue are usually most severe at certain points where a wire rope is subject to great stress. Cutting short lengths of the rope from the drum end will change where the rope is stressed over pulleys and load points, increasing its service life. This approach requires that the initial length of rope be slightly longer than needed for the hoisting or rigging application. Another approach is to change the rope end for end.
- Forming a Loop in Wire Rope: When forming a loop in wire rope, the length of the short end or dead end should be at least 35 to 50 times the diameter of the rope, and a thimble should always be used in the loop. The number of clamps to be used and their spacing should conform to the rope manufacturer's recommendations. In general, the number and spacing of clamps (clips) varies with the diameter of the rope. For example, two clamps spaced at 75 mm (3 inches) for 13-mm (1/2-inch) diameter wire rope, and six clamps spaced at 150 mm (6 inches) for 25-mm (1-inch) diameter rope would be typical. Retighten cable clamps after 1 hour of running time. If a cable or wire rope is used at or

about its full safe working load, it is very important to install the correct number of cable clamps (clips) at the proper spacing and in the proper direction. Efficiency ratings for wire rope in termination (end) loops is based on the breaking strength of the wire rope. To construct an end treatment, use the turn-back amount recommended by the wire rope manufacturer and apply the first clip one base width from the dead end of the wire rope. Apply the U-bolt of the clip over the dead end of the wire rope with the live end resting in the saddle of the clip. (A memory aid to maintenance personnel: You cannot saddle a dead horse.) Tighten nuts evenly, alternating from one nut to another until reaching the recommended torque. Apply the second clip as near to the loop or thimble as possible and tighten the nuts evenly, alternating from one nut to another until the recommended torque is reached. However, if more than two clips are required, turn the nuts on the second clip firmly but do not tighten to the required torque. Then add all the additional clips, evenly spaced between the first and second clips, before tightening all clips to the specified torque. Test the assembly with a load equal to or greater than the load expected in the application. Check and retighten nuts to the recommended torque. Inspect all end terminations (loops) periodically.

- Elastic Stretch: Wire rope will stretch when a load is applied. The amount of stretch is estimated by the relationship: (change in length) = (change in load times the length) divided by (wire rope cross-sectional area) times (steel modulus of elasticity). Make certain the units of all items in the relationship are consistent when estimating the stretch.
- Sizing and Cutting: Great care is taken in the manufacture of wire rope to produce a balance of the tension loading among the wire strands. This balance will be lost if insufficient sizings are applied before cutting a wire rope. Sizing is the process of winding needle wire around the wire rope. When a wire rope is cut in the warehouse or in the field, place sizing on both sides of the point where the cut is to be made. If this is not done, the balance will be disturbed or one or more strands will slip back and others will carry more than their portion of the load.

Scaffolding: Tubular scaffolding (usually made of steel tubing) is a manufactured product similar in design to toy construction sets. The scaffolding can be built from the ground up by putting pieces together. The pieces can be adjusted for irregular ground and when firmly placed provide good support to a working platform.

Wood scaffolding is sometimes required. A competent carpenter who is familiar with contruction regulations and any applicable building code regulations should construct wood scaffolding. Wood scaffolding needs to be strong enough to withstand the expected loading and secured against slipping or overturning. Loose boards should not be allowed to project beyond their supports. Screws should be used instead of nails in tension areas to hold scaffolding or falsework. All screws and nails should be driven in flush. Scaffold horses should support evenly and should be nailed to the platform that they are supporting.

Swinging or suspended scaffolding is used for bridge maintenance work when there is no base for erecting tubular or wood scaffolding. Boatswain's chairs or work baskets are used in access areas that are hard to reach.

Metal Scaffolding Assembly Cautions:
- Select a platform that is long enough to safely hold the workers and materials needed to do the job. Before scaffolding is used, hoist it about 0.3 meter (1 foot) off the ground and apply a test weight equal to four workers for a period of 5 minutes.
- Attach guide rail supports to the platform. Spacing between the supports should not be more than 2.4 meters (8 feet). Use as many supports on each side as necessary. End supports should be 600 mm (12 inches) minimum to 460 mm (18 inches) maximum from each end of the platform.
- Attach wood or aluminum toe boards. Toe boards that are 100 mm (4 inches) high may be used, although 140-mm (5.5-inch) boards are recommended.
- Insert guardrails and mid-rails using U-bolts. Guardrails should be at least 1,067 mm (42 inches) high, with the mid-rails an equal distance between the platform and the guardrail.
- Attach each end to guardrail and toe boards.
- Slide stirrups onto each end of the platforms and attach them with U-bolts close to the end guardrails and the toe boards.
- Install a hand winch or motorized cable-climbing machine if necessary.
- Securely fasten all clamps, bolts, and safety knots before using the scaffolding. Place hooks directly above the scaffolding. As the angle at which the wire rope is used increases, the load that it can hold decreases. The hooks should be attached to a sound portion of the bridge.

Motorized Cable-Climbing Machine: A motorized cable-climbing machine can be attached to each end of the platform to move the platform to different positions. The platform can be moved to different heights and to different lateral positions with a simple control lever. Installation points for which maintenance supervisors should be watching are as follows:
- Attach the cable-climbing machine to the stirrups by means of two stirrup connecting bolts on each stirrup. Use one machine at each end of the platform.
- Install the electric-powered cable or air hose, whichever is the power source.
- Reeve the machine according to the manufacturer's recommendation.
- After the machine is reeved, loop and secure the bottom free end of the wire rope with a cable clip to prevent the rope from running through the machine.

Hoisting: A hoist is a machine for raising or lowering heavy or bulky items.

Hand Signals for Rigging and Hoisting: In rigging work, communication must be maintained among workers involved in all aspects of moving the load. A person on the ground should direct the operator of the hoisting equipment in how to move the load for safety and efficiency using standard hand signals. Basic rules for hand signaling are as follows: (1) only one person should be responsible for transmitting hand signals to the operator (for maximum safety, however, it is helpful if all personnel in the area know and understand the standard hand signals); (2) the person giving the hand signals must stand in full view of the hoist operator; and (3) the person giving the hand signals must have no other duties or function to perform while engaged in signaling. Hand signals are not effective unless the signal is plainly visible to the operator.

Safe Operation of Hoisting Equipment: Many serious accidents result from unsafe operation of construction and weight-handling equipment. Untrained operators or failure of the operators and other workers on the job to recognize and respect the dangers inherent in power-driven equipment contribute to these accidents. Familiarity or long acquaintance with the equipment can also contribute to carelessness and indifference to the hazards present. Bridge maintenance supervisors should be alert to the following cautions when a crane is being used as a part of bridge maintenance activity:

- Test both equipment and rigging under load before hoisting and swinging.
- Move heavy loads slowly maintaining control at all times. Loads should be moved in one direction at a time. The critical moment in lowering a heavy load occurs when the load is checked (stopped). At that moment, the equipment must withstand the inertia of the load as well as its weight.
- Loads should be guided and prevented from swinging sideways with tag lines tied to the load and held by one or more workers on the ground.
- Never lift loads during strong or gusty winds.
- When hoisting equipment is in operation, the operator should not be permitted to perform any other work. In addition, the operator should not be permitted to leave the controls while a load is suspended from a crane or derrick.
- Never permit operation of equipment without some means of signaling or warning nearby workers of the operator's intentions.
- Never work or walk under skips, buckets, or other operations where falling objects are a hazard.
- Never enter or stand in a confined area when loads are being raised or lowered.
- Never ride on a load or a sling.
- Never grab or hold onto a cable or rope as it is being pulled through a set of sheaves.
- Never operate a crane near power lines without proper and adequate supervision.
- Never walk in front of a moving crane.
- Watch out for swinging booms.
- Never feel for matching holes in girders with fingers or hands. This is a good way to lose one.
- Avoid an angle of inclination or obliqueness on the legs of a sling that puts hazardous or undue stresses on the sling. As the legs of the sling become more nearly horizontal, the load stress in the sling cable or wire rope increases dramatically.

Hooks: The strength of standard hooks should be taken from the manufacturer's recommended standards for safe loading. If its standard is unknown, a hook should be measured and the safe loading estimated by a professionally qualified structural engineer. Hooks should always be inspected before use. At least annually, inspect the hook and the threads for cracking (some disassembly may be required) by magnetic particle, dye penetration, or both methods if the hook is used in severe loading conditions. Do not use hooks that are bent or disoriented. Do not use hooks with cracks, nicks, or gouges. These defects may be repaired by grinding lengthwise following the contour of the hook, providing that the reduced dimension is still within the required structural cross section. Never repair a hook by heating, welding, burning, or bending. Never side-load, back-load, or tip-load a hook.

Eyebolts: Always inspect eyebolts before using them. Never use eyebolts that show signs of wear, damage, bends, or elongation. Never machine grind or cut eyebolts. Never exceed the rated structural load limits of an eyebolt. Never cut an eyebolt to seat the shoulder against the load. Always countersink the receiving hole or use a washer to seat the shoulder of the bolt. Always screw eyebolts down completely for proper seating. Always tighten nuts securely against the load.

Slings: Slings are used to hold material while it is being hoisted. Loads on slings should always be within the rated capacity of the sling. The sling should have suitable characteristics for the type of load, hitch, and operating environment. Consideration should be given to the angle of width that may affect the lifting capacity of the sling. Diameter of pins and hard shapes may also affect the capacity of the lift sling. Slings should be protected from cutting and sharp edges. Abrasion padding should be placed between the sling and the load. Slings should not be twisted or tied in knots to shorten them. Slings should not be pulled out from under loads that are resting on them, nor should vehicles be driven over them. Damaged slings should not be used. All personnel should stand clear of suspended loads and should not ride on slings or suspended loads. Avoid snatch or shock loading. Slings should be stored in a clean, dry place. Avoid storing slings near heat sources or places that are not ventilated. Slings should be inspected periodically for evidence of any cutting, tearing, abrasion, or heat damage. Written inspection reports should be prepared for each 30 days of use. Slings removed from service should be destroyed. Slings can be tested by pulling the slings at twice their rated capacity for a period of 15 seconds. Good practice in using a sling includes using the proper sling for the job and starting and stopping the lifting operation slowly. When selecting a sling for a particular job, determine the load to be lifted, decide what hitch shall be used, select the lifting device with adequate capacity to lift the load, make certain there is enough room to lift and maneuver the load, select the length of sling, and use the rated capacity of the sling to define load limits.

Blocking: Blocking may be used to support a load before or after it is lifted. Maintenance supervisors should take note to see that wood blocking is placed under slings lifting heavy steel beams to prevent the slings from being cut; that blocking or cribbing is always used to secure necessary height under jacks (jacks should not be set on a post or strut where they might kick sideways under strain); that when placing two sling legs on a hook, the angle from the vertical of the outermost leg does not exceed 45 degrees and the total included angle between the two legs does not exceed 90 degrees.

Block twisting can occur when the wire rope twists slightly or unlays when a load is applied. This twisting can be reduced by reducing the length of wire rope, reducing the load applied, increasing the sheave size that increases the amount of separation between wire rope parts and improves stability, restraining the twisting block with a tag line, and using rotation-resistant wire rope.

3.1.8.13 Safety Review

The dangers inherent in bridge maintenance are often made more severe by the need to work at great heights above the ground or water. Personnel need special protection, and they often need special tools, special equipment, and special materials-handling procedures. Maintenance supervisors need to be particularly alert for any conditions that might reduce safety of the work environment. The following items are suggested for inclusion in a checklist of things to keep in mind to promote safe bridge maintenance:

- Is every person in the crew fully aware of individual responsibilities and assignments before work starts?
- Are all manila ropes on the site fresh and in good condition with the inside bright and clean?
- Are all manila ropes coiled loosely, hung above the floor, and stored under cover when not in use?
- Are thimbles used to attach manila rope to a ring, hook, or pulley block?
- Are all wire ropes properly lubricated?
- Have thimbles been used to form all loops in wire ropes?
- Is all wire rope on-site in good condition without excessive numbers of broken wires?
- Are personnel and loads always transported to the work area separately?
- Is scaffolding (tubular or wood) properly constructed and erected?
- Are work platforms long enough to safely hold the worker and the materials for the job?
- Are platforms equipped with toe boards, guardrails, and mid-rails?
- Are all ladders fast at the top (if not, will ladders be held at the bottom)?
- Are all wood ladders unpainted?
- Do swinging scaffolding units have a lifeline for each worker and are they being used properly?
- Are boatswain's chairs, safety harnesses, or safety nets used when workers are working at heights in excess of 8 meters (25 feet)?
- When working from bucket trucks (cherry pickers), are personnel using the safety belts provided with such equipment?
- Are cranes correctly located with respect to operator visibility and utility lines, and has the responsibility for directing the operation and giving hand signals been properly assigned to only one person?
- Are slings properly formed to safely carry the loads for which they are to be used?
- Does the foreman assigned to supervise the crew appear to be constantly aware of potential problems that can arise from poor safety practices?
- Have all crew members been briefed on the following equipment safety rules?
    → Never leave a piece of equipment when its engine is running.
    → Never stand in the line of a moving load.
    → Never be careless when getting on, off, or working around moving equipment.
    → Never operate a defective piece of equipment or permit it to be operated.
    → Do not permit unauthorized personnel to ride moving equipment.
    → Never stand within the angle formed by a line carried over a block.
    → Never use or permit equipment to be used for any purpose other than that for which it was designed.
    → Never attempt to lubricate, clean, repair, or refuel any piece of equipment while it is in motion or running.
    → Always keep all guards over moving parts of the equipment. Never remove them except for repair or adjustments, and then only when the power is off.
    → Never operate a piece of equipment in a careless or thoughtless manner.
    → Always insist on proper maintenance.
    → Never overload construction equipment.

3.1.9 Movable Bridges
3.1.9.1 Types of Movable Bridges
The three basic types of movable bridges are (1) vertical lift, (2) bascule, and (3) horizontal swing. All movable bridges are used to cross navigable waterways where some feature of the site makes construction of a bridge with sufficient clearance above the water for navigation either technically or economically not feasible. Bascule bridges are sometimes further subclassified as (1) the Scherzer type (a rolling lift); (2) the Rall (a rolling lift); (3) the Chicago (a simple trunnion); and (4) the Strauss (a modified simple trunnion). For a high-volume vehicular traffic route, many transportation agencies prefer bascule bridges because the break in the roadway floor of a double-leaf bascule is located ahead of a vertical plane passed through the axis of the trunnion. As the leaf of the deck trunnion bascule forms a traffic barrier the instant it begins to lift, this barrier remains throughout the entire time of opening and closing [7]. Counterweights are used to balance the movable spans and their attachments in any position. A small positive dead-load reaction at the supports results when the bridge is seated (closed). Provisions must be made for unbalanced conditions of the machinery and power equipment [8]. A swing span of unequal lengths will typically be balanced by counterweights.

Movable bridges have mechanisms to align the bridge and roadway, and to fasten them securely in position so that horizontal and vertical displacement is prevented under traffic. End-lifting devices are used for swing bridges, and span locks for bascule bridges. Span locks are also provided in swing and vertical lift bridges normally left in the open position. These aligning and locking devices need to be regularly inspected and properly maintained to ensure that they function when needed. Cases have arisen where welders have had to cut several millimeters off the end of a span to close a swing bridge for vehicular traffic.

Two traffic gates are generally provided at each approach roadway to a movable span bridge. The first acts as a warning device, and the second functions as a physical barrier to stop vehicles. Even though bascule bridges provide a traffic barrier by the very nature of how they operate, traffic gates are usually still provided. These traffic barricades and warning devices need to be kept in proper operating condition at all times for which there is any possibility that the movable span will be opened. Emergency operation plans need to be in place in advance of any traffic barrier failure.

Houses are typically provided for machinery and operations. These need to be constructed and maintained to meet applicable building code regulations as well as all applicable bridge regulations for such facilities.

Traffic signals are generally mandatory for movable span bridges and are supplemented with warning bells. Navigation lights and audible navigation signals are also standard features and must be installed and maintained according to U.S. Coast Guard regulations. Applicable regulations of the transportation agency having jurisdiction over the movable span bridge must also be followed. The traffic signals require the same type of maintenance that would be applied to roadway signals.

Stairways, platforms, walkways with railings, and elevators are typically provided on the structure to allow safe and convenient access to various bridge components during bridge operations and maintenance. A movable span bridge is constantly subjected to large stress reversals and load variations by the traffic rolling across it. Therefore, many structural and mechanical components need to be continuously serviced and maintained, including the lifting or moving machinery, trunnions, counterweights, lighting, bridge seats, and controls.

3.1.9.2 Structural Maintenance

All of the maintenance processes and practices that apply to bridge traffic wearing surfaces, bridge decks, and substructures and superstructures on bridges without movable spans apply to bridges with movable spans. However, there are additional aspects of movable spans that are unique in their maintenance needs.

3.1.9.3 Machinery and Equipment Maintenance

To make certain that the movable span is properly balanced, the mechanical parts of the bridge must be inspected. The inspection team must check and document the function of all components and accessories [8]. The team should be familiar with the mechanical aspects of movable spans and be experienced in the inspection process for movable spans. Generally, the machinery should be checked for proper lubrication, any unusual noises, possible looseness in all the shafts and bearings, proper alignment, and the manner in which parts are secured in place. The movable span should be checked first in the closed position, then also as it is moving to the open position. Any unusual motion, deviation, or abnormal sound may indicate a mechanical problem. The teeth of the open gear should be examined for scratches, indentations or ridges, and any indication that the metal is overstressed. Trial openings should be made as needed. However, no trial opening for inspection is to be made concurrently with an opening for the passage of vessels, since operator attention in that circumstance may be diverted from a vessel that is negotiating the channel. Auxiliary standby power plants should be started and checked thoroughly in addition to the primary plant.

Many agencies require measurement of gear teeth to determine the extent of wear and establish a probable rate of wearing. Span or vernier tooth calipers can be used to measure gear teeth wear. To check the wear, the original design and manufacture dimensions must be on file with the original construction and installation documents. Span calipers are usually 610 mm (2 feet) long and can provide measurement across many teeth on the gear. Span calipers can be placed in any position along the tooth and then directly measure several tooth faces. Vernier tooth calipers are used to measure tooth thickness on individual teeth along the pitch line. Vernier tooth caliper measurements of circular tooth dimensions are converted into chords.

The inspection should also cover the electrical aspects and electrical components of the operation. Normally, the mechanical and electrical operations complement each other, and the inspection of these two areas should be the result of a coordinated and preplanned effort. Inspection of the electrical system should be thorough and include items such as wiring, conduits, motors, and lights. Special care should be taken to identify worn or broken lines that may be hazardous. Other conditions that could be potentially hazardous should also be identified.

Submarine cables carrying power and control circuits should be examined in areas above the water line. The underwater lines should be inspected by divers following the occurrence of unusually high water or at any time when there is indication of possible damage.

The inspection should be extended to traffic gates, barriers, and signal systems for highway and marine traffic. Likewise, fenders and dolphins should be examined for possible damage from marine traffic. Timber sheathing, wales, and piles should be inspected for decay, damage from marine borers, and tightness of bolts and cables. The overall setup of fender systems should be checked to ensure that they are properly maintained.

Typical problems associated with movable span bridges for which a maintenance response is required include worn machinery, ropes, bearings, and gears; defective or deteriorated controls; inefficient brakes; poorly aligned parts and unbalanced conditions; broken, frozen, or cracked parts and supports; missing or loose bolts; poor lubrication; exposed wiring; and motors with too much or too little power.

3.1.9.4 Inspection and Maintenance of Specific Parts and Components

Bearings: Bearing clearance should be measured at the four opposite locations and checked with the allowable values. The pivot bearings of swing bridges are generally difficult to inspect because the span must be made inoperable and then jacked up slightly to expose the bearing device. A fiberscope can then be slipped between the bearing surfaces, and the image broadcast and recorded on a suitable video monitor. Bearing defects may be similar to deficiencies discussed for bearing devices in other sections and include accumulation of dirt and grit, scratches, and lack of proper lubrication. To correct defects, the bearing may be removed for repair and refinishing, or may be completely replaced.

The inspection should check the tolerance for a given shaft diameter or journal. It is not necessary to always remove the bearing cover to examine the surface of the shaft or bearing. However, when this becomes necessary, it should be done carefully and under fully qualified maintenance supervision, noting that the movable span will not be operating during this maintenance activity. In some cases, the shaft may exert an upward force on the bearing, making it difficult to replace the device and the bearing cover.

Ropes: Wire ropes are typically used in vertical-lift bridges. Examination should emphasize the potential danger of cracked or broken wires. This problem may arise especially at the sockets where dirt accumulates and where corrosion can occur. The ropes should also be checked for flat spots or wear. Where defects imply section loss, a professional structural engineer should determine the remaining capacity of the wire rope.

The tension force should be equally divided in the ropes. This can be verified rather simply by lifting the span about 1.5 to 1.8 meters (5 to 6 feet) off its bearing, and shaking each individual wire rope suspending the lift span. All of the ropes should vibrate at about the same frequency, which will indicate similar stress levels in all of the support ropes.

Seating: All movable bridges should be properly seated when in the closed position. This will assume that the loads are distributed as intended by design and will also prevent bouncing of the span as traffic moves across the bridge. Any vertical misalignment of adjoining movable spans will also increase impact effects and is a potential cause of serious damage.

Usual Repairs of Bascule Bridges: Repairs of main structural systems including the deck, supporting girders or trusses, substructure, and fender systems are essentially similar to stationary bridges. However, when the repairs include the moving leaf, attention must be given to the adjustment of the balance. For example, a relatively small weight change at the end of the movable span can affect the bridge balance a great deal.

Bascule bridges usually have normal-weight concrete, but there are exceptions where the counterweight has been constructed from heavyweight concrete containing heavy aggregate or steel punchings. Because adjustments of the counterweight are normally needed following initial construction, balance blocks are placed on the exterior of the counterweight or in a chamber provided in the counterweight for the purpose of holding balance weights. The blocks may consist of steel members or concrete blocks that can be added, removed, or shifted to restore the needed balance.

Professionally qualified structural engineers can compute the bascule bridge balance with analytic procedures considering the exact weight and center of gravity of each contributing component of the counterweight system. This involves assumptions with regard to possible variations in unit values; the variability of dimensions; and items such as paint, rivets, and bolts. Thus, adjustments may be necessary and need to be provided for in the structural engineer's design.

Strain gauge balancing with portable electronic testing instruments is a more precise method of ensuring the bridge is balanced when the equipment and personnel trained in its use are available.

> *Bascule Bridge Notes:* Check the center locks on double-leafed spans and note whether there is excessive deflection of the center joint or vibration on the bridge. Inspect the locks for fit and movement of the leaf (or leaves). Check for lubrication and loose bolts. Check the lock housing and its braces for noticeable movement (this can be accomplished by observing the paint adjacent to the housing for signs of wear). Check the differential vertical movement at the joint between the two leaves for adequate clearance. Check the front live-load bearings to determine whether they fit snugly. Also observe the fit of tail locks at rear arm and of support at outer end of single leaf bascule bridges.
>
> Check the bumper blocks and the attaching bolts for cracks at the concrete bases. Check the counterweight well for excessive water. Check the sump pump for mechanical functioning, the concrete for cracks, and the entire area for debris. Check the brakes, limit switches, and all stops for excessive wear and slip movement. Note if the cushion cylinder ram sticks or inserts too easily. Check the shaft or trunnion bearings for excessive wear, lateral slip, and loose bolts. Check shear locks for wear. Excessive movement should be reported and investigated further.
>
> On rolling lifts, check the segmental rim and girder and the track plate and girder for the following conditions: extension of the rim plates and track plates in either direction; wear and poor fit of toothed rim and track plates, including cracking in the corners of the slots or in the rim plates, and fractured teeth; distortion of rim and track plates including curling edges and separation of plates from their supporting girders; cracking at the fillets of the angles forming the flanges of the segmental and track girders; cracking of the concrete under the track; and looseness between walking pinion gear and top rack. Check top rack for lateral movement when bridge is in motion. On heel trunnion (Strauss) bascules, check the strut connecting the counterweight trunnion to the counterweight for fatigue cracks. On several bridges, cracking has been noted in the web and lower flanges near the gusset connection at the end nearer the counterweights. Such cracking would be most noticeable when the span is opened. The rack and pinion should be inspected for gear wear, cleanliness, and corrosion.

Usual Repairs to Lift Bridges: Vertical-lift span bridges have counterweights placed in the structural system. Balance blocks are placed inside the pocket at the time of construction and are used to balance the span. Because the counterweights may be exposed, the concrete will eventually be subjected to deterioration such as spalling, resulting in a change in the bridge balance. Unlike a bascule bridge, the balance of a vertical-lift bridge changes as the bridge operates. This is caused by the shifting weight of the wire ropes from one side of the counterweight rope sleeves to the other. A secondary counterweight or a device such as balance chains can offset the shift in the weight of the ropes.

Vertical-Lift Bridge Notes: Check span and counterweight guides for proper fit and free movements. Span guides are usually castings attached to suspended span chords, which engage a T-section attached to the tower. Counterweight guides are angles or tees attached to the tower and engaging grooved castings attached to the counterweights. These grooved castings must be inspected closely for wear in the grooves. Check cable holddowns, turnbuckles, cleats, guides, clamps, and splay castings. Check the motor mounting brackets to ensure secure mounting. Check alignment and wear of cables, drums, and sheaves. Note if cable is running properly in sheaf grooves. Recommend replacement of all frayed or worn cables. Look for any obstructions to proper movement of cables through pulleys, etc. Check spring tension, brackets, braces, and connectors of power-cable reels. Check the travel rollers and guides, brakes, limit switches, and stops. Since the machinery room is usually under the main deck, check the ceiling of the machinery room for leaks or areas that allow debris and rust to fall on the machinery. Survey lift bridges, including towers, to check both horizontal and vertical displacement. This should identify any foundation movements that have occurred.

Swing Bridges: When a swing bridge is in the closed position, wedges or similar devices are used to support the outer ends of the span. These are designed to induce a dead-load reaction at the piers when the span is closed. The wedges also serve to align the roadway of the swing span vertically with the approach roadway since the ends of the swing span tend to deflect downward under the cantilever action. These items must be checked and, if necessary, adjusted for proper swing span operation.

Swing Bridge Notes: Check the wedges and the outer bearings at the rest piers for proper adjustment. If they are not adjusted, there will be excessive vibration of one span or uplift when a load comes upon the other span. Check the live-load wedges and bearings located under the trusses or girders at the pivot pier for proper fit. Check the teeth of all gears for wear and for proper alignment. Check the meshing of gears to determine if any gears are climbing one upon the other.

On center-bearing swings, check the center pivot, the housing, the tracks, and balance wheels for fit, wear, pitting, and cracking. Check for proper and adequate lubrication.

On rim-bearing bridges, inspect the center pivots, rollers and roller shafts, and guide rings or tracks for proper fit, wear, pitting, and cracking. Check for proper and adequate lubrication.

Other Repairs: The mechanical, electrical, and hydraulic components in a movable span bridge create a need to routinely inspect and maintain numerous other units such as the following: mechanical gears, bearings, bushings, shaft journals, bearing housings, shafts, trunnions, reducers, couplings and associated accessories, brakes and clutches, locking devices, tracks, emergency drives, electrical motors and controls, and hydraulic components. If any unit of the mechanical, electrical, and hydraulic systems goes out, the bridge is likely to be inoperable for either navigable waterway traffic or vehicular roadway traffic.

3.1.9.5 Inspection Items

The following inspection items for movable span bridges are listed with related corrections to defects [7].

Cables: Counterweight cables as well as up-haul and down-haul cables on lift or bascule spans should be inspected carefully for wear, damage, corrosion, and evidence of inadequate lubrication. To properly inspect cables, old lubricant must be removed. After inspection, the cable should be lubricated again. Check for any binding at the travel rollers and guides. Check the piers for rocking when the leaf span is lifted.

Counterweights and Attachments: Check the counterweights to determine if they are sound and are properly affixed to the structure. Further check temporary supports for the counterweights that are used during bridge repair operations.

Where steel members pass through or are embedded in the concrete, check for any corrosion of the steel member and for rust stains on the concrete. Also, look for cracks and spalls in the concrete.

Check for debris, animals, and insect nests in the counterweight blocks. Where cable counterweights or balance chains are used, check the links, slides, housings, and storage areas for deterioration, adequate lubrication (where applicable), and protection. Determine if the bridge is balanced and whether extra weights are available as needed. A variation in the power demands on the motor according to the span's position is an indicator of balance problems. Paint must be periodically removed from the lift span proper. Otherwise, the counterweights will eventually become inadequate as more layers of coating are added.

Drainage: Check to determine if the counterweight is properly drained. On vertical lift bridges, be sure that the sheaves and their supports are well drained. Examine any portion of the bridge where water can collect.

Piers: Check the piers for rocking when the leaf is lifted. Rocking could be an indicator of a serious substructure problem and should be reported at once for a more detailed investigation and subsequent repair as needed.

Warning Devices: Check the operation of safety gates, traffic barriers, and warning signals. Be sure they function properly and that the warning signals give sufficient notice to permit vehicles to clear the automatic gates. Determine if the safety gates and barriers are sound and well maintained. Note any deteriorated areas at bolts and other connections. Note if these need to be replaced or repaired. Note the locations of the safety gates in relation to the warning lights, signs, and bells, and the bridge opening itself. Check the location of warning lights to determine if motorists can easily see them.

Machinery: On all movable bridges, the machinery is so important that considerable time should be devoted to its inspection. The items covered and termed as machinery include all motors, gears, tracks, shafts, linkages, overspeed controls, brakes, and any other integral part that transmits the necessary power to operate the movable portion of the bridge. A machinery specialist or a movable bridge specialist should make an inspection of the following items.

Check the alignment of all gears, locks, and other interlocking mechanisms. Check the adequacy of the lubrication of all movable parts, particularly where meshing or contact occurs between the movable parts. Check the schedule of lubrication to determine if the frequency of lubrication is sufficient. On live-load bearings, check the wedge (lock) linkages for loose knee pins and excessive play. Note the closing and releasing of the wedge locks or pin locks for proper functioning. Check all gears for cracks including the teeth, spokes, and hub. Inspect all shafts for twisting, strain, and play within bearings. Check the keyways on the shafts and gears for looseness. Check the keys for looseness also. Check the braces, bearings, and all the housings for cracks, especially at welded points. Inspect the concrete for cracks in areas where machinery bearing plates are attached. Note the tightness of bolts and the tightness of other fastening devices used. Check all brake devices for proper functioning. Check to see whether stops are used and needed.

Motors and Engines: Motors and engines should be inspected by mechanical specialists. Inspection should include checking the belt drive for wear and slippage if that is the type of drive; checking the condition of all belts and replacing any that need to be replaced; if a friction drive is used, checking for wear and uneven bearing areas; if a direct drive is used, checking all bracing and bearings for tightness; if a liquid coupling is used, checking the fluid to determine that the proper quantity is used and looking for leaks.

Control House Notes: Consult with the bridge operator to ascertain if there have been any unusual developments in the operation of the bridge. Note where the control panel is located in relation to the roadway and the waterway. Note if the bridge tender has a good line of sight to approaching boats and vehicles. Note if the structure shows any cracks. Determine if it is windproof and insulated. In some cases, only control boxes are provided, without a bridge tender. Note this situation and check the security system. If controls are separate, note description of bridge tender's house or shed, and include its condition as well as information about the control house. Note if alternate warning devices such as bullhorns, lanterns, flasher lights, or flags are available. Also note if all U.S. Coast Guard, U.S. Army Corps of Engineers, and local instructional bulletins are posted. Check for obvious hazardous operating conditions involving the safety of the operations and maintenance personnel. Check for any material accumulations that may be readily combustible.

Check controls and electrical panels on movable structures. An electrical specialist should be available for this part of the inspection. Check controls while the bridge is opening and closing. Look for excess play and sparks. Check electrical cabinets for loose wires, heaters, and bunched-up wires. Note debris or material hidden in cabinets. Inspect the electrical system of the bridge including the wiring, the conduits, the motors, and the lights. Check for worn or broken lines. Check for any existing hazardous condition. Check for rusted out or mismatched members. Determine whether the controller is outdated or if parts need replacing. Determine whether electrical interlock is working. Check whether panel doors are secured. Note whether the bridge tender has any complaints about the panel. Check the span's speed-control resistor banks for overheating.

Main and Submarine Cable Notes: On the main cables, note the condition of the power lines coming to the bridge. Where high-voltage lines extend to a transformer in the control house, check that the main lines or cables are fully insulated and out of public reach. Where transformers are on a power pole near the bridge, check the rigidity of the pole and inspect the condition of guy lines, ground line, and cable to bridge. Check the transformer in the control house, if any, for bracing, insulation loss on high-voltage cables, leaks, and cable protection. If a cable is attached to the bridge, check anchors, clips, concrete bases, and insulators or armor and check for attached growth or debris. Determine if the line or transformers should be replaced or relocated. Check the lightning arrestor device for signs of distress.

Submarine cables should be labeled for size and number of conductors contained. For each cable indicate the number of conductors used, the number of spares available, and the number of conductors that have failed. This should equal the total number of conductors in the cable. Note if the cable is protected from boats and the public, and if it is behind the fender system. Note if the cable is kinked, hooked, or exposed either above or below the water. Note if the ends are conditioned and protected from moisture. Check cables at tidal areas for excess marine or plant growth. If the cable has been spliced, note conditions of the box seal. Inspect clamps and securing clips.

Auxiliary Power Notes: Operate auxiliary power or crank and note condition and reliability. On double-leaf bascules, note whether both sides have auxiliary power systems. On hand-crank systems, determine whether standing platforms are free of grease and debris. Determine the number of persons needed in a cranking operation and whether a portable generator-powered mechanical device can replace the personnel needed to operate the bridge.

3.1.10 Maintenance Work Zone Traffic Control
3.1.10.1 Control for Work Zones

Whenever bridge maintenance work is done on or near the roadway, drivers are faced with changing and unexpected traffic conditions [1]. These changes may create hazards for drivers, workers, and pedestrians unless protective measures are taken. Standard and uniform traffic control is especially important through work sites. Unusual conditions are the norm in bridge maintenance work sites, so traffic is particularly dependent on traffic control devices to guide and direct it safely and efficiently through what would otherwise be hazardous areas. The constantly shifting and changing nature of maintenance activities requires frequent readjustments of traffic control devices to handle the new situations. Thus, it is important that the traffic controls used and the way they are placed conform to the state DOT manual on uniform traffic control devices (in most cases, the Federal Highway Administration's *Manual on Uniform Traffic Control Devices*). Traffic control procedures at work sites are intended to fulfill several purposes:

- Warn motorists and pedestrians of the hazards involved and advise them of the proper manner in which to travel through the area.
- Inform the user of changes in regulations or additional regulations that apply to traffic passing through the work site area.
- Delineate areas where traffic should not operate.

(Also see chapter 2.)

Planning Traffic Control: Traffic control plans should be developed for every project that requires workers or equipment in traveled lanes or adjacent to traveled lanes. Traffic may traverse a project by detour, by temporary roadways, or by passing through the work operation. Care in the layout of the work zone, in the use of delineation and warning devices, and in the control of the actual work is necessary to minimize the impact of these operations on the safety of both motorists and workers. The following are some steps that are useful in planning and maintaining a safe work zone:

- Provide detour alignment and surfaces that will allow traffic to pass smoothly around the work zones, when practicable.
- Use long tapers for lane drops or for squeezing traffic flows in transition areas.
- Adequately maintain all traffic control devices and use pavement marking materials that can be removed when traffic patterns change.
- Use roadway illumination and warning lights as necessary, with steady burning lights preferred to flashing lights to delineate the travel path throughout and around a work

area. The very short "on" time of flashing lights does not enable motorists to focus on the light and may not add significantly to a driver's ability to make a depth perception judgment. Flashers should be limited to marking spot hazards or occasional short lengths of straight-line delineation.
- Use flashing arrow boards, cones, delineators, reflective drums, or lightweight barricades as a means of channeling traffic.
- Use rumble strips to slow traffic down when appropriate.
- Remove signs and markings from the job site when they are no longer needed.
- Use energy dissipaters (attenuators) to protect workers and traffic when appropriate.
- Use transports to carry slow-moving equipment between work sites.
- Remove equipment completely off the roadways and shoulders at night, on weekends, and whenever equipment is not in operation.
- Hold conferences and job site discussions well off the traveled way and never in the median or on a narrow bridge.
- Protect personnel exposed to traffic hazards during the performance of their duties, including by requiring use of safety vests by all workers.
- Consider limiting working hours to off-peak traffic times.
- Park workers' private vehicles well off the shoulder.

Human Factors and Traffic Control: Proper management and guidance of traffic passing through maintenance operations on the highway is a complicated task. It requires full cooperation of drivers and workers. Driver attitudes, habits, and capabilities need to be understood and taken into consideration. Studies indicate that drivers do not quickly or easily change their habits. Thus, it is important that the traffic control fit the driver as much as is practicable. Most of the time, drivers perform at a level below their full capability. Traffic control for work zones needs to be designed for a "below average" driver who is tired, ill, or otherwise impaired. Because of their advancing age, elderly drivers have reduced capabilities including daylight glare vision difficulties, an increased level of night blindness, slower reaction times, poor hearing, reduced ability to concentrate, and increased susceptibility to medicinal reactions and interactions.

Decisionmaking: Once information is presented to a driver, the driver must process the information and take appropriate action. A typical human behavioral model includes the following processing steps: perception, recognition, decision, and reaction. The entire process may take a significant amount of time, especially considering the speed at which the vehicle is traveling. The time required for any given situation and any given driver varies greatly. More time is needed if
- the situation is unfamiliar (it may be familiar to a worker by virtue of having done this type of job many times before, but the driver may not remember ever driving through a similar type operation before);
- there are several possible choices;
- the problem posed is complex; or
- the driver is impaired in any way (tired, angry, under the influence of alcohol or other drugs, myopic, frail and elderly, etc.).

Typically, the perception-reaction time may be less than 1 second for an alert, clear-headed, physically fit driver who is familiar with the situation and faces only one of two choices, such as stopping for a flagger signal or slowing and proceeding. On the other hand, the perception-reaction time might be 5 seconds or more if the driver is in any way impaired or faces a complex work zone traffic control situation, such as trucks crossing the road, a crossover to opposite lanes, merging to reduced lanes, and a flagging or traffic signal operation.

Condition Response: Drivers tend to act as creatures of habit (i.e., through conditioned responses). Normal driving habits include (1) maintaining uniform speed for a given highway situation, especially in high-speed areas; (2) traveling in a particular lane and being guided by lane lines or construction joints for that lane; and (3) assuming the right-of-way unless otherwise instructed by signs or signals. Extra effort is required to change the habitual behavior of drivers to adapt to maintenance work zones with temporary signs, detour routes, unexplained speed limits, and equipment and people in the roadway.

Expectancy: Through their driving experiences, drivers build expectations of how the roadway environment will be laid out and how they can navigate it. Traffic control that reinforces what a driver can reasonably expect will produce a safer and more efficient work zone. Standard traffic control devices, standard messages, uniform application, etc. in work zones build on positive aspects of driver expectancy.

Accommodating the Driver: Drivers make their own decisions based on information that is available to them. They cannot be directly controlled. Persons responsible for traffic control in work zones need to use drivers' general habits and natural tendencies in a positive way.

- Provide clear concise information. Recognize that information transfer is reduced when the information presented is not enough for a quick decision; is too much to be comprehended quickly; is confusing, ambiguous, or misleading; is not in the place a driver is most likely to look; or is outdated, incorrect, or no longer applies to the work zone.
- When appropriate, use redundancy to increase the likelihood that critical information is received. Redundant messages help because the information missed in one form may be noticed in another form, the percentage of drivers who understand the information is increased, and any potential ambiguity is reduced.
- Help drivers make proper choices by identifying alternatives; minimizing the number of alternatives; separating alternatives along the roadway so that no more than two need to be considered at one time (i.e., space out decision points); giving adequate advance warning; providing only pertinent information; and allowing time (distance along the road) for an appropriate choice to be made.
- Avoid unexpected situations that are a major factor in driver confusion and poor driver performance. The best traffic control procedure is one that offers no surprises for drivers.

Use of Typical MUTCD Layouts [1, 9]: The MUTCD and most state traffic control manuals offer a variety of suggested layouts for highway work zone traffic control. All personnel involved in planning and executing traffic control for work zones need to remember that these layouts are only intended to be samples illustrating how traffic control might be laid out incorporating the principles contained in the manuals. Furthermore, these layouts are only prepared for "typical, normal" conditions. Implementation of the typical layouts needs to be adjusted for the actual work zone roadway site. Perhaps other devices not shown in a typical layout should be added for an improved message communication. Perhaps the spacing of signs or markings should be changed to improve the ability of the driver to interpret the area. Perhaps a different taper length is needed to improve the workability of the layout. Judgment is required in adapting the typical layouts to a proper design for a particular work site. The following items are factors to be considered in developing the site-specific layout in accordance with MUTCD principles:

- Drivers may not perceive or understand one or more of the devices placed in the traffic control zone. Thus, some redundancy may be required to protect motorists and workers.
- Consideration must be given to what is apt to happen if the motorist does not get the required information and does not take proper corrective action. The higher the degree of hazard, the more emphatic should be the delivery of the message and the greater should be the physical protection.
- Where possible, a recovery space should be provided in case the traffic control system fails.
- When the guidelines for traffic control cannot be met in one area, this deficiency should be compensated for by exceeding the minimum guidelines in another area.

Installation and Removal: Installation and removal of work site traffic control zones creates situations that are often far more hazardous than operation of the completed zone. Workers placing signs and channeling devices must be in the roadway at points of high conflict without the full protection of the devices being placed. Also, the placement operation creates an unexpected situation for the motorist. To minimize these hazards, it is essential that the installation proceed in an orderly fashion with an emphasis on safety.

Planning: To reduce the degree of exposure of both the public and the workers, the installation operation should be completed as quickly as possible.

Assign Responsibility: Responsibility for work site traffic control must be clearly established, especially if more than one organization is involved. One specific individual should be designated as the person in charge.

Coordination: Coordination includes providing advanced publicity about the planned work, scheduling operations, and notifying emergency services such as police and fire departments.

Assemble Equipment: All devices that are required by the traffic control plan should be inventoried and assembled before starting the work. Signs should be cleaned, barrier lights checked, and all devices inspected to ensure they will perform properly. Devices should be packaged on trucks so they will be taken out in the order in which they are needed. Additional devices may be needed during the installation process, such as extra cones and an arrow board.

Training and Instruction: All crew members should be trained in their tasks with particular emphasis on safety considerations. In addition, to ensure that all crew members know their assignments and to ensure an efficient and speedy operation, the supervisor in charge of the installation should review the installation process with the crew before going to the field. A chalkboard or large chart board for drawing out the operation is useful. If a new or different procedure from the last installation is to be used, or if there are new workers in the crew, this final review before starting is especially important.

Installation Procedures: The traffic control truck should park in a safe location to allow crews to disembark and unload the devices. The truck carrying the devices may also serve as an advance warning system by using its flashing or rotating beacon lights during the time that the first warning devices are placed. To protect the crew, the device truck should be located upstream of the crew. This is a bit clumsy when the signs or devices are unloaded from the rear of the truck, but it provides essential protection for the crew.

Devices should be installed in the direction in which traffic flows—that is, downstream. The first device placed is the first advance warning sign. The installation should then proceed in sequence through the work zone.

On high-speed roads, a backup, shadow, or protection vehicle should be used. This vehicle should first be positioned on the shoulder some 30 meters (100 feet) or more behind the device truck when the first signs are placed. The shadow vehicle uses a truck-mounted attenuator and special lights or an arrow board to warn traffic. When the crew needs to encroach upon the roadway, the shadow vehicle moves out into the traveled lane with the crew.

When closing a lane, tapers are laid out in a straight line starting at the shoulder. Each channeling device is then placed in sequence moving downstream. When placed by hand, the devices should be moved laterally out from the shoulder with the worker looking toward traffic as he or she moves out into the lane to place the device. When channeling tapers are installed, each device placed is positioned 300 mm (1 foot) further into the lane being taken. For recurring installations, it is helpful to mark the locations of channeling devices on the pavement using small painted spots.

Whenever possible, traffic control devices should be removed in a reverse sequence to that used for installation. This requires moving backwards or upstream through the traffic control zone.

Inspection: Concern for uniformity and consistency using the requirements of the MUTCD cannot be overemphasized. Things to be alert for in inspection of control devices include improper barricades, poorly trained flaggers, dirty barricades or signs, improper spacing of signs, insufficient advance warning signs, poor protection for workers, nonconforming signs, insufficient reflectivity, and inadequate lighting. Persons in charge of a bridge maintenance work zone should drive through the zone in both directions during day and night hours to check that all hazards to the traveling public, pedestrians, and work crews on the job have been corrected or are properly marked according to good traffic control practice. The effectiveness of the initially installed traffic control scheme should be immediately observed and evaluated to determine if any modifications are needed.

Flagging: Flagging of traffic is provided at maintenance work sites to help protect the work crew and to enhance traffic safety, either by stopping traffic as needed or by maintaining continuous traffic flow past the work activity at reduced speeds. Flagging is a very hazardous activity. Flaggers assume a vulnerable exposed position with great responsibility. Too often a person assigned to flagging is inexperienced, poorly trained, and not motivated to do a good job.

Flaggers should be trained in the importance of their job and in the proper way to conduct a flagging operation. Flaggers need to know the correct ways to stop traffic, slow down traffic, and keep traffic moving. They should know where to position themselves in the roadway and what actions they should be prepared to take in different situations. If there is more than one flagger, they should know how to coordinate their actions. Flaggers should have instruction on how to deal with the public courteously, explain delays, provide assistance if necessary, and deal with situations where drivers ignore their instructions.

Maintaining Traffic Control Zones: Once traffic control zones are established, they should be observed frequently to ensure proper operation. Every inspection or observation should be logged into an inspection or supervision record for future reference in case of an accident. A periodic drive-through during both day and night should be conducted. Any deficiencies should be noted, reported, and corrected. Traffic behavior through the work zone should be observed to ensure that drivers are interpreting the traffic controls as intended. Damaged devices, skid marks, frequent brake applications, and erratic maneuvers by drivers may indicate a potentially hazardous situation that needs to be corrected.

When accidents occur in the work zone, immediate action should be taken to prevent a secondary accident and provide necessary first aid and emergency services. Once immediate emergency needs are satisfied, the circumstances of the accident and the condition of the work zone need to be carefully documented. If a video camera or 35-mm camera (preferably both) is available, a photographic record should be made of the location, position, and condition of all traffic control devices, roadway conditions, and the position of all equipment. There should be a safety engineering evaluation of the work zone to see if modifications are needed to prevent a recurrence of a similar event.

3.1.10.2 Temporary Structures [1]

The sudden failure or major damage of a bridge can require an immediate temporary structure to carry the traffic. A location where a local detour would require unreasonable out-of-the-way travel to bypass the damaged bridge is an example of a situation that might require a temporary bridge. Temporary bridges are more likely to be required for smaller, low-capacity spans, such as spans of 12 to 15 meters (40 to 50 feet). Short-span bridges are called live-load structures since the weight of large vehicle loads far exceeds that of the structure itself. Large and long-span bridges are referred to as dead-load structures because their massive structural weight is large compared to vehicular loads on the bridge. Temporary bridges are not usually feasible as bypass structures for a dead-load structure, except in military operations.

Temporary Bridge Strategies: The most economical or expeditious strategy should be selected from the available options. The following are some alternatives that work in the appropriate situations:
- Construct a runaround using a culvert or culverts to handle dry-weather stream flow. If the bridge repair can be completed in a dry-weather period, this type of temporary structure is a viable option.
- Construct a one-way wood or salvaged rolled steel beam superstructure on pile bents with automatic traffic signals to control traffic. Sufficient time needs to be allowed for traffic from one direction to be stopped, and all vehicles that might be already moving across the bridge to clear before traffic is released to cross the bridge from the other direction. Thus, this method is typically only applicable to relatively low-volume roads.
- Erect a prefabricated bridge such as an Acrow or Bailey. This may be the most expeditious alternative.

Advance Planning: If a highway agency has a large number of bridges in its jurisdiction, some advance preparation is possible. For example, an inventory can be maintained of used steel I-beams that can be used in building a temporary bridge, or it may be feasible to stock an Acrow or Bailey bridge for emergency use.

Traffic Control: It is always good public policy to keep the traveling public well-informed. This is particularly true in the case of a sudden change in a traffic pattern because of a bridge failure. Public service announcements using the radio and press are a good way to inform the public of detours. Signs should be posted at intersections well in advance of any temporary bridge to allow vehicles with heavy loads ample opportunity to either turn around or select an alternate route.

Prefabricated Bridges: Two well-known manufacturers of steel truss bridges that can be quickly assembled in the field are (1) Bailey Bridges, Inc., of San Luis Obispo, California, and (2) Acrow Corporation of America, of Carlstadt, New Jersey. The prefabricated bridges of the two companies are quite similar in principle, but the components and parts are not interchangeable. The similarities of the two bridges make it easy for a crew trained to erect one system to adjust quickly and be able to erect the other system.

Key features of both systems include standard preassembled steel components that are easily erected at the site using light equipment and unskilled labor. These standard components are stocked by the manufacturers and can be easily transported to the site. These bridges can also be found all over the United States and other parts of the world making availability from other sources quite likely.

Temporary bridges come in a range of widths and can accommodate spans up to 100 meters (300 feet). They are extremely versatile because they can be disassembled and used at other sites. This versatility, however, often results in the use of a stronger temporary bridge than a particular site requires because the width needed is not available in inventory. One county has 10 of these prefabricated bridges, which are used on a semipermanent basis. These bridges are durable. Some models can be launched into place from one end.

3.1.11 Developing Issues
3.1.11.1 Nonmetallic Repair of Steel Bridge Components

Traditionally, steel beams and steel bridge components have been repaired with steel components, welded, bolted, and riveted. Recently, several agencies have experimented with composite materials that can be chemically welded to damaged steel components. In particular, the Florida DOT has used this technique to make repairs to both steel and concrete bridge damage in the marine environment. As materials science and engineering research and applications continue to develop in areas appropriate to bridge maintenance, it is important that maintenance engineers monitor progress in this technology.

3.1.11.2 Nonmetallic Bridges

As carbon filament composite materials advance from aerospace structural applications to automotive structural applications, experiments are underway to evaluate the potential to design and construct small span bridges completely with nonmetallic composite materials. If these materials prove to be feasible, they will have a significant impact on maintenance of small span bridges. Nonmetallic bridges may potentially be completely resistant to deck corrosion damage from salt and de-icing chemicals. They may also potentially be replaced with prefabricated segments and sections during the repair process.

3.1.12 REFERENCES

1. Federal Highway Administration. *Bridge maintenance training manual.* Report FHWA-HI-94-034. Washington, D.C.: Federal Highway Administration, 1994.

2. Techniques enable flatcar bridges to attain higher load ratings. *Roads and Bridges.* Des Plaines, IL, August 1995, pp. 42–43.

3. How Mississippi uses underwater inspection. *Better Roads.* Park Ridge, IL, May 1990, pp. 22–23.

4. Underwater inspection team saves funds. *Roads and Bridges.* Des Plaines, IL, November 1990, p. 52.

5. DeLozier, F. James. Bridge abutments: An on-site point. *Better Roads.* Park Ridge, IL, January 1997, pp. 22–23.

6. Creswell, J. S., D. F. Dunlap, and J. A. Green. *Studded tires and highway safety—Feasibility of determining indirect benefits.* Report 176. Washington, D.C.: National Cooperative Highway Research Program, Transportation Research Board, 1977.

7. White, Kenneth R., John Minor, Kenneth N. Derucher, and Conrad P. Heins, Jr. *Bridge maintenance inspection and evaluation.* New York: Marcel Dekker, 1981.

8. Xanthakos, Petros P. *Bridge strengthening and rehabilitation.* Upper Saddle River, NY: Prentice Hall PTR, 1996.

9. Federal Highway Administration. *Manual on uniform traffic control devices, Part VI*, 3rd rev. Washington, D.C.: Federal Highway Administration, September 3, 1993.

10. Manning, David G. *Waterproofing membranes for concrete bridge decks.* Synthesis of Highway Practice 220. Washington, D.C.: National Cooperative Highway Research Program, Transportation Research Board, 1995.

11. Sprinkel, Michael M., and Mary DeMars. *Gravity-fill polymer crack sealers.* Transportation Research Record 1490. Washington, D.C.: Transportation Research Board, 1995, pp. 43–53.

12. Weyers, Richard E., Jerzy Zemajtis, and Rick O. Drumm. *Service lives of concrete sealers.* Transportation Research Record 1490. Washington, D.C.: Transportation Research Board, 1995, pp. 54–59.

13. Doody, Michael E., and Rick Morgan. *Polymer-concrete bridge deck overlays.* Transportation Research Record 1442. Washington, D.C.: Transportation Research Board, 1994.

14. Sprinkel, Michael M. *Polymer concrete bridge overlays.* Transportation Research Record 1392. Washington, D.C.: Transportation Research Board, 1993, pp. 107–116.

15. Baun, Mark Douglas. *Steel fiber-reinforced concrete bridge deck overlays: Experimental use by Ohio Department of Transportation.* Transportation Research Record 1392. Washington, D.C.: Transportation Research Board, 1993, pp. 73–78.

16. Markow, Michael J., Samer M. Madanat, and Dmitry I. Gurenich. *Optimal rehabilitation times for concrete bridge decks.* Transportation Research Record 1392. Washington, D.C.: Transportation Research Board, 1993, pp. 79–89.

17. Van Lund, John A. *High-load elastomeric bridge bearings.* Transportation Research Record 1541. Washington, D.C.: Transportation Research Board, 1996, pp. 8–17.

3.2 BRIDGE MANAGEMENT AND BRIDGE MAINTENANCE MANAGEMENT

3.2.1 Introduction

Surveys of U.S. bridges continue to indicate that more than 200,000 bridges have deficient capacity or lack functional performance on the roadway they serve; more than 100,000 are posted for restricted vehicle weight less than the legal load limit for the roadway they serve. More than 5,000 are completely closed. Every year, about 150 to 200 bridge spans either partially or completely collapse. The cost of repairing all bridge deficiencies has been estimated to be more than $90 million [1].

The source of many bridge deficiencies can be found in inadequate maintenance. Damage resulting from salt-contaminated debris buildup, plugged scuppers, leaky joints, and failed paint systems accelerates deterioration and compounds repair requirements. If left unrepaired, the damage ultimately imposes a severe limitation on the structure's operational capabilities. The longer maintenance is delayed, the worse the bridge situation will become.

Structures not classified as deficient must be included in a systematic maintenance program. This practice precludes the possibility of premature replacement or rehabilitation. A bridge maintenance management system is an integral and valuable tool in developing and managing the necessary databases to effectively maintain a bridge [5, 6].

3.2.1.1 Bridge Maintenance Management Process

A systematic approach to bridge maintenance will provide a planned procedure for performing routine major or minor maintenance on all bridges. It will also provide priorities on bridge replacement that should optimize overall life-cycle costs of bridges as a major component of the highway network. The process should organize bridge maintenance into units including inspecting, planning, budgeting, scheduling, performing, reporting, and evaluating, all coordinated by a management information system.

Planning: Planning for bridge maintenance management involves the selection of objectives and the determination of the policies, programs, and procedures to be used to achieve those objectives. Of all management functions, planning is the one that has the most productive effect on the use of available labor, equipment, and materials. Inadequate planning is often the root cause of many of the difficulties in maintaining bridges that generate criticism of the status of bridge maintenance.

Planning should consider the entire bridge system and bridge maintenance needs over an extended period of time. Planning must also account for everything that affects or can affect the bridge system. Planning must begin by analyzing the bridge database consisting of the inventory, history, plans, and maintenance requirements to define priorities, needs, and quantities. The bridge maintenance work plan is basic to the system. Once the work plan is developed and budgeted, it is the base from which specific work orders are developed and scheduled.

Budgeting: Budgeting is the process by which the funds and resources to implement the plans are obtained. Government agencies are accustomed to reviewing line item budgets that break down the budget requests into defined classes such as "pay of personnel," "equipment purchases," etc. A performance-based budget that relates work to be accomplished with the required resources of labor, equipment, and materials and their associated costs is far superior because alternative budgets can be associated with needs.

Work-estimating procedures that are fairly standard are used to develop a budget. The maintenance plans evolved in the planning process indicate the amount of work required. Using the work estimates, costs are compiled and the budget developed. The plan may need to be revised to match the available budget. One key to effective budgeting is to collect and manage cost data within the bridge management system, and for this purpose, NCHRP Synthesis of Highway Practices 227 provides a comprehensive assessment of available practices and procedures [18].

Scheduling: Scheduling is the process of laying out future work. A good scheduling process develops work schedules on at least three levels: organizational, supervisory, and foreman (or working). Scheduling is a tool to help achieve project objectives without exceeding the budget. Scheduling accounts for labor, equipment, and materials. Schedules must be more detailed at the working level than at higher administrative levels. The workers' schedules are the final approved and budgeted work plans. The foreman's schedules are very specific as to location, date, time, and assigned crew. The work schedule reflects seasonal requirements. Since emergency work is always a possibility for public works agencies, the scheduling process also requires flexibility so emergency conditions can be met without undue strain on the organization.

Performing Maintenance: Performing is the process that is concerned with actual completion of the work in the field. The work should be done in a manner that conforms with prescribed quality and quantity standards. Materials, equipment, and labor are used as indicated in performance standards, or as nearly so as is practicable. Variations occur, but if these are consistently larger or smaller than the standards, either the standards are in error or procedures other than those prescribed are being followed.

The schedule is generally translated into performance through a work order system. Two types of work orders are usually applied: one for in-house forces and one for contract activity. The tasks specified in the work order should be planned in detail by the crews who will do the work. This may involve decisions about the specific work methods and the provision of resources as well as the actual performance of the work.

Reporting: Reporting represents the primary means of communication between those performing the work and those managing it. Work completed and the resources used are reported as they occur. This informs management that planned and scheduled maintenance goals have been met and provides data to account for funds budgeted. Obviously, standard reporting forms and methods that are easy to prepare and complete with the correct information generate the best data. Work completed must be entered into the inventory, database, and historical record to provide an updated record of bridges in the system.

Evaluating: Evaluating is the means by which the quantity and quality of work is measured and is the basis upon which management can exercise control actions. Provisions should be made so that those most involved with the actual work—the foremen, for example—regularly receive information so they can judge their own performance. Reports are provided to each administrative level appropriate to the needs of that level. Evaluations are based on comparisons between like crews or personnel performing like tasks with due account taken for extenuating factors such as individual work site specifics (i.e., compare apples to apples and oranges to oranges). Comparisons are also made against standards so the standards may be adjusted to make expectations reasonable and so planning and budgeting efforts are more accurate. Evaluation is also the system element that measures the effectiveness of all other elements and provides the means by which bridge maintenance and work performance can be improved. A bridge maintenance management system can supply the manager with the means to more effectively and efficiently accomplish the maintenance program. At the same time, the system will increase control of operations and provide credibility to the program since the manager will have information with which to respond to requests and to justify budgets.

3.2.1.2 Federal Bridge Management System Requirements

General minimum standards have been developed to guide the development and implementation of bridge management systems. These requirements are intended to ensure that agencies can share data, concepts, successes, and failures as well as help each other improve in the continuous quality improvement sense.

- Data must characterize the severity and extent of bridge element deterioration.
- Data must include estimating costs of bridge maintenance actions.
- Statistics must be available to support the estimates of user costs.
- A history of conditions and actions taken on each bridge, excluding minor or incidental maintenance, must be available.
- A computer model for network-level analysis and optimization must be resident that includes procedures to predict deterioration; identify feasible actions to improve conditions; estimate the cost of actions; determinate the least costly maintenance, repair, and rehabilitation strategy; perform optimization across the network; generate reports as needed for planning; monitor the status of maintenance actions; and update the database.

3.2.1.3 PONTIS

Many state DOTs have adopted a particular computer-based bridge management system called PONTIS in their efforts to satisfy federal bridge management requirements. PONTIS supports a series of bridge management activities involving information gathering, information interpretation, prediction of bridge condition, cost accounting, decisionmaking, budgeting, and planning. It consists of a set of interconnected computer modules that address each of these functions systematically. PONTIS combines modules that predict deterioration, provide costs, and compare feasible actions. As inspections are reported and repairs are made, PONTIS is updated and refined with additional cost, deterioration status, and feasible action data. If precise historical data are not available, the operation can be started with engineering judgment values and national bridge inventory system data to generate initial predictive models that can be adjusted and refined as precise data on the bridge network become available.

PONTIS, like all computer-based bridge management systems, relies on significant mathematical assumptions to create predictive models [7, 8]. While it is not necessary for maintenance engineers to be experts in the mathematical logic that underpins these models, they do need to understand the strengths and weaknesses of the estimates and projections produced by the bridge management system.

3.2.1.4 System-Level vs. Project-Level Management Decisions

PONTIS was designed to provide assistance in making system-level management decisions for groups of bridges in an area. The system helps a manager determine and support funding for maintenance needs to keep bridges in good condition. Bridge managers need a tool that demonstrates to those that make final decisions on funding programs that it costs less to maintain bridges in good condition than to defer maintenance and let bridges deteriorate. Other approaches have been developed for system- or network-level optimization of bridge management decisions, but many of them require a more sophisticated methodology than PONTIS [9–12]. At the system or network level of a state agency with a large number of bridges, it is not normally necessary to have a detailed methodology to prioritize needs.

Simplified approaches to bridge management optimization have been developed that offer promise for agencies with networks having small numbers of bridges [13, 14].

PONTIS was not designed to provide information for decisions needed to manage a specific bridge or for "project-level" decisions. Some management systems can provide project-level support. In these systems, bridge maintenance activities such as cleaning, patching, and joint sealing have been identified and performance standards have been developed. The performance standards include the resources (personnel, equipment, and materials) needed to perform a specified maintenance activity. The activities needed and the urgency required must be identified for an individual bridge. It is important for maintenance managers to keep in mind that any computer-based bridge maintenance management system that will generate project-level decision guidance requires a database with a greater level of detail on all of the bridges in the bridge network than does a system like PONTIS.

The Pennsylvania Department of Transportation (PennDOT) maintenance management system (called MORRIS) includes performance standards for 76 maintenance activities. The system also tracks personnel availability, equipment, and materials available to each bridge crew. PennDOT bridge inspectors identify maintenance needs for each bridge and give each need a priority. After the maintenance needs are input, the system develops bridge maintenance work orders for performing maintenance on specific bridges.

Since no two bridges are exactly alike, it is helpful to have the maintenance worker visit the bridge before the repair is planned to evaluate the requirements. Management systems are tools to assist managers. As the models and logic upon which they are based are refined, the results will more accurately reflect a specific bridge. However, there will always be a need for engineering judgment in the process.

Bridge management systems are not just applicable to large agencies. Local agencies can benefit from simplified computer-based bridge management systems to assist in setting priorities among their wide range of bridge needs [2].

Good planning at the system level should facilitate good planning at the project level. One agency determined that working on weekends was the most efficient way to perform major deck replacements on a highway route that was a critical regional commuter link. Thirty-three spans were completed in 22 weekends with minimal interference to necessary traffic flows [3].

3.2.1.5 System Interfaces

In some agencies, the bridge maintenance management system is a part of the maintenance management system, and in others bridge maintenance management is a part of the bridge management system. Regardless of how the total management information system is organized, bridge maintenance management systems must interface with all other management information systems to avoid repetitive collection of data and duplication of data processing that provides input to the management information systems. Bridge maintenance management needs to interface with the maintenance management system; safety management system; pavement management system; bridge management system (if this system is not directly integrated with the bridge maintenance management system); and management system containing the fiscal and inventory control data. While all agencies are working toward complete integration of these database management systems, much of the integration has to be achieved by human interaction rather than by automated computer communication links.

3.2.2 Planning and Scheduling Bridge Maintenance

The key to successful execution of any bridge maintenance project is proper planning and scheduling of the work. Increased needs combined with reduced funds means that maximum advantage must be taken of every available hour and resource. This can only be done when a small part of the available hours are spent planning and scheduling the work. Planning and scheduling must be done prior to beginning the project and continue during execution of it.

If a bridge maintenance management system is available, it may be used to aid in the planning and scheduling process. Getting the work done can be divided into two areas: job layout and work performance. The first involves planning how the job is to be accomplished. The second involves coordination and efficiency of the work procedures.

3.2.2.1 Long-Range Planning and Scheduling

The planning and scheduling of bridge maintenance is performed on several levels. The worker may not be directly involved in the long-range planning. The general steps in the process are as follows:

(1) Needs are identified for all the bridges that are included in the planning region. These needs include procedures performed on a regular basis and procedures performed on an as-needed basis. A ballpark cost estimate is developed and a budget is prepared.
(2) After the budget is approved, a more detailed estimate is developed to determine the personnel and equipment requirements to accomplish the work. This resource estimate is compared to the personnel and equipment available. Special needs for certain skills or equipment that may not be available within the agency are also evaluated.
(3) If either personnel or equipment are lacking (including personnel with special skills needed), a decision is made regarding whether to contract out part of the work or to add to the in-house resources.
(4) Assignments are made for each bridge maintenance crew.

After crew assignments are made, schedules can be developed and work orders can be generated. In most agencies, the worker is responsible for and involved in developing the crew schedule. The scheduling process involves determining the urgency of the work; when in the year the work can be performed; when necessary approvals or permits can be obtained; and when the necessary support, equipment, and materials will be available. Realistic scheduling improves efficiency. Workers should be challenged, but if too little time is allotted to a job, short cuts and omissions will be encouraged or the schedule will not be taken seriously.

3.2.2.2 Work Orders

Work orders are prepared in advance to provide details of each job. The work order contains a description of the particular type of work to be performed, the exact location of the work, names of all workers assigned to the project, and the list of equipment and materials available or needed to do the work. The work order may be prepared by the bridge management system based on performance standards and condition data from the inspection, or it may be prepared manually by someone in the maintenance division after the assignment is made and the job layout is completed.

3.2.2.3 Job Layout

The worker (or a bridge maintenance engineer) usually performs the job layout after the job is assigned to the crew. This involves visiting the bridge to determine detailed work requirements. Factors that must be considered in preparation for each bridge maintenance project include the following:

- Traffic, employee safety, and environmental conditions.
- How the work will be performed.
- Job assignments for the crew.
- Equipment and materials required.

To prevent unnecessary delays, it is important to preplan every project. General guidelines can be created for the procedures that need to be followed for each factor of the bridge maintenance project. Each agency will have its own detailed process.

Traffic Control: Any project that will place workers or equipment on or adjacent to the roadway requires traffic control procedures that conform to uniform agency practices. This includes the correct use and placement of signs and flaggers. Adequate traffic control is often the most important safety factor of the crew's protection. Improperly planned traffic control can place the agency in a vulnerable position with respect to potentially costly liability judgments.

Environmental Considerations: Environmental factors are becoming a major source of delays in conducting bridge maintenance work. Typical concerns are air, water, or soil pollution and disposal of toxic materials, such as salvage material containing lead-based paint or creosote.

Employee Safety: The high cost of employee injuries and OSHA penalties have become a significant management concern of all transportation agencies. Some agencies require that fall protection schemes or confined space occupancy plans be developed and submitted for approval prior to beginning work that places workers in conditions associated with potential falls or confined spaces. Likewise, some agencies require higher level management approval for any bridge maintenance activity in which workers will be exposed to toxic materials or excavation with any potential for embankment collapse. Each agency needs to integrate worker and management safety training with any approval process to determine a reasonable system of checks and balances that properly considers worker safety in the planning process yet allows work to proceed in an orderly fashion.

Work Procedures: It is necessary to determine exactly what work must be performed. The worker should visit the site and make this determination. While at the site, the worker should determine the exact location of the work and the total units of work needed. Storage of equipment and materials at the site is also a consideration.

If performance standards are available, they can be compared to the work to be done to determine whether the standards are applicable or need to be modified. If the information in the performance standards is applicable, it can be used to help determine the resource requirements of the job with modifications as necessary. If performance standards are not available, a written description of the work to be performed should be prepared. Usually, a simple outline listing the items of work to be done and the order in which they will be accomplished is sufficient.

Job Assignments for Personnel: Performance standards aid in determining the number of workers required and either the classification of personnel or the skills that will be needed. If performance standards are not available, required skills for each item of work in the outline of work previously prepared should be noted. The knowledge of the actual work that is to be done and the conditions at the work site should temper the requirements. Variations may be required because of traffic conditions, unusual complexity of the job, or weather factors. These assignments should be compared with the capabilities of the crew, then the work should be assigned to the appropriate individuals in the crew.

Whenever possible and appropriate, individuals should be assigned specific tasks in accordance with their capabilities, the requirements of the job, and the available training opportunities. Specific precautions or instructions should be given to the employees at the time of assignment. If required skills are not available in the crew (e.g., a certified welder), arrangements should be made with an upper-level manager to obtain the necessary assistance.

Equipment Requirements: Performance standards aid in determining the type and amount of equipment required to properly perform the work. Modifications may be necessary if unusual work requirements are determined when work procedures are initiated, if personnel qualifications do not match standards, or if the equipment available is different from the equipment identified in the standard. If performance standards are not available, units of equipment required for items of work in any work procedure outline should be noted. Backup units or alternative plans may be necessary so equipment failures will not unduly delay completion of the work. This is especially important for critical tasks in the maintenance process or when reopening the bridge to traffic is critical.

Tools that might be required during the maintenance operation must also be considered and compared to the complement of tools readily available to the crews. If additional items are required, arrangements should be made to have them available and ready for use. Nothing can be more time consuming, inefficient, and destructive to morale than to have a crew inactive while one member drives back to a staging area for a vital tool. As anyone in the public sector knows all too well, an idle crew on the job site can also be damaging to an organization's public image. Every crew is obviously expected to carry a complement of standard tools.

Material Requirements: The amount of materials needed for the job can also be estimated from the performance standards, or estimated from past experience with similar jobs. If any materials requiring a special order are needed, the necessary paperwork should be prepared. A reserve of the materials should also be taken to the job site in case the estimated amounts are understated. Other materials that might be needed, such as wood for forming or aggregate for backfilling at approaches, should also be included. As is the case for other resources needed to perform the work, delay caused by a lack of proper preparation to secure materials is inexcusable.

3.2.2.4 Short-Term Scheduling Procedures

Once the job layout has been performed, the work can be programmed in the short-term schedule. Schedules are "roughed in" several months ahead of time, and as the time to do the work gets closer, the job layout is performed and scheduling becomes more precise.

When scheduling work for the month, it is important that maintenance activities be coordinated as much as possible. This includes such considerations as repairing concrete on several bridges in the same area at the same time, cleaning and painting several adjacent bridges, or ordering material for several jobs at the same time. As scheduling moves to a weekly level, it will become more detailed and should include the following items:

- Check to see if any work scheduled for the present week will be carried over into the new workweek because of emergencies, bad weather, or delays.
- Review the work scheduling to see what projects are scheduled to receive the highest priority.
- Review the monthly schedule to see what projects are scheduled and if adjustments are needed.
- Determine the employee days, equipment, and material required for the week.
- If manpower is still available, alternative or fill-in projects should be identified.
- Complete the schedule by assigning workers and equipment to the specific projects.

Obviously, inclement weather may disrupt the work schedule. It is important that other work be planned that can be performed in bad weather so that time will not be wasted. In many states, repairs in the winter months are generally limited to emergency patching, bridge cleaning, some joint sealing, and other emergency work. Personnel training and equipment repair are well suited to times when weather interferes with the actual conduct of maintenance. Upper-level management should examine the possibility of using bridge maintenance personnel in other areas of activity not as adversely affected by winter weather conditions.

3.2.2.5 Graphical Methods of Scheduling

One of the graphical scheduling methods commonly used in the construction industry can be used to plan the bridge maintenance work schedule. Two common methods that are easy to learn, understand, and use are the critical path method (CPM) and bar charts. These methods can be effective in developing and communicating the schedule process and also in using the schedule process to control and manage bridge maintenance. However, it is important that all personnel—from the crew supervisor to the high levels of maintenance management—have a common rudimentary understanding of how the CPM diagram or the bar chart is developed and what it tells a person. The potential positive effect of either method is limited if only the person developing the charts understands them.

3.2.3 Performing the Work

3.2.3.1 Job Execution

The daily routine of any job consists of preparing for the job at the yard and moving to the job site, performing the work at the job site including occupying and leaving the site, and cleaning up after the job at the yard. The daily routine can contribute to coordination and efficiency of performance of the work at the job site if the routine is properly and carefully planned.

Yard Preparation: Before leaving for the job site, the crew is first assembled to determine if changes must be made in job assignments because of absences or other reasons. Following the assignments, the crew should be informed of the tasks each is to perform, including those related to safety. This allows each crew member to know what his or her assignment is and who will act as support to whom. The crew then gathers the required equipment and material needed. Once the necessary resources (personnel, equipment, and materials) are assembled, the crew is informed of the staging location and proceeds to the work site.

If one job takes several days, the process is much simplified after the first day. Depending upon local rules, regulations, and agreements, it may be possible and even economical for the crew or some members of the crew to report directly to the work site. In any event, time getting ready is not wasted.

At the Site: When the crew arrives at the work site, the crew leader should review the safety procedures to be followed on the job and ensure that safety-related assignments are fully understood. Proper traffic control is then established by placing signs, cones, or other devices. Flaggers are stationed, if necessary. Once proper traffic control has been established, the crew can proceed with the work. Traffic control items should be periodically checked, and the site should be kept as clean and uncluttered as possible while the work is underway.

Leaving tools and materials lying around the work site is hazardous and unprofessional. When a job is complete, tools and excess materials should be properly stowed before leaving the site. Once the site is clean, clear, and again suitable for traffic flow, the traffic control devices can be removed. Opening the site to traffic requires that signs and traffic control devices be removed.

Start at the work site and proceed away from it upstream in the direction of oncoming traffic with the most distant sign from the work site removed last.

Yard Cleanup: Certain tasks must be performed after the crew returns to the yard. These tasks include storing tools and materials in their proper locations, preparing equipment for the next day, and reporting the work accomplished. To save time the next work day, all equipment should be serviced, and fuel and oil added as needed. Any problems encountered with the equipment during the day should be reported. The final and perhaps most important task for the crew lead is to report the work accomplished during the day and ensure that work for the following day is planned with the resources available to perform it.

3.2.3.2 Personnel Resources

To ensure that the bridge maintenance job can be performed properly, the crews should be made up of personnel who have been trained to perform the specialty tasks that are required. These personnel may include the following:
- Carpenters to build formwork or to do necessary woodworking.
- Welders for steelwork.
- Concrete and masonry workers.
- Special equipment operators (e.g., crane operators).

Specialists such as these will be able to perform most of the skilled labor tasks in bridge maintenance. The remainder of the crew can be composed of general laborers who can assist the specialists and who can perform other required tasks. In addition, the crew should have the necessary equipment and stockpiled materials to ensure that the scheduled work can be performed.

3.2.3.3 Equipment

Each bridge maintenance crew should have as full a complement of tools and equipment as possible so no time is lost attempting to obtain items from other sources. The scheduling process will often reveal if necessary items are continually lacking and causing delays to bridge maintenance projects. These delays are often far more expensive than the cost of acquiring proper equipment and tools. Some suggested items in equipping a bridge maintenance crew are as follows:
- Flatbed truck with a winch and an A-frame (or other lifting equipment).
- Pickup trucks, as needed.
- A variety of air-powered tools.
- Air compressor.
- Small concrete mixer or mortar mixer.
- Oxygen and acetylene welding and cutting equipment.
- Heavy-duty hydraulic jacks—180 kn to 450 kn (20 ton to 50 ton) capacity.
- Portable arc welding unit with electric service outlets.
- Heavy-duty electric drill fitted with an electromagnet.
- Small, portable high-pressure water pump.
- Sandblasting equipment.
- Hand tools for steel, carpentry, concrete, and mechanical work.
- Staging (scaffolding).
- Spray-paint outfit.
- Tow cable and chains.
- Radio equipment.
- Chain saw.
- Heavy-duty chain hoist.
- Miscellaneous survey equipment (tape, level and rod, theodolite or total station, etc.).

3.2.3.4 Materials

Each bridge maintenance crew should have a small supply of materials, especially for emergency repairs. Many materials can be accumulated from salvaged materials or materials leftover from new bridges.

- Timbers for blocking and cribbing—usually salvaged material.
- Assorted bridge planks.
- Steel decking.
- Assorted I-beams, angles, channels, and plates.
- Steel reinforcing bars.
- Sheet piling.
- Timber and steel piling.
- Cement, mortar, masonry sand, and aggregates.
- Epoxy.
- One gallon of penta or creosote and brushes.
- Paint, primer and finish, paint brushes.
- Nails, spikes, bolts, nuts, washers, drift pins, and lag screws.

3.2.4 Reporting Bridge Maintenance Accomplishments
3.2.4.1 Why Report?

New personnel often question the importance and urgency of reporting work accomplished. After all, the job got done and that is what counts. Maintenance engineers and managers need to communicate the importance of reporting as a means of building an information base to help do things better and more efficiently. The following list includes some of the results of reporting that might be of interest to maintenance personnel:

- Develops a historical record of maintenance and repair on each bridge.
- Maintains a record of regular periodic and special expenditures as a basis for developing and justifying future budgets.
- Maintains a current record to establish cost-to-performance relationships.
- Provides a source of information to enable maintenance engineers and managers to develop maintenance trends, and to establish the need for additional cost controls or work item controls.
- Develops a source of information for public relations, for accomplishment reports, and for the defense of tort liability claims.
- Provides a record of cost to compare with budget cost estimates.

3.2.4.2 Report Requirements

Each state has specific reporting requirements. In general, maintenance managers need much common information. Five general classifications are used to describe the types of information any reporting system should include:

Who: Indicate who performed the work by specifying the crew or individual in charge of the work. This permits the work to be charged to the proper department and permits future verifications and follow-up in case of discrepancies or claims. The actual coding of the "who" identifier will vary with state requirements.

What: Report the activity number that has been assigned to the specific type of work performed. The amount of work performed should also be recorded. The performance standards for each of the activities indicate how the amount of work is to be measured. The report of "what" and "how much" is used to evaluate crew performance, the suitability of standards, and project progress and is also used for budget comparisons.

When: Report the date (or dates) on which the work was performed. This information is helpful in determining when work should be scheduled in future years and is required for scheduling periodic maintenance, particularly preventive maintenance.

Where: Report the location of the bridge by reference to a route or a milepost (or both) and to the bridge number. This information correlates the work to the repair history of the bridge. In addition, this information allows sorting by road types and locations.

How: Report the resources that were used and the process used to get the job done. The hours of labor, types of equipment, and type and amount of materials used are included. This permits computing the cost of performing the work and provides resource utilization data. This information also permits managers to determine the monthly resource needs for use in future scheduling.

3.2.4.3 Reporting Procedures

The actual reporting of the work performed has many variations. Many states already collect the major portion of the information needed to create the comprehensive database for an integrated bridge management and bridge maintenance management system. Some states use separate forms for bridge maintenance and other states include bridge maintenance reporting in the general reporting for all highway maintenance or for the personnel/payroll reporting system. Whatever administrative data processing system is used, maintenance field personnel will be relied upon as the source of the information. Generally the person in charge of the crew is also charged with reporting data. Errors can result if more than one person from a crew is responsible to input data.

As the report information is minimized, accuracy and reliability of the information will increase. The goal is to ask only for what is needed to conduct the necessary maintenance management functions and to ask that any given data item be reported only once. Thus, if the payroll system collects information on hours worked by each employee, the equipment management system collects information on equipment unit use, and the materials management and inventory system collects information on materials use, then these databases should be interconnected with the bridge maintenance management system so that data can be transferred into it for compilation into bridge maintenance reports.

With the move to larger and more integrated transportation management information systems attempting to incorporate all information needs into one management information system, there is a tendency to attempt to use one universal reporting form to collect all data for the entire department. Maintenance engineers and maintenance managers should analyze any such attempt carefully with respect to the reporting capabilities of maintenance personnel and the burden this type of reporting will place on them. Information reporting overload often leads to reporting errors or unrealistic reporting. (For background material on this process, consult NCHRP Report 344, *Maintenance Contracting*.)

3.2.5 Contract Maintenance for Bridges

All state DOTs contract for some portion of their maintenance work. A bridge maintenance program requires purchasing and procuring a wide variety of materials, supplies, equipment, and services. Routine purchasing and procurement is usually handled with procedures established at the department level. Contracting represents a specialized area of procurement that is becoming more common in all maintenance work. Contracting for maintenance should be based on the agency's need to be more efficient and effective, not on some arbitrary policy decision. A thorough maintenance management analysis in making a decision on whether or not to contract for bridge maintenance should consider the following factors. (The discussion presented here on contract maintenance applies equally to contracting for roadway maintenance.)

- Limitations on in-house staff.
- The need for specialized equipment not currently in the agency inventory.
- The need for specialized personnel in conjunction with limitations on acquiring such personnel or limitations on the potential to train current personnel to reach the needed skill level.
- The need to cover peak work loads when the agency resources are geared to an average work load significantly below the peak load.
- The potential to obtain required services at lower total system cost.
- The need to conduct emergency maintenance for which the agency does not have the capacity to respond.
- Legal restrictions on the amount of work that can be performed by agency forces, legal restrictions on contracting, or employee agreements restricting contracting.

Contracting is a formal process governed by state laws and department policies and regulations. It is common for contracts of different types and of different amounts of money to entail different restrictions on the agency's ability to enter into them. Before bridge maintenance engineers and managers propose entering into a contract for bridge maintenance services, they should consult with higher level management and, if needed, department legal staff to ensure that a contract can be drawn up which will meet the functional need of the maintenance group and will conform to any contracting restrictions on the agency. This is especially important when any portion of the bridge maintenance is funded with federal reimbursement funds or when the maintenance activity could be interpreted as rehabilitation work subject to federal guidelines.

State DOTs are almost universally required to submit any contract for bridge maintenance to a competitive bidding process. Some agencies do have a minimum contract amount under which the contract is not subject to competitive bid or legally defined emergency conditions under which the agency is freed from any competitive bidding process requirement. Maintenance engineers and managers should become aware of these restrictions and any conditions under which the restrictions are not applicable.

3.2.5.1 Lump Sum Contracts

Lump sum contracts are only suitable if the amount and scope of work can be defined precisely, as might be the case in the complete replacement of specific elements of a structure. There must be reasonable assurance that unexpected conditions, such as deterioration of adjoining elements, will not cause delays, increase materials requirements, or cause other problems. Unanticipated difficulties can also arise when some portion of the work is not defined in sufficient detail, thereby causing a situation where the agency and the contractor cannot agree on the intended interpretation of the plans and specifications. If the contract is written to allow some flexibility in the amount of work required for a single price, the contractor will have to assume the worst case to protect its interests, which tends to increase the contract cost. Thus, it is recommended that lump sum contracts for bridge maintenance only be used for projects for which all features can be easily and conveniently described in the plans and specifications.

One variation of the lump sum contract allows some flexibility in the quantity of work performed. This variation solicits a lump sum bid for an element of work but specifies that a variable number of units of that work element may be required to be completed. For example, perhaps a contract for the complete replacement of damaged or defective piles in a wood trestle bridge specifies a minimum and maximum number of piles to be replaced. The exact number is to be determined as the work progresses. The maximum and minimum amount of work must be specified in the contract for the protection of both parties. The price in contracts of this kind is a single lump sum for all labor, equipment, and materials to do one element of work, complete and in place. This lump sum price may also be used for some items in a unit price contract.

3.2.5.2 Unit Price Contracts

If the amount and scope of work can be defined within reasonable limits, say 10% or so, the unit price contract is usually the best choice for bridge maintenance contracts (as it is for bridge and highway construction contracts). Unit price contracts are the most common contracts used by transportation agencies for construction projects. In unit price contracts, the contractor receives payment for the actual amount of work done and the contracting agency retains a reasonable degree of control over the extent of the contract. Frequently, some well-defined items within a unit price contract are more appropriately handled as a lump sum.

For bridge maintenance projects, special modifications can be made to a unit price contract to increase its flexibility. This is done to make the contract more convenient to administer and to eliminate sources of controversy between agency personnel and the contractor. An example of such a modification in bridge maintenance might be a contract to repair, rather than to replace, a concrete deck. The contract could specify that removal of the deteriorated concrete and its replacement will be paid for in one of three ways, depending how work progresses. First, the agency could pay for removal and replacement of the concrete with no reinforcing steel exposed. Second, it could pay for removal and replacement of deteriorated concrete with the top layer of reinforcing steel completely exposed and cleaned. Third, it could pay for removal and replacement of the full-depth slab. Each item would be paid for on a square meter (or square foot) basis for all work and materials required. In this example, a complete survey of the deck condition would be required so that the contract documents could provide a reasonably precise estimate of the quantity of each type of deck repair required.

Whenever possible, unit price contracts should be used for bridge maintenance work instead of other types of contracts. A limited number of labor and equipment hours can, and probably should, be included as bid items to account for unexpected conditions. This will permit some flexibility within the contract; however, excessive use of these bid items should be avoided since they are likely to raise costs and have other undesirable effects similar to a cost reimbursement contract, as described below.

3.2.5.3 Cost Reimbursement Contracts

If it is impossible to define the amount and scope of work required, it may be necessary to resort to a cost reimbursement contract. In such contracts, the contractor is reimbursed at a predetermined rate for labor, equipment, and materials used. Cost reimbursement contacts require a great deal of inspection and record keeping on the part of the agency to verify the contractor's charges. If federal participation will be requested, discussions prior to executing the contract must be held to ensure that the contract will be approved for reimbursement. A common difficulty in obtaining federal reimbursement for work performed under a cost reimbursement contract (also on work done by agency personnel under force account) is the lack of a complete audit trail for all work done because of typical deficiencies in the record-keeping system.

Another key problem with many cost reimbursement contracts is that there may be a perceived, if not actual, loss of competition. A rigorous and detailed inspection based on good specifications is about the only way complaints of reduced competition can be deflected. Cost reimbursement contracts have been successfully used as an adjunct to force account projects where highly skilled personnel or specialized equipment is needed for only a part of the force account project.

Cost reimbursement contracts have also been used to contract for a variety of maintenance services over a fixed time period (6 months, 1 year, 18 months, etc.). Examples of the types of work that have been included in such contracts include the following:
- Remove and replace concrete curb.
- Remove and replace concrete deck or portions of a deck.
- Remove concrete or block slope protection.
- Repair damaged reinforcing steel or structural steel.
- Repair miscellaneous painted areas.
- Remove, repair, and replace damaged railing or balustrade.
- Remove or replace bridge sidewalks.
- Repair scupper and drain systems.
- Repair pier, pier caps, or abutments.

Contracts have been awarded to the lowest bidder on the basis of specified quantities of labor (foremen, laborers, welders, concrete finishers, painters, various equipment operators among 15 categories of labor); equipment (various sizes and types of trucks, front-end loaders, cranes, concrete mixers); and traffic control and management items (cones, flashing arrow boards, signs, barricades, steel plates). Crew size furnished by the contractor is determined by the size of the project but typically ranges from five to eight persons. Except in emergencies, the contract typically allows a 7-day notice to the contractor of specific work that is to be performed, including recommended equipment and labor requirements. This type of contract permits maintenance agencies to use contractor resources to offset reduced agency staffing levels and to smooth out demand for maintenance services in peak work load periods. The following represents some precautions to be noted before implementing fixed-time-period cost reimbursement contracts for bridge maintenance services:
- Items and quantities of items used in the bidding process should be carefully selected. Unbalancing of bids can result if items are shown that are rarely used or if unrealistic quantities are indicated.
- Every effort should be made to encourage competition between contractors. If a single contractor becomes too entrenched in a district, prices can become unreasonably high.
- The state's representative should be particularly well trained and experienced. The representative should have hands-on experience in bridge maintenance and should also be well-grounded in contract management and administration.

Cost reimbursement contracts should be used only when a thorough engineering and management evaluation indicates they are the most cost-effective means of conducting the needed bridge maintenance.

3.2.5.4 Negotiated Contracts

In some instances, it may be necessary to resort to a negotiated contract for bridge maintenance. Negotiated contracts are used when they are the best way to get the best value for the price paid. They are commonly used to procure professional engineering services. However, there may be circumstances when services other than engineering may need to be procured through a negotiated contract. If only one feasible source exists for some needed maintenance service, a negotiated contract is the only logical agreement. Even when there is more than one source, if there is some important difference among the possible competitors—such as experience records or safety records—a negotiated contract may be the appropriate agreement.

3.2.5.5 Cost Comparison of Contracting vs. Performing Work In-house

Cost of Contracting: When deciding whether to contract or when estimating the total cost of a project to be done by contract, the incidental costs of contracting should be included. Incidental costs may include the following:

- Administrative costs of preparing requests for a proposal or bid documents and plans.
- Costs of selecting the contractor or vendor, particularly for negotiated contracts.
- Administering the contract while the work is being done.
- Inspecting the work.
- Field record keeping.
- Consultations with the contractor.
- Verifying charges, approving final results, and approving invoices for payment.

In-house Cost: When comparing contracts to in-house work, there is a tendency to compare the salaries of the agency employees with those of a contractor, while ignoring the hidden cost to the agency of having the employees on the payroll (overhead, etc.). Other costs beyond salaries to perform work in-house are as follows:

- Overhead costs such as employee benefits, work space, administrative support (payroll, etc.), and tools.
- Cost of keeping employees on the payroll during bad weather, during winter months, or in other times when the work load does not require the full work force.
- Cost of purchasing equipment and keeping it operational.
- Liability costs associated with the maintenance operation, such as injury to employees and the traveling public.

3.2.6 Quality Control and Quality Assurance of Bridge Maintenance Operations

Quality Control/Quality Assurance (QC/QA) is an important part of supervision. QC is normally performed within the group or the maintenance crew. It includes performing quality work safely, on schedule, and within budget. QA is performed from outside the group. QC is the responsibility of the bridge maintenance worker. (For background material, consult NCHRP Report 376, *Customer-Based Quality in Transportation*. For guidance in establishing QC/QA, consult NCHRP Project 14-12, *Final Report,* at <www.nap.edu/readingroom>; select the heading "Transportation," then go to NCHRP Web Document 8.)

3.2.6.1 QC at the Work Site

There are two general classifications of items to be reviewed at the work site. The first is how well the work is being performed in terms of quality of results and the amount of work being done, or productivity. The second is how the work is being performed in terms of safety to the workers and the public.

While the quality review would often be considered the most significant and the most difficult to review, it is often far easier to perform than the safety review. Even if performance standards have not been developed, standard practices, formal or informal, are usually available. However, safe practices and total crew motivation to perform the work safely are almost always different matters since small errors of omission or commission can produce unfortunate or even tragic results. The general attitude of a crew toward satisfactory or unsatisfactory work practices can be revealed in a number of ways that would only be noticed by someone looking specifically for such indications. In addition, it is often tempting for an agency to overlook departures from the best practices because of expediency, or to avoid cautioning or reprimanding a foreman who attains high performance and productivity through the use of shortcuts or questionable practices that carry relatively high risks of causing personal injuries.

3.2.6.2 Technical Site Review

If performance standards have been developed for the work, the reviewer should use them as a guide. State construction standards are also a good resource for basic repair methods and procedures. The person conducting the review should be thoroughly familiar with such documents before visiting the site to take full advantage of his or her available time there.

When bridge maintenance crews are asked to perform work for which plans, sketches, or engineering drawings have been developed to guide the work, workers should be thoroughly familiar with these before visiting the site. It is best if field personnel are consulted during the preparation of any such documents so that the workers are familiar with the project from its conception. This provides an opportunity for the project to be developed in a way that considers the capabilities of the crews that may be assigned to the work.

In agencies where performance standards have not been developed or plans are not needed for a bridge maintenance project, the work is done by standard practices, with or without written instructions. Thus, a worker must use a base of experience or knowledge of satisfactory practices to evaluate the progress of work at the site. In smaller agencies with a stable work force and relatively light work loads, this process can be satisfactory for as long as those conditions prevail. Unfortunately, as more bridge maintenance work is required and if turnover of employees is frequent, a lack of written standards can become a significant problem.

Even if standards or plans are available, there are a number of specific items related to quality control that would not usually be shown on such documents and that should be reviewed by the worker. Items that workers are supposed to keep in mind are best laid out in checklists by topical area.

3.2.6.3 Traffic Control Site Review

Almost all bridge maintenance activities interfere with normal traffic flow to some extent. While all traffic control is based on the MUTCD or the state manual on traffic control, it is not reasonable to expect the worker to be familiar with all of the MUTCD. A traffic control checklist should be developed for the worker to monitor aspects of traffic control at the site. This checklist should include factors the maintenance engineer or maintenance manager considers important to quality control of traffic safety and protection of workers.

3.2.6.4 Site Review of Tools and Equipment Use

All work related to highway maintenance is inherently associated with some level of risk and danger because of the proximity to traffic flow and the nature of the tools, equipment, and materials used. This is especially true for bridge maintenance work because of the restricted areas in which the work must be performed and because of the necessity to do a lot of handwork. Every precaution that is practicable should be taken to ensure that the use of tools and equipment presents the least risk possible to workers and the public. A checklist of key points regarding the use of tools and equipment should be available to supervisors for their use when visiting work sites.

3.2.6.5 Site Review of Rigging and Climbing Practices

Among the inherent risks and hazards of bridge maintenance is the difficulty of performing work at a significant height above the ground or the water. This requires that personnel be protected when working at such heights and that tools, equipment, and materials be safely and efficiently transported to the work level. When visiting sites involving activities that use cranes, staging, ladders, worker lifts, or other devices, the worker should be particularly alert to practices and procedures that create unnecessary hazards. A supervisor making a site visit should have a checklist available of the significant issues regarding climbing and rigging practices.

3.2.6.6 Budget Monitoring

A major control task is monitoring budget expenditures and accomplishments. Two major elements of this task are controlling funds to keep spending within budget limits and reallocating work. The first requires continuously current information on actual expenses. Necessary changes or reallocations can then be based on accurate information.

A method of controlling the budget is redirecting resources. This determination applies primarily to labor and equipment use where scheduling has the most impact. Adjustments may include the following:
- Change work scheduling.
- Add or reduce labor.
- Add to or reduce the equipment fleet.
- Reduce peak demand for labor and equipment.

3.2.6.7 Schedule Monitoring

Methods of developing and monitoring work schedules for bridge maintenance crews are quite straightforward and require little additional comment. Interruptions of bridge maintenance activity for emergency repairs or because of bad weather tend to be frequent enough that such schedule interruptions are usually ignored. However, an important part of quality assurance is to make certain that the schedule is updated to properly account for interruptions. The schedule must be realistic and it must be current. Completing maintenance assignments on schedule has the following direct impacts on performance:
- Efficiency can be measured by success in meeting schedules.
- Support and equipment has to be scheduled ahead for the next assignment and adjustments influence other activities; therefore, they should be minimized.
- Interruptions tend to be the same over a year's time so that they cancel out when comparing performance between years or between crews.

3.2.6.8 Methods of Quality Assurance

QA is performed from outside the maintenance crew. Some agencies have formal QA programs that are developed to measure the crew performance. The QA evaluation may be generated by the bridge management system. Components of the evaluation could include the person-hours, equipment, and materials used in accordance with performance standards; acceptable condition ratings at the next inspection cycle; and accidents or injuries related to crew assignments.

If there is no formal QA program, QA is probably taking place on an informal basis. The disadvantage of informal QA is that it is often subjective and rarely quantified. Therefore, it may be applied based on the experience of the person making the evaluation. The items that are checked are often based on recent problems. A carefully developed QA program that everyone understands, that is considered fair, that provides objective evaluations, and that provides quantitative results will have the best results. (It is again suggested that NCHRP Project 14-12, *Final Report,* be consulted.)

3.2.7 Bridge Inspection

3.2.7.1 Inspection Process

Bridge maintenance crews and bridge inspection teams need to work in close coordination. The inspectors are the primary source of identified bridge maintenance needs.

A thorough, well-documented inspection is essential to determine bridge maintenance requirements and make practical recommendations on suggested actions to correct the deficiency or preclude the development of bridge defects or deficiencies. Regular inspections should be considered a primary maintenance responsibility. Besides searching for defects that may exist, inspections should also look for conditions that may indicate possible future problems.

Regular inspections are conducted every 2 years under the federally mandated inspection program. The emphasis of this program is establishment of rehabilitation and replacement priorities. As bridge management systems become fully operational, these data management systems will be able to generate maintenance and repair needs and establish short-range and long-range budgets.

3.2.7.2 NBI Program

The 1968 Federal-Aid Highway Act established the National Bridge Inspection (NBI) Program, which requires states to inventory and inspect all structures on the federal aid highway system. The program has been expanded to include all bridges on public roads that are not on the system. The federally mandated Structure Inventory and Appraisal System especially benefits states that implement a detailed system to inspect bridge maintenance and to provide the data needed (current condition of the bridge) to implement a bridge maintenance management system.

3.2.7.3 Performing the Inspection

The following make up the basic content of a bridge inspection.

Waterway or Stream Channel: Anticipating problems and taking adequate protective steps will avoid or minimize the possibility of serious difficulties in the future. Some important conditions to investigate are as follows:
- Adequacy of the waterway opening under the structure. Sand or gravel deposits may reduce the size of the bridge's waterway opening.
- Soundness and performance of the existing bank and shore protection.
- Possible waterway obstruction. Debris or plant growth can contribute to scour and present a possible fire hazard.
- Streambed erosion and scour around piers, abutments, and under the bridge.

A channel profile record for the structure provides valuable information on the tendency toward scour, channel shifting, degradation, or aggradation. The record should be revised as significant changes occur. These indications can help predict when protection of pier and abutment footings may be required.

Abutments and Piers: When inspecting concrete abutments and piers, specific types and locations of defects should be particularly noted. Those likely to occur most frequently are as follows:
- Spalling and deterioration of concrete at the water line.
- Deterioration of concrete under bridge bearings.
- Cracks in abutments, especially at the corners where the wings join the face of the

abutment. These cracks should be observed over a period of time to determine if they are growing. When a crack enlarges, this indicates that movement is taking place in the abutment or pier. Movement can be an indication of subsurface problems.

Timber abutments and piers are subject to specific types and locations of defects. Those that should be investigated are as follows:
- Decay that usually occurs at the ground line, the water line, or at bolt holes.
- Missing or broken bracing.
- Broken or cracked piles resulting from ice or debris collisions.
- Decayed, cracked, or crushed pile caps.
- Tipped or rotated abutment caps.

Bearings: All bearing devices, both fixed and expansion types, should be examined to make certain that they are functioning properly. Unexplained changes in the bearings can indicate serious problems in other portions of the structure. Pier or abutment movement is an example of such a problem. When problems exist, the following inspections should be made:
- Check anchor bolts for damage and nuts for tightness.
- Check anchor bolt nuts on the expansion bearing for proper setting to allow designed movement.
- Check expansion bearings for indications of proper movement.
- Check for dirt and debris in and around bearings.
- Check for excess bulging, splitting, or tearing in elastomeric-type bearings.
- Check grout pads and pedestals for cracks, spalls, or deterioration.
- Check condition of roller nest bearings.

Bearings should be examined carefully after any unusual occurrences, such as heavy traffic damage, earthquake activity, and battering from debris during flood flows. Any bearings, fixed or expansion, that show signs of distress or malfunction should be reported promptly to the appropriate official.

Beams or Stringers: Beams or stringers can be fabricated from timber, steel, or concrete. Each material presents specific maintenance problems for which inspection is needed.

Timber Beams: Defects commonly found in timber beams or stringers are as follows:
- Split, broken, or decayed stringers.
- Lack of surface treatment that allows large longitudinal cracks to develop that can extend the full length of the beam.
- Crushing of the stringer at bearing areas, which normally indicates serious decay problems and reduces live-load-carrying capacity.
- Loose bridging or diaphragms between timber stringers.

Steel Beams: Defects commonly found in steel beams or stringers are as follows:
- Rust below expansion joints.
- Rust on the beam caused by moisture coming through deck cracks.
- Paint deterioration such as peeling, blistering, or cracking.
- Looseness of beam connections.
- Cracking and corrosion around rivets and bolt heads in built-up girders.
- Cracks in welds and base metal.

Concrete Beams: Defects commonly found in concrete beams are as follows:
- Disintegration of the structural slab forming the horizontal section of a T-beam.
- Inoperative bearings from freezing of sliding plates.
- Damaged T-sections that expose reinforcing steel to corrosion.
- Cracked beam ends.

Any of the defects mentioned with respect to concrete beams are particularly significant when found in a prestressed beam. If an open crack is found in a prestressed-concrete beam, the appropriate official should be notified immediately.

Trusses: Trusses are nominally classified in three broad categories based on their position with respect to the roadway. They are categorized as high trusses, pony or half through trusses, or deck trusses. Inspection of any truss should begin by sighting along the roadway rail (or curb line) and along the truss chord members to determine any misalignment in either the vertical or horizontal planes. Each truss member should be checked. Inspection should include the following:

- Observe the truss alignment and grade.
- Check truss span bearings and expansion plates, ensuring freedom of movement.
- Check straightness of compression members.
- Examine truss and bracing members for traffic damage.
- Check all upper and lower lateral bracing members for damage and proper functioning.
- Examine the paint and determine the extent of corrosion, particularly around bolt and rivet heads. Connection details are especially susceptible to corrosion.
- Check the pins at the connections and make certain nuts and keys are in place.
- Check for loose, rusted, or missing rivet heads and bolts.
- Examine tension members for cracks, particularly at connections.
- Check for loss of steel thickness from rust.

Decks: All decks should be examined for skid resistance to determine if a hazard exists. A check should be made to ensure that decks drain properly and that there are no areas where water can pond. Drains and scuppers should be open and clear because poor deck drainage can contribute to deterioration. Drain outlets should not discharge water where it can be detrimental to any parts of the structure, cause fill and bank erosion, or spill onto a traveled lane below.

Timber Decks: Timber decks should be examined for decay at their contact surfaces where they bear on the stringer, and between layers of planking or laminated pieces. The following conditions may also create maintenance problems: loose nails or spikes; openings between planks over abutments and over piers that may allow dirt to sift through; and split, worn, broken, or decayed planks.

Steel Decks: Steel decks should be checked for corrosion and welds that are not sound. A check should also be made for the following conditions: dirt collected in open-grid decking on the top of stringers, deteriorated paint, and loose welds where a steel deck is fastened to the stringers.

Concrete Decks: Concrete decks should be examined for cracking, leaching, scaling, potholing, spalling, and other evidence of deterioration. Each defect should be evaluated to determine its effect on the structure. Any evidence of deterioration in the reinforcing steel should be inspected closely to determine its extent. Decks that are treated with de-icing salts are especially apt to be affected. Cracking of concrete allows moisture and chemicals to penetrate to the reinforcing steel, which then rusts, expands, and causes spalling of the concrete.

Wearing Surfaces: Asphaltic concrete or other wearing surfaces on a deck can hide deck defects. The surface must be examined very carefully for evidence of deck deterioration. Cracking, breaking up of the surface, or excessive deflection can indicate such deterioration. Any evidence of water passing through cracks in slabs should be noted. Areas where deck deterioration is suspected may require removal of small sections of the surface to facilitate a more thorough investigation.

Approaches: Approaches are an important adjunct to a bridge. Approaches should be level with the bridge deck. If the transition between the approach and the structure is not smooth and even, additional impact loads of substantial energy can be introduced onto the bridge, which can cause extensive and serious structural damage over extended time periods. Approach pavement conditions should be checked for unevenness, settlement, or roughness. Cracking or unevenness in an approach slab may indicate a void under the slab caused by fill settlement or erosion. The joints between the approach pavement and the abutment back wall, which are designed for thermal movement, should be examined to determine if there is adequate clearance and a proper seal. The shoulders, slopes, drainage, and approach guardrail should also be evaluated as part of the inspection of the approaches.

3.2.7.4 Inspection Reports

The NBI inspection report varies among states. However, it always includes condition and inventory information in a standardized record-keeping system suitable for computer data files. The information is provided to the Federal Highway Administration and is updated on a regular basis. It is included as part of the national Structure Inventory and Appraisal System that is used to establish eligibility for and priority for bridge rehabilitation or replacement.

The inspection report includes numerical ratings that represent the condition of each bridge element. The meaning of this rating varies somewhat among the states and among the bridge elements. Normally the rating ranges from "1" to "9," with a "9" meaning new condition and a "1" meaning the bridge is currently closed. In between, typically "8" means "good," a "6" or a "7" means that minor repair or maintenance is required, a "4" or a "5" means that major repair is needed, a "4" usually signals that major repair is urgently needed, and a "2" rating means that the bridge should be closed to traffic. Since the ratings are interpreted somewhat differently among the states, each state should have clearly expressed definitions of each rating for use within its jurisdiction.

Inspection reports include written descriptions, sketches, and photographs to provide details of the findings. The report should also provide the location of problems and estimates of their extent.

3.2.7.5 Inspection Performed by Maintenance Supervisors

Bridge maintenance crews should also function as inspectors whenever they are in the field working. It is important to look for defects that might represent a potential safety hazard or a defect that will cause problems at some time in the future. It is much easier and more cost-effective to correct the problem while the crew is already at the site than to go back at some future date.

Maintenance crews may spend more time at the bridge site than the inspector does. When cleaning and preparing for a repair, they may discover or expose problems that the inspector did not see. The best results can be achieved when the bridge maintenance personnel and the inspectors work together as a team.

3.2.7.6 Resource Estimating

The condition inspection reporting system should include information that can be used for estimating the resources necessary to maintain or rehabilitate a bridge. The accuracy of estimating is improved when it is performed in the field in conjunction with the inspection. The estimate made in the field is a "quantities" estimate that can be used for computing a cost estimate later in the office. Working together, the engineer and the supervisor can anticipate specific procedures that are needed, and the accuracy of the estimate will be improved. When preparing estimates, engineering office personnel are hindered by a lack of familiarity with a specific site and the special circumstances likely to be encountered. They are, therefore, limited to the use of statistical averages for determining time and resource requirements. Field estimate data eliminate this problem. The obligation for the engineer and the supervisor to perform the maintenance activity within the estimate is easier to fulfill when the field estimate procedure is used.

3.2.7.7 Identifying Fracture-Critical Bridges

A fracture-critical bridge must have one or more fracture-critical members (FCMs). An FCM is a tension member or component whose failure will produce a sudden collapse of the structure. It is important to know if a bridge is fracture critical when evaluating damage or performing repairs.

The portion of a member in tension is being pulled apart, which causes any cracks to grow, and fracture is the final result. Hangers, suspension cables, and some truss members are normally in axial tension. Maximum tension is in the bottom flange at mid-span of beams. A continuous span is typically in its critical tensile stress in the top flange over an interior support. High stresses may concentrate in tension members at points where the cross-sectional area changes or where there is a member discontinuity.

Redundancy: Redundancy is the ability of other members of a structure to help carry the load when a member becomes weak or fails. Lack of redundancy in a bridge makes it more susceptible to sudden collapse when a member fractures. Three different forms of redundancy may result from a particular design approach: load path, structural, and internal redundancy.

Load path redundancy is associated with a minimum number of structural members supporting the bridge deck. A bridge with less than three trusses or girders supporting the deck is not considered redundant, and therefore, it is classified as fracture critical. There can be degrees of redundancy depending upon girder spacing, stiffness of the deck and framing system, and other interdependencies of the structural elements. For some bridges, it may be necessary to have a professionally qualified structural engineer conduct a bridge capacity analysis to predict the failure scenario.

Structural redundancy refers to the support provided by the cantilever effect created after a continuous member is weakened. This effect occurs only on interior spans with members that are continuous across supports on both ends. Thus, there must be a minimum of three continuous spans to have a structurally redundant span, which for three spans would only be the center span.

Internal redundancy relates to crack propagation through the cross section of a member. Many members are composed of several parts. A crack in one part must start again as a new crack in each separate part to completely penetrate the cross section in an internally redundant member. Built-up members with plates attached by rivets or bolts, reinforced-concrete members, cables, and members composed of several separate sections all have internal redundancy. Rolled steel members and members with built-up sections from welding do not have internal redundancy.

Many agencies define fracture-critical members in terms of load path redundancy, but structural redundancy and internal redundancy should also be considered in the evaluation. Examples of fracture-critical spans are spans supported by two or fewer single web girders, trusses, suspension spans, cross girders, caps, and tie members on tied arch spans. Spans supported by four or less pin-and-hanger assemblies also qualify as fracture-critical spans.

Fatigue and Fracture: It is very important for the bridge owner to identify a crack or flaw before the member fractures. Physical characteristics make certain members more susceptible to fracture. The magnitude of the total stress and the number of times a member is stressed contribute to fractures. Certain design details can contribute to the beginning of a crack. Residual stresses in the member itself can also increase the tendency to crack. An inspector's efficiency in identifying FCM problems is significantly enhanced by an understanding of fatigue and fracture.

A fracture requires a driving force. This force normally results from the load on the bridge. The force on a particular member cross section is called stress and may take the form of compression, tension, or shear. Compression squeezes or pushes down on a member and tends to help resist crack growth. A crack in a compression member is rare, and such cracks would not be expected to show evidence of growing. Since tensile forces pull the member apart, cracks in tension members are a serious concern. Cracks in tension members can be expected to grow perpendicular to the direction of the tensile force and produce fracture. Shear is similar to tension except that the force is in a sliding or slicing direction across the cross section so that shear tends to tear the material. Some cracks may grow as a result of shear stresses. Shear forces act at 45° to the direction of the force. Bridge members may be resisting only one of these forces, or a member may have to simultaneously resist a combination of these forces.

A fracture may be the result of an overload when the member is stressed beyond its capacity of the material yield point. This is rarely the case for bridges designed to carry standard legal loads. More often, repeated loads that do not exceed the legal limit cause the cracks. Repeated flexing and stressing of the material at a point below the yield limit produces an "internal working" of the molecular bond within the material that gradually reduces the strength of the material in what is called fatigue failure. One load is a cycle. A cycle must subject the member to a certain magnitude change in stress or stress range before it is significant in causing fatigue cracks. Bridges that carry a large volume of heavy loads are more likely to experience fatigue failures than low-traffic-volume bridges.

Fatigue crack initiation is related not only to the number and size of the stress cycles, but also to the degree to which design details affect the fatigue resistance of a member. Stresses concentrate at locations where the rigidity of the member changes. To be most effective in their respective activities, both bridge designers and bridge inspectors need to be aware of the effect of these design details. Welds, bolt holes, rivet holes, notches, copes, flame cuts, and grinding marks all contribute to stress concentrations. However, bolt or rivet holes cause stress concentrations to a lesser extent than some of these other items. Service flaws that can contribute to fatigue include collision damage, damage from improper straightening, or cross-section area loss from corrosion.

> *Material Considerations:* The two types of fracture are ductile and brittle. When ductile fracture occurs, the material stretches before it separates into two parts. The fracture is slow, and often there is time to prevent a disaster. A brittle fracture occurs very rapidly and is of particular concern to the bridge owner. Certain members are more likely to fail by brittle fracture. Because they tend to break rather than bend, members composed of thick plates are more likely to experience a brittle fracture than members made of thinner plates. Cold temperatures reduce metal ductility and increase the likelihood of a brittle fracture. Modern bridges are designed and specified to be constructed of steel with minimum toughness to resist brittle fracture. Old bridges were not required to meet such specifications and may need to be tested to determine the steel toughness, analyzing the bridge's potential for a fracture failure.
>
> *Design Considerations:* Fatigue cracks initiate at locations in steel members where the rigidity of the member changes. Usually these locations result from a design engineer attempting to save material to reduce bridge weight and save money. Cover plates added

to rolled section beams to avoid specifying a larger size rolled section change the rigidity in ways that increase susceptibility to fatigue cracks along the edge of the cover plate. If the designer uses a thin-web rolled section and increases the beam rigidity by adding stiffener plates to the web, this encourages fatigue crack development. Fatigue cracks may develop from in-plane bending when the load is distributed from the floor directly to the member. Fatigue cracks may develop from out-of-plane bending when the floor load is transferred to the member through a secondary member. This produces a twisting action that places excess stress levels on thin parts of the member that are not well suited to resisting the forces. Bridge inspectors should report all cracks to professionally qualified structural engineers who can evaluate the nature of the cracking and determine how critical it is with respect to needed bridge maintenance.

Loads on the Structure: The loading rate on the structure can also be an important factor in whether the fracture will be brittle or ductile. Static loading (dead load) is least likely to produce brittle failure. Dynamic loading (live load) is far more likely to produce brittle or sudden failure. Since bridges undergo a combination of static and dynamic loading, it is important to identify bridges where the dynamic loading is exceptionally high. Bridges that receive heavy pounding loads resulting from low approaches or poor vertical alignment are candidates to watch carefully.

Repeated loads that produce high-stress cycles cause fatigue cracking. Large loads relative to the average design load create stress cycles that cause fatigue damage. A particular design detail may be capable of carrying a limited number of cycles caused by a very heavy but legal load using the structure. When the number of stress cycles exceeds the limit, cracking will occur at locations in the structure that can be predicted quite reliably. Thus, a knowledge of the loading history of a bridge is helpful in evaluating the probability of fatigue cracking.

Crack Initiation and Propagation: Most cracks in steel bridges occur at predictable locations. Cracks occur at areas of stress concentration. Normally, they originate at a flaw. The flaw is often associated with a weld. When a fatigue crack caused by in-plane bending grows to a size visible to an inspector, typically at least 80% of the service life of the member has already been destroyed. A small crack has been growing beneath the surface in a semi-elliptical pattern, and after it reaches the surface of the steel, it must penetrate through the paint before it is visible to the inspector. Sometimes a rust stain will allow an inspector to detect a crack earlier. Nondestructive testing (NDT) can help verify the existence and extent of a crack after it has been initially found. However, NDT is not very effective in general inspection to find cracks initially. Indications of cracks are generally found by visual inspection.

3.2.7.8 Testing Existing Bridge Components

To properly plan a bridge maintenance repair, it is sometimes necessary to obtain more information on the condition of the existing material than can be obtained from a visual inspection. Tests are performed to provide the information required. While the maintenance supervisor normally would not perform the tests, maintenance personnel should know when they are needed and what the results of the tests mean.

Reinforced-Concrete Corrosion Survey: Several tests are available to investigate concrete element deterioration from reinforcing steel corrosion.

Delamination Survey: Sounding the surface of a reinforced-concrete deck can determine the presence of delaminations (internal cracks caused by corrosion of the reinforcing steel). A grid is laid out on the deck surface. The surface is sounded (usually by a drag chain), and the delaminations are noted. The areas containing delaminations are marked on the surface and mapped for a report of the survey. The amount of delamination is computed as a percent of the surface area. Concrete spalls are not included.

Cover Depth Over Reinforcing Steel Using a Cover Meter: Devices are commercially available that use a magnetic field to detect the presence of reinforcing steel within concrete. If the size of the steel bar is known, the devices can estimate the depth of concrete over the bar. An estimate of the depth of concrete cover is needed if part of the surface is to be removed in any maintenance procedure. It is helpful to check the precision and calibration of the instrument at one location on the deck by exposing a reinforcing bar and comparing the reading to the actual depth. This is especially helpful because the concrete may contain magnetic particles that disturb the instrument measurements. The exposed reinforcing bar can be used later as a half-cell test bar ground connection.

Chloride Content: Powdered samples of concrete are collected with a concrete drill at several increments between the concrete surface and the level of the reinforcing steel. The chloride (salt) content of the powder is measured in kilograms per cubic meter (pounds per cubic foot) with a portable kit in the field or sent to a laboratory for testing. The threshold chloride contamination is 16 kg per cubic meter (one pound per cubic foot).

Corrosion Potential Survey: This procedure determines the potential for reinforcing steel corrosion by measuring the electrical potential of the reinforcing steel. Electrical measurements are made by placing a probe, connected to a half-cell corrosion detector, on the deck surface at predetermined points based on a grid established on the concrete deck surface. The surface is usually wetted for a better electrical contact. This probe is grounded to an exposed reinforcing bar in the deck that is in contact with the reinforcing steel bar mat being tested, and an electrical connection is made. The half-cell corrosion detector reads the electrical potential of the steel at the predetermined points, and these readings are recorded. A corrosion potential survey is not recommended if the deck contains epoxy-coated reinforcing steel or galvanized reinforcing steel (epoxy-coated bars are insulated from each other, and readings on galvanized bars merely indicate the electrical potential of the zinc coating). Also, this test cannot be used if the concrete is overlaid with a dielectric material.

Corrosion Contour Map: Corrosion tests are typically tied to a 1.3-meter (4-foot) or smaller rectangular grid established on the deck. The test findings are recorded at the proper location on a sketch of the deck, and contours are plotted to show the areas that have delamination, chloride contamination, and active corrosion.

Newer Corrosion Tests: Concrete deck deterioration from salt contamination is a continuing major expense in bridge maintenance. Research and development efforts to find faster, more reliable, and more precise methods of detecting and quantifying corrosion damage continue to make bridge maintenance management more effective. One such effort is the development of a rate of corrosion measurement that is based on determining the polarization potential of the reinforcing steel. Corrosion current is calculated from a simple equation and expressed in terms of mill-amperes per square meter (or square foot) of the reinforcing steel area. The test should be conducted at locations of highest corrosion potential (peak negative values) as determined by the half-cell test. Corrosion rate tests should not be carried out where epoxy-coated or galvanized reinforcing steel is used.

A second such test is the permeability of concrete test. This test determines the relative permeability of the concrete (or a concrete overlay). The permeability is indicated by the electrical charge passed through the concrete, expressed in coulombs.

Noncorrosion-Related Concrete Tests: A number of tests can be conducted on a reinforced-concrete deck to determine characteristics that may be useful in planning bridge maintenance.

Cores: Cores can be drilled out of the deck to provide a sample of the hardened concrete. These cores can be tested for compressive strength. However, since most deck problems are related to durability rather than strength, cores are rarely tested for compressive strength. Cores can be used for petrographic analysis, air content analysis, materials compatibility tests, or chemical contamination tests. Since coring is moderately expensive and destructive, cores are typically taken only when other evidence indicates further investigation is warranted.

Alkali Silica Reactivity: Some aggregates react with cement and create a gel in the hardened concrete. Over time, this gel expands, causing cracking and disintegration of the bond between the concrete ingredients. A uranyl acetate and ultraviolet light test is available to determine presence of the gel. Little can be done to prevent this problem on existing bridges, except to do everything possible to avoid using reactive aggregates in future construction and repairs.

Tests for Special Problems: Tests are available that are usually considered too expensive to be routinely used in testing for routine maintenance conditions. However, one of them might be used for special situations.

Ultrasonic Pulse Velocity: Ultrasonic pulse velocity measurements measure the time of transmission of an ultrasonic pulse of energy through a known distance of concrete. The velocity of the pulse is proportional to the dynamic modulus of elasticity (or hardness), which in turn is an indicator of concrete strength. The test evaluates homogeneity and determines crack location. The results can be affected by many factors including variations in aggregate and the location of reinforcing steel. The results obtained are quantitative but they are only relative in nature; thus, they need to be correlated with corings to yield absolute values.

Radiographic (X-ray) Inspection: Radiographic inspection can be used to locate cracks, reinforcing steel, and internal voids within the concrete. Up to 200 mm (0.65 foot) of concrete can be penetrated. The method is not destructive but requires access to the back of the bridge element. It is very expensive and must be used with great care because of the potential radiological health hazard.

Computer-Assisted Tomography: Computer-assisted tomography scanning uses a nuclear source to develop a cross-sectional view of a member. It yields information on the size and location of aggregate, cracks, voids, density, and extent of corrosion. This method is not destructive and can be used to scan members up to 1 meter (about 3 feet) thick. It is very expensive, gives no direct measure of strength, and poses a significant potential health risk to the user.

3.2.7.9 Tests for Steel Members

Various testing methods are available for evaluating steel members that are suspected of having problems. When maintenance work is being planned on steel bridge members, it is important to determine the strength of the steel, the ingredients in the steel, and the existence and location of flaws or cracks in the steel that cannot be seen by visual inspection.

Coupon Samples: A small coupon sample is taken from an area of the steel member where its location will not cause problems (as determined by a professionally qualified structural engineer). The sample can then be tested for tensile strength and steel ingredients (for load capacity and weldability). This test is destructive, so it is used sparingly.

Dye Penetrant Test: This test is used to identify and enhance surface cracks in steel members. It is simple and inexpensive. A photograph will provide a permanent record.

Magnetic Particle Testing: This method locates surface cracks in steel by an induced magnetic field. The magnetic particles are fluorescent and suspended in a slurry. The magnetic field attracts the particles to discontinuities in the steel surface. This method is quick and low cost although only applicable to surface defects.

Ultrasonic Testing: This method uses sound waves to locate cracks or flaws inside steel members. It is commonly used on welded splices, cover-plated ends, and pin-and-hanger details. It is most effective in identifying cracks that are perpendicular, rather than parallel, to the direction of the sound wave. It is not destructive and can be used to measure the member thickness.

Radiographic (X-ray) Inspection: Radiographic inspection locates cracks by using a film cassette with an x-ray or gamma ray source placed on opposite sides of the member. It produces a permanent record of the crack. Up to 350 mm (about 14 inches) of steel can be penetrated. This method is expensive, hard to use, and poses a health risk to the operator unless extreme care is taken in its application.

Acoustic Holography: This method locates cracks using an array of ultrasonic transducers to produce a multidimensional picture and a permanent record. The test is expensive and somewhat experimental.

3.2.7.10 Tests for Timber Members

Wood is one of the oldest building materials used for bridges. In spite of this long history, testing methods are still under development to add to those currently available.

Pointed Probe: A pointed probe (e.g., an ice pick) can be used to subjectively measure the quality of existing timber.

Increment Bores: Increment bores are used to obtain samples of the interior cross section of the timber member. Since decay starts on the interior of a treated timber member, this test is needed to determine if a member should be removed as part of a repair operation.

Advanced Timber Testing: The following are two of several efforts to develop additional testing capabilities for timber bridge members:

(1) Sonic Pulse Velocity Testing: This method indicates relative timber strength and section loss as a single value based on a transmitted pulse velocity that is proportional to density and the modulus of elasticity. Correlation of results with samples of known strength is needed to yield an absolute value. Some success has been reached in making measurements from a boat to test timber at a depth of 1 meter (about 3 feet) below the water line.

(2) Handheld Moisture Meters: Moisture meters are available to measure the moisture content of solid wood, including glue-laminated members. These meters may be either conductance meters or dielectric meters. They can provide a rapid estimate of moisture content. They also can infer strength based on electrical parameters measured, but these inferences may or may not be reliable. Measurements must be compared to a calibration curve to obtain an indirect measure of moisture content. Preservatives (e.g., creosote) and adhesives (in glue-laminated members) can significantly affect readings.

3.2.7.11 Load Tests

Most analytical methods of measuring the capacity of a bridge member predict the stress that will be produced in that member from a certain weight vehicle. The prediction is based on a simplified application of structural theory combined with experience factors. To allow for unknown field conditions, the analytical methods are usually somewhat conservative. Sometimes very expensive repairs can be avoided by using a more precise method to measure the actual stress in a member by applying a test load to the bridge. Strain gauges are attached to the member at prescribed points determined by a professionally qualified structural engineer, and the strain produced by different loads is measured. The strain readings are converted to stress levels to yield a precise estimate of the load capacity. Two recent reports comparing load testing to analytical methods suggest that actual stresses and strains are typically appreciably less than those predicted by standard analytical methods [15, 16].

3.2.8 Preliminary Site Visit to Plan a Bridge Repair

State DOTs follow different procedures in responding to bridge inspection findings. If maintenance or repair needs have been identified, a bridge inspection supervisor or local (or assistant) bridge maintenance engineer will review the report, decide if an engineer should visit the site, and determine who should perform the work. The options vary significantly with respect to what work is contracted and what work is conducted in-house. Decisions are based on the degree to which a particular state has special crews that perform bridge maintenance or whether this type of maintenance is generally a contract process. The site visit provides valuable input to prioritizing bridge maintenance needs.

After the assignment has been made, someone must visit the site to collect information to plan the repair. Depending upon the state and the type of work to be done, the supervisor may be totally responsible for the site visit or the supervisor may work with an engineer (or other person with extensive bridge repair experience and knowledge). The basic information needed to plan the repair is considered first, followed by special considerations for each component of the bridge.

3.2.8.1 Basic Information Required

It is important that the person who performs the preliminary site visit understand the limits of his or her knowledge and experience. If structural members are damaged, a professionally qualified structural engineer should be involved in the inspection. Some types of damage may look similar to past cases, but the type of stress or the type of structure can cause the repair to require different safety procedures.

Quantities: If maintenance and repair quantities have been provided, they should be verified in the field. Since some time may have passed since the inspection was done, conditions may have changed enough to change the estimated quantities required.

Location: The exact location of the damage should be verified so that the crew can locate it without any difficulty. This will contribute to efficiency on the job.

Traffic Control: The roadway should be studied in both directions to verify that any traffic control plan developed under the guidance of the MUTCD and the state manual is appropriate and applicable for both worker safety and motorist safety.

Staging Area: A place is needed near the site to safely store equipment and materials.

Repair Work Not Specified: Check for other damage not included in the work order. It is more cost-effective to do all needed maintenance and repairs in one trip. However, this must be balanced against any backlog of more critical repairs needed elsewhere, and additions to the original work order need to be agreed to by maintenance engineers and managers at higher levels of management.

3.2.8.2 Channel Damage

Problems with the channel have the potential to change rapidly if a major storm occurs. Debris may direct the water force toward a vulnerable portion of the bridge substructure. Damaged riprap may leave an abutment footing unprotected. To avoid a potential catastrophe, channel repairs should be treated with appropriate urgency. Major floods can provide an opportunity to assess the potential for bridges to resist scour under extreme water flows [4].

Cause: The cause of channel problems should be identified and corrected, if possible, as part of the repair.

Type of Repair: Determine if the problem is caused by local scour or if it is related to a change along a significant length of the channel (which might be best solved with a bridge modification).

Size of Material: To determine what size of material needs to be used to correct a scour problem may require an analysis of specific characteristics of the stream flow by a professionally qualified hydraulics engineer.

Materials and Equipment: The equipment you have available may dictate the method or repair you can choose. The type of material available may be a significant factor in determining how much material is needed.

Permits and Damage to Adjacent Property: Evaluate if special permits will be needed or special precautions will be necessary for environmental control. The potential negative impact of any repair on adjacent property needs to be evaluated.

3.2.8.3 Approach Damage

Approaches can present special problems because roadway maintenance crews (who may not be aware of the effects of approach roadway damage on the bridge) often maintain the approaches. NCHRP Synthesis of Highway Practice 234 provides a comprehensive examination of the problems associated with settled bridge approaches and the maintenance treatments appropriate to their repair. [17]

Settlement: Settled approach roadways to bridges are often merely built up with asphaltic patching material. Bridge maintenance personnel should determine if the settlement is related to any scour problem behind the abutment.

Pavement Pressure: If joints in a concrete approach roadway become filled with incompressible materials, the pavement will push on the bridge abutment when it expands in hot weather. In this case, a relief joint will need to be installed in the approach roadway to prevent damage to bridge bearings, bridge joints, etc.

Repaved Approach Roadway: When an approach roadway is repaved, any overlay or additional depth of pavement should not extend across the bridge because it will change the dead load of the bridge. Neither should there be a bump left at the end of the bridge because it will create dynamic loading on the bridge.

Drainage and Debris: Check any maintenance of the bridge approach to ensure that surface water from the bridge deck will continue to drain away from the bridge, and that trash does not accumulate at the transition from the approach roadway to the bridge.

3.2.8.4 Deck Damage

The deck is exposed to many elements, including de-icing salts, abrasives, and mechanical wear from vehicular traffic. Since the deck provides the riding surface for traffic, damage that can be tolerated elsewhere on the structure is a major problem on the deck. Thus, repairs to the deck are sometimes seen as "emergency" conditions. The goal of bridge maintenance is to prevent deck problems that require an emergency response.

Preventive Maintenance of Concrete Deck: Good maintenance of a concrete deck requires more information than a biennial bridge inspection typically provides. Reinforcing steel corrosion causes most concrete deck problems. Once the steel corrosion starts, extensive rehabilitation is required for permanent repair. Thus, the most cost-effective process is to perform concrete bridge deck maintenance before spalls and potholes begin. If tests indicate no corrosion of the steel and no delamination of the concrete, and if the salt contamination of the concrete is minimal, then annual flushing of the deck or resealing may be the most cost-effective maintenance possible. If tests indicate any of the above conditions exist, repair—and perhaps removal and replacement—of the deck may be required.

Temporary Concrete Deck Repairs: If the level of salt contamination in the concrete and the areas of delamination are extensive, it may be more cost-effective to replace the deck than to repair it. In this case, patching with a material and process that will least interfere with traffic and last as long as possible may be best while a deck replacement is being programmed and budgeted.

Timber Deck: Unless the damage to a timber deck is isolated, it is generally better to replace the entire deck than to replace a few planks at a time. This is especially true for laminated timber decks.

Orthotropic Bridge Decks: Cracks of a minor nature in an orthotropic bridge deck will not require restriction of traffic other than during the repair period. After the cause of the crack has been determined and corrective action taken, the crack can be repaired with a welded cover plate. These bridges generally have an abrasive wearing course that has to be removed and replaced over the work area.

Steel Grid Decks: Steel open-grid decks usually require little maintenance. Since their failure exposes raw steel to corrosive elements, the welds or rivets that join the grid end and hold down the steel grid decks should be replaced if broken. Broken welds should be removed by grinding and then welded again. Grid sections with severe corrosion and section loss should be replaced. Steel grid decks tend to become slippery when wet or frost covered. Studs that are about 8 mm (about 5/16 inch) in diameter and about 10 mm (about 3/8 inch) high may be welded on intersections of the cross members, providing a grip surface that overcomes slipperiness.

Joint Repairs: To properly protect the bridge components below the deck, the joints should be waterproof. To be waterproof, the joints must be sealed. If a bridge deck joint is not sealed because it is an older joint design, consider remodeling the joint to accommodate an effective sealer. Some agencies are retrofitting bridges to eliminate as many bridge joints as possible. Bridge joints should only be eliminated if a professionally qualified structural engineer has determined that stresses will not be transferred to other parts of the structure, thus creating additional problems. If a bridge joint fails, it may create a problem for traffic using the bridge, and emergency repair may be needed until a full joint repair can be scheduled.

Rail Repairs: Vehicular collisions are the primary cause of bridge rail damage, and the cost of repair may be recovered if the errant vehicle can be identified. It is important to check other parts of the structure to make certain that the rail impact did not cause problems elsewhere on the bridge.

3.2.8.5 Superstructure Damage

Superstructure repairs often require more time and resources than repairs to other parts of the bridge. Planning is important to ensure that these resources are available when needed and are used efficiently. The site visit is essential to effective planning.

Access: Access is always an important consideration for repairing superstructures. Depending upon the access to the superstructure, it may be necessary to consider rigging a platform from the bridge, erecting a scaffold from the ground, working from a barge, or using mechanical platforms.

Lifting and Support: Repair of superstructure damage often requires lifting and supporting parts of the bridge. An engineer should design this procedure. However, the feasibility of the procedure needs to be verified during the site visit. Think about the number and size of jacks required, shoring requirements, jacking and shoring locations, foundation and terrain for shoring, and the opening and condition of deck joints.

Steel Beams: Any crack or fracture in a steel beam, girder, or truss should be considered a sign of serious distress, and immediate corrective action should be taken. In some instances, the only action required may be to drill a hole at the end of the crack to control further crack growth. Cracks in steel beams usually occur at welded areas. Indiscriminate field welding or making improvised attachments to beams is discouraged unless a professionally qualified structural engineer has analyzed the beam's stress conditions and recommends welding as an appropriate solution.

Trusses: Most truss bridges are made of steel, although some old bridges with wrought iron trusses are still in use. Repair normally consists of replacing a damaged member or strengthening weakened members by adding steel reinforcing plates. The type (tension or compression) and magnitude of the stresses on each member must be determined before beginning repairs. A professionally qualified structural engineer should assist in this assessment and in the development of the repair procedures.

Timber Beams: Certain types of cracks in timber bridge beams can result in a loss of load-carrying capacity and may require immediate repair or replacement before traffic is allowed to return to the bridge. Using U-bolts fitted around the beam and extending through the deck can sometimes temporarily repair timber beams with a longitudinal crack. A new beam may also be fastened alongside the damaged beam by this method. Normally all the stringers that have decayed must be replaced.

Reinforced-Concrete Beams: The concrete deck is normally an integral part of the beams. Since they carry the load together as one unit, damage or repairs to one affect the other. Repairs may be cosmetic or structural. Cosmetic repairs are appropriate only on small areas where the reinforcing steel is not damaged. To perform structural repairs to concrete beams, the bridge should be supported by shoring. Spalling of the deck surface above the beam can affect the strength of a concrete T-beam. Corrosion of the main longitudinal reinforcing steel in the bottom of simple span beams can cause the concrete cover to spall, exposing the reinforcing steel to further corrosion. A crack at the end of the beam that extends diagonally upward away from the bearing is an indication of shear failure.

3.2.8.6 Substructure Damage

Substructure repairs are often unique; therefore, a general list of things to look for during a site visit for a typical substructure repair is difficult to compile. However, most of the considerations for superstructure repairs also apply to substructure repairs. Access to the repair area certainly needs to be checked, and lifting a support may also be necessary.

Cracks: Before a method is identified for repairing a crack, it must be determined if the crack is active (moving). A moving crack should be sealed with a flexible material. A passive crack can be sealed with a rigid material. If possible, determine if the crack is a full-depth crack. This is hard to do in an abutment.

Wet or Dry Repair: The method of repair and its associated cost may be quite different depending upon whether the repair can be performed in a dewatered area or whether it must be done underwater.

Column Repairs: Column repairs often cover a deeper problem with a jacket or layer of concrete. Determine if the repair is only intended to protect the column against additional damage or if it is intended to replace the column's lost load-carrying capacity. A corroding steel pile, a concrete pile with corroding reinforcing steel, or a decaying timber pile may all continue to deteriorate under a concrete repair jacket. This could result in serious consequences if the jacket is not designed to carry the column load.

3.2.8.7 Emergency Damage

When damage to a bridge presents an immediate safety hazard, priorities are different than on other site visits. The objective is to assess the urgency of the situation and to begin to take appropriate actions.

Close or Restrict Use of the Bridge: A standard procedure to implement an immediate bridge closing should be available in every agency so it can quickly be put in place once personnel making an on-site visit determine that the bridge needs to be closed. The procedure should outline who and what agencies must be contacted, what signs must be erected, how to establish a detour, and how to use the mass media to inform the public. A professionally qualified structural engineer should be contacted if there's a question about whether the bridge should be closed or only restricted in the way vehicles are permitted to use it.

Emergency Repair: After the immediate safety issues are resolved, the repair urgency is determined and the maintenance and repair process can proceed according to the priority set on the repair. After the type of repair is determined, the personnel and equipment needed can be mobilized and the necessary materials can be located while the repair details are worked out. Special provisions for working at night and lodging the repair crew near the site may also be necessary.

3.2.9 REFERENCES

1. Federal Highway Administration. *Bridge maintenance training manual*. Report FHWA-HI-94-034. Washington, D.C.: Federal Highway Administration, 1994.

2. How to maintain local bridges at the least cost. *Better Roads*. Park Ridge, IL, May 1989, pp. 29–30.

3. Weekend work eases major deck replacement. *Better Roads*. Park Ridge, IL, December 1994, pp. 25–26.

4. Mueller, David S., Mark N. Landers, and Edward E. Fischer. *Scour measurements at bridge sites during 1993 upper Mississippi river basin flood*. Transportation Research Record 1483. Washington, D.C.: Transportation Research Board, 1995, pp. 47–55.

5. American Association of State Highway and Transportation Officials. *Guidelines for bridge management systems*. Washington, D.C., 1993.

6. Organization for Economic Co-operation and Development. *Bridge management*. Paris, France, 1992.

7. Mohamed, Hosny A., A. O. Abd El Halim, and A. G. Razaqpur. *Use of neural networks in bridge management systems*. Transportation Research Record 1490. Washington, D.C.: Transportation Research Board, 1995, pp. 1–8.

8. Hawk, Hugh. *BRIDGIT deterioration models*. Transportation Research Record 1490. Washington, D.C.: Transportation Research Board, 1995, pp. 19–22.

9. Ravirala, V., D. A. Grivas, A. Madan, and B. C. Schultz. *Multicriteria optimization method for network-level bridge management*. Transportation Research Record 1561. Washington, D.C.: Transportation Research Board, 1996, pp. 37–43.

10. Vitale, Jeffrey D., Kumares C. Sinha, and R. E. Woods. *Analysis of optimal bridge programming policies*. Transportation Research Record 1561. Washington, D.C.: Transportation Research Board, 1996, pp. 44–52.

11. Farid, Foad, David W. Johnston, Bashar S. Rihani, and Chwen-Jinq Chen. *Feasibility of incremental benefit-cost analysis for optimal budget allocation in bridge management systems*. Transportation Research Record 1442. Washington, D.C.: Transportation Research Board, 1994, pp. 77–87.

12. Lu, Yun, and Samer Mandanat. *Bayesian updating of infrastructure deterioration models*. Transportation Research Record 1442. Washington, D.C.: Transportation Research Board, 1994, pp. 110–114.

13. Sanders, David H., and Yuqing Jane Zhang. *Bridge deterioration models for states with small bridge inventories*. Transportation Research Record 1442. Washington, D.C.: Transportation Research Board, 1994, pp. 101–109.

14. Grivas, Dimitri A., B. Cameron Schultz, David J. Elwell, and Anthony E. Dalto. *Span-based network characterization for bridge management*. Transportation Research Record 1442. Washington, D.C.: Transportation Research Board, 1994, pp. 123–127.

15. Saraf, Vijay, Andrej F. Sokolik, and Andrzej S. Nowak. *Proof load testing of highway bridges*. Transportation Research Record 1541. Washington, D.C.: Transportation Research Board, 1996, pp. 51–57.

16. Kim, Sangjin, Andrej F. Sokolik, and Andrzej S. Nowak. *Measurement of truck load on bridges in Detroit, Michigan area.* Transportation Research Record 1541. Washington, D.C.: Transportation Research Board, 1996, pp. 58–63.

17. Briaud, Jean-Louis, Ray W. James, and Stacey B. Hoffman. *Settlement of bridge approaches.* Synthesis of Highway Practice 234. Washington, D.C.: National Cooperative Highway Research Program, Transportation Research Board, 1997.

18. Thompson, Paul D., and Michael J. Markow. *Collecting and managing cost data for bridge management systems.* Synthesis of Highway Practice 227. Washington, D.C.: National Cooperative Highway Research Program, Transportation Research Board, 1996.

4.0 EQUIPMENT SYSTEMS
4.1 INTRODUCTION

The maintenance equipment used by state transportation agencies to conduct their maintenance activities varies with each organization, the geographic region served and its associated climate, and general state policies governing public capital investment in extensive equipment fleets. Maintenance equipment could include a wide range of equipment, tools, instruments, and technologies. This chapter, however, will be restricted primarily to trucks and field equipment.

General equipment management organization varies with each individual state. However, there are typically two common management structures: (1) a subdivision of the maintenance structure under the state maintenance engineer and (2) a separate equipment management division within the department of transportation with a state equipment engineer or manager as the principal manager.

Most states operate garage facilities as part of their maintenance facilities and operations. Other states have garage facilities available to them through a separate state agency or as a separate organization within their transportation agency that is neither a part of nor the responsibility of their maintenance structure. Some states have field mechanics for field service of equipment problems as part of their maintenance field organization.

All states have formal equipment replacement programs and criteria established for the replacement of maintenance trucks and field equipment. Criteria for replacement range from a combination of age and mileage, mileage only, age or hours used, present value for sale and recovery of initial capital investment in comparison to replacement cost, and annual maintenance cost (basically a cash stream analysis on a life-cycle cost basis). Most states use "present condition" of equipment to be replaced as part of their replacement criteria.

In most states, the actual purchasing of maintenance equipment is handled by another state agency but is processed on the request and recommendation of the transportation agency's maintenance organization. In many states, the specifications used for the purchase of maintenance equipment are prepared by either the maintenance organization or in cooperation with the maintenance organization. Most states receive specific appropriations for the purchase of trucks and equipment.

Requests and recommendations for types of maintenance equipment normally originate in the state transportation department's district or division offices and are then reviewed by central office personnel. Most maintenance equipment is acquired for use in specific districts or divisions. Some specialized equipment is acquired for central office control because it requires statewide use to generate a level of use that justifies ownership.

A maintenance truck is the most common item of maintenance equipment. The types of attachments, allied equipment, and associated modifications purchased with maintenance trucks vary widely among the states. Some states supplement their state-owned maintenance truck and equipment fleets with rented equipment on an as-needed basis. The second most prevalent item in the maintenance equipment inventory is the tractor (used in right-of-way maintenance operations). As with trucks, the attachments and modifications purchased with tractors vary greatly among the states. A recent survey of all U.S. state and Canadian provincial DOTs found the 10 most needed pieces of equipment to conduct their maintenance responsibilities to be (in order): dump trucks, snowplows, loaders, spreaders/salters, pickup trucks, motor graders, impact attenuators, mowers, excavators, and street sweepers [1].

All states have some type of cost record system for their maintenance trucks and equipment, but the extent of the system and the level of detail varies. Most states have some type of preventive maintenance program for their trucks and equipment. Training programs for truck drivers and equipment operators are provided in most states.

Most states have an equipment management system (EMS), either in the developmental stage or moving into the mature implementation stage. Integration of an EMS with the agency's maintenance management system (MMS) and bridge management system (BMS) will greatly enhance the agency's ability to determine what is the optimal equipment fleet to support the functional mission of the maintenance division. EMSs are intended to monitor such things as equipment use levels, equipment operating costs, fuel consumption, and equipment maintenance as well as to assist in developing replacement criteria.

Most states have fabricated or developed specialized maintenance equipment in their own shops. For instance, the Nevada DOT modified an existing striping unit to change to waterborne paints for improved environmental control, the West Virginia DOT modified mowing units to truck-mounted boom units, and the Minnesota DOT developed a remote-controlled mowing unit for increased operator safety on steep slopes. The Montana DOT (and several others) modified various truck units to use chemical de-icers and salt brines directly on the pavement for snow and ice control [2]. Many DOTs are also cooperating in joint efforts to conduct research and development seeking new, improved equipment systems to support the maintenance function. These cooperative efforts across an entire region having similar maintenance activity needs offer the potential to create a large enough market, and the purchasing power, to interest equipment manufacturers to provide equipment tailored to maintenance agency needs. Estimates of the annual expenditures of all state agencies combined for the purchase of new maintenance equipment have ranged as high as over $200 million. Efforts to optimize the functional return on this large investment are an important part of an overall budget management strategy to achieve maximum return on the public's transportation investment. An evaluation of new equipment systems for precise distribution and placement of de-icing materials on pavement has found that these newer equipment systems use less chemicals (which is better for the environment and reduces materials costs) and provide more effective snow and ice control than older equipment designs (which increases traveling public safety) [3].

4.2 EQUIPMENT APPLICATION

The application of equipment to a maintenance activity will be largely governed by the standards defined in each agency's maintenance management system. However, some typical categories of equipment that are assigned to different kinds of maintenance jobs are listed below to provide an apppreciation for how the transportation agency's equipment inventory is so extensive and varied compared to the equipment inventory of a contractor who specializes in a limited scope of work.

Job	Equipment
Hauling	Trucks, Trailers, Scrapers
Ditching and Diking	Bulldozers, Drag Lines, Hydraulic Excavators, Graders, Rotary Ditchers, Backhoes
Road Grading	Graders, Bulldozers, Tractor Drags, Maintainers
Truck Loading	Tractor End Loaders, Drag Lines, Hydraulic Excavators, Beltloaders
Mixing Road Materials	Graders, Special Mixing Machines
Dirt Moving	Bulldozers, Tractor Scrapers, Drag Lines, Front-end Loaders, Backhoes
Excavation	Bulldozers, Power Shovels, Drag Lines, Tractor Scrapers, Front-end Loaders, Hydraulic Excavators
Snow Removal	Snowplows, Rotary Brushes, Snow Blowers, Graders, Front-end Loaders, Bulldozers, Trucks, Belt Loaders
Scarifying and Shaping	Graders with Attachments, Tractors with Attachments, Rippers, Hydraulic Excavators
Backfilling	Bulldozers, End Loaders, Backhoes, Trenchers
Compaction	Crawler Tractors, Flat-Wheel Rollers, Rubber-Tired Rollers, Sheeps Foot Tamping Rollers, Pneumatic Rollers
Stockpile Maintenance	Bulldozers, Drag Lines, End Loaders
Pipeline Laying	Tractor Winches, Bulldozers, Tractor Side Booms, Backhoes, Drag Lines
Watering Gravel or Fill	Water Truck with Pressure Pump
Flushing Pipe	Water Truck with High-Pressure Pump and Hoses
Crushing Rock or Gravel	Jaw Crusher, Gyratory Crusher Rolls
Spreading Asphalt for Treatments	Distributor Trucks and Tanks, Trailer-type Distributors
Heating Asphalt	Steam Generators, Heating Kettles
Spreading Rock Chips	Dump-bed Truck with Spreader Box Attachment, Self-Propelled Spreaders with Trucks
Traffic Striping	Truck-Mounted Striping Unit or Hand-Operated Unit, Shadow Vehicles
Pavement Joint Sealing	Trucks, Trailer-Mounted Sealant Applicator Unit, Shadow Vehicles, Trailer-Mounted Air Compressor with Lances, Joint Router, Attenuators

Note: Bridge maintenance equipment for typical operations is noted in the bridge maintenance section because of the equipment's specialized application and the frequent use of contracting as a means of administering bridge work.

When using equipment in the vicinity of moving traffic, maintenance personnel safety is always an issue for maintenance engineers and managers. Technological developments have resulted in warning equipment systems being added to inventory; these warning systems have the potential to reduce accidents significantly when personnel are operating equipment. One development is "radar emulation," in which a fake radar pulse is emitted in advance of the maintenance work and makes drivers with radar detectors think they are being tracked by police units [8]. A second development is a device that tracks the speed of vehicles approaching work zone protective devices to predict a vehicle impact in advance and sound an alarm to workers before the vehicle actually intrudes into the workers' area [9]. If such devices are to be effective, they must be maintained in a high state of readiness by persons assigned to maintain such equipment.

Originally laser control of grade and alignment of equipment was considered applicable only to paving machines associated with the construction process. However, laser beam control systems and sonic beam control systems are becoming flexible enough and sufficiently varied in the types of equipment units to which they can be applied that maintenance equipment used in pavement repair, ditch grading, and slope repairs are candidates for these equipment control systems [10].

Following the lead of the construction industry, maintenance is beginning to incorporate multiple-use equipment in areas beyond trucks. One example is the versatile small front-end loader known as the skid steer [11–13]. While these units are typically thought of as a front-end loader, a wide range of attachments are available for them, and they can be hauled to the job site on a trailer behind a truck, making them a preferred choice for many repair activities. Flexibility in maintenance functions with smaller crews requires maintenance agencies to review profit-motivated organizations for examples in their equipment inventory that can be adapted to support public efficiency.

4.3 MAINTENANCE OF EQUIPMENT

Equipment maintenance involves considering innovations, not necessarily high-tech ones, that reduce the cost of maintaining a vast equipment fleet [4]. For example, one agency found that low-cost tires with a thin recap provided better wear and service than more expensive new tires in urban service. Another agency found that corrosion damage to trucks could be significantly reduced in a snow and ice belt area by resealing all units with silicone sealants at new delivery before putting them on the road.

Extending the life and improving the serviceability of engines on equipment units can sometimes be accomplished simply by having the proper oil filter, transmission filter, fuel filter, etc. [5]. Persons responsible for equipment maintenance should coordinate with persons responsible for purchasing, making sure that the filters purchased for routine equipment maintenance meet the manufacturer's specifications for the equipment. As motor oils continue to develop into lubricating fluids that reduce internal engine friction and release less pollutants during the combustion process [7], they must be coupled with the correct filters.

Most agencies are planning for the day when their truck and automobile fleet will be fueled with some alternative to gasoline and diesel fuel [6]. This alternative will require more planning on the fuel storage and dispensing side of the operation but will likely reduce the demands for routine spark plug maintenance and the frequency of oil changes.

4.4 ACQUISITION AND REPLACEMENT

One of the critical elements of any equipment replacement program is the depreciation period, or life, of a particular equipment item. Unlike for a private sector organization, the life over which an equipment item is considered useful to an agency is not bound by Internal Revenue Service regulations on equipment depreciation. Thus, a particular agency should develop depreciation periods for each equipment item on the basis of criteria that match their equipment philosophy and their budget philosophy, considering annual maintenance cost for the item, downtime caused by loss of the item in service, equipment maintenance personnel requirements, recoverable sale value of the item, initial capital cost of replacement, any increase or decrease in equipment maintenance costs associated with replacement of the item, etc. The following typical depreciation periods for various maintenance equipment items are only intended for general comparisons but are also useful in evaluating if an agency's depreciation schedule for some item may be inconsistent with the depreciation periods it uses for other classes of equipment.

Type of Equipment	Depreciation Period (years)
Air Compressor, Truck-Mounted	7
Air Compressor, Wheel-Mounted	7
Air Tools	5
Automobile	4
Batcher, Measuring	6
Batcher, Weighing	5
Bin, Aggregate	10
Blower, Portable	5
Bucket, Clam or Drag Line	6
Bulldozer, Tractor Attachment	7
Concrete Saw	4
Conveyor Belt, on Wheels	5
Crack-Filling Machine	8
Crane, Crawler	10
Crusher, Rock, Portable	8
Crushing and Screening Plant	8
Cultivator, Motor-Driven	5
Distributor, Bitumen	8
Drill, Core	7
Drill, Drifter	5
Drill, Wagon	7
Dryer, Aggregate	8
Dryer, Elevator	8
Engine, Gasoline or Diesel	7
Excavator, Telescoping Boom	10
Finishing Machine, Concrete	5

Type of Equipment	Depreciation Period (years)
Generator, Electric, Gasoline, or Diesel	7
Gradation Control Unit	8
Grader, Motor	8
Grader, Pull	10
Guardrail Straightener	10
Heater, Aggregate, Revolving	8
Heater, Bitumen Kettle	8
Heater, Tank Car	8
Hoist, Bucket, Truck-Mounted	5
Hoist, Drum, with Power	7
Joint-Cleaning Machine	4
Loader, Belt, Blade Feed	6
Loader, Chain Bucket	6
Loader, Scoop, Crawler	7
Loader, Scoop, Wheel Tractor	6
Magnet, Road	5
Maintainer, Self-Propelled	6
Maintainer, Drawn	5
Mixer, Bituminous	7
Mixer, Concrete	7
Mixing Plant, Bituminous	8
Mower, Reel	6
Mower, Rotary	7
Mower, Sickle Bar	5
Mud Jack	6
Nuclear Density Unit	3
Pavement Breaker	5
Paver, Bituminous	8
Spray-Paint Outfit	8
Hammer, Pile	10
Hammer, Sheeting	7
Pump, Asphalt	6
Pump, Water	6
Road Mixer, Digging Rotor	6
Road Mixer, Pug Mill Type	8
Road Roller	10
Roller, Pneumatic	10

Type of Equipment	Depreciation Period (years)
Roller, Sheep's Foot	10
Roller, Trench	10
Roller, Vibratory	10
Rooter or Ripper, Heavy	8
Scarifier, Rotary	5
Scraper, Drawn	7
Scraper, Self-Propelled	7
Screen, Vibrating	8
Screening and Loading Plant	8
Seed-Gathering Machine	4
Shovel, Crawler	10
Shovel, Truck Mounted	10
Snowplow, One-way	10
Snowplow, Rotary	10
Snowplow, V	10
Spraying Machine, Insect	5
Spreader, Drawn or Attached	6
Spreader, Self-Propelled	5
Sprinkler, Water, Truck-Mounted	7
Stump Cutter	7
Subgrade Finisher	4
Sweeper, Rotary	5
Tank, Bitumen Storage	10
Tank, Bitumen, Wheel-Mounted	10
Tank, Water, Skid-Mounted	10
Tractor, Crawler	7
Tractor, Wheel	6
Traffic Line Marker	7
Trailer, House	7
Trailer, Platform	7
Tree Mover, Truck Mounted	8
Truck Tractor, with Semitrailer	7
Trucks, Multiuse	5
Trucks, Snow Use Only	10
Vibrator, Pneumatic	5
Wagon, Semi, with Tractor	7
Washing and Screening Plant	8
Welding Outfit, Electric	5

Chapter 4 Equipment Systems

4.5 REFERENCES

1. Dump truck tops state equipment choices. *Better Roads*. Park Ridge, IL, August 1997, pp. 18–20.

2. Adapted machinery helps state DOTs. *Better Roads*. Park Ridge, IL, August 1997, pp. 22–23.

3. Fleege, Edward J., and Robert R. Blackburn. *Spreader equipment for anti-icing*. Transportation Research Record 1509. Washington, D.C.: Transportation Research Board, 1995, pp. 22–27.

4. Technology report: What's new in trucks. *Better Roads*. Park Ridge, IL, August 1997, pp. 24–27.

5. What you should know about filters. *Better Roads*. Park Ridge, IL, December 1994, pp. 18–19.

6. Easier ways to switch to alternatively fueled vehicles. *Better Roads*. Park Ridge, IL, December 1994, pp. 20–21.

7. Motor oil change will be mandated soon. *Better Roads*. Park Ridge, IL, April 1997, pp. 31–32.

8. Radar emulation sounds an alarm. *Roads and Bridges*. Des Plaines, IL, January 1997, pp. 44–46.

9. Intrusion alarm goes high tech. *Better Roads*. Park Ridge, IL, February 1997, pp. 15–16.

10. Machine control: It's more than just lasers. *Roads and Bridges*. Des Plaines, IL, May 1997, pp. 54, 56, 58, 61.

11. Skid-steers: The 90's market. *Roads and Bridges*. Des Plaines, IL, January 1996, pp. 14, 16.

12. Skid-steers: The contractor's Swiss army knife. *Roads and Bridges*. Des Plaines, IL, April 1996, pp. 44, 46–48, 50, 52.

13. Round 'em up. *Roads and Bridges*. Des Plaines, IL, April 1997, pp. 22–24, 67–68.

5.0 MAINTENANCE RESEARCH AND DEVELOPMENT

5.1 ADMINISTRATION

Research of interest and benefit to the maintenance engineer and maintenance manager can be conducted through many mechanisms. More important than the mechanism used to administer the research is that the research be proposed and funded. In any case, the role of maintenance must continue to grow in transportation research. Almost all research in structures, materials, design, operations, and communications technology directly affects maintenance if the research produces data that can be implemented. Consequently, there is a self-interest on the part of the maintenance community to be involved in transportation research; there is also a professional and societal responsibility to be involved, to guide research expenditures into avenues of inquiry that will make a lasting life-cycle cost improvement in the transportation system.

The first and most obvious step in advancing maintenance research is for maintenance engineers and maintenance managers to encourage field personnel to communicate their concerns about methods, materials, and processes that do not "work right" to their supervisors and to other people in their own organization, generating ideas that might be appropriate for study.

Then maintenance engineers and managers should inquire within their own agency how an idea gets submitted to the AASHTO review process for consideration as an annual NCHRP effort. Whenever any idea, or even part of one, from your maintenance organization becomes an NCHRP project, this should be communicated to all personnel, and a communication liaison should be established to keep your organization informed about progress on and the results of the research.

Some agencies have their own research program. If your maintenance personnel have an idea that is worthy of study, analysis, and development, but your agency but does not have sufficient interest in competing to win NCHRP funding, maintenance engineers and managers should interact with persons who manage and program the local research fund to promote the maintenance project. In cases where the research idea might interest agencies in adjacent states, a joint effort can be arranged. This joint effort can be direct or it can be administered through NCHRP as a "pooled resource" project outside AASHTO's regular national priority system. One example of such a multistate effort is the advanced concept maintenance vehicle that the Iowa DOT and the Minnesota DOT are jointly designing.

The Minnesota DOT brings maintenance research needs from field personnel into a research program funded with a small fraction of the total maintenance budget and managed by the maintenance division. Ideas are forwarded from field-level personnel, evaluated and refined, and submitted for prioritization with respect to the funds available. Then the projects that can be pursued each year are initiated. (The program is known as the Statewide Maintenance Operations Research Program [1].) Field forces that have participated in the process are kept informed of the progress of their ideas and any projects that arise from the process. This innovative approach to creating a modest in-house research program that focuses on maintenance needs pays dividends in several ways: (1) projects that fulfill a research and development need to improve maintenance but would not ordinarily be competitive for research funds in a large-scale program get initiated; (2) maintenance personnel who have a keen intuitive sense about what needs to be done but ordinarily never get to participate in the research and development process become active participants; (3) field personnel who actually implement the results of research efforts are stimulated to be more observant and more analytical about potential ways to improve methods, materials, and equipment because they see the direct application of research results in which they or their colleagues were active participants.

Finally, agencies need to encourage and to assist maintenance engineers and maintenance managers to serve on technical and administrative committees of AASHTO and the Transportation Research Board, bringing a maintenance community perspective to research dissemination and evaluation as well as increasing the interest of a broad range of researchers in conducting maintenance research.

5.2 METHODS

Maintenance engineers are always looking for better ways of doing things; it is just the nature of their personality and the work of their people. In Minnesota, the Office of the Legislative Auditor published a report on their administrative examination of snow and ice control across all levels of agencies in the state. This report compiles the best practices from a public administration view [2]. This approach is not normally the type of review associated with research, but it does present a fresh viewpoint that can stimulate thinking about what is good and what might be studied to make maintenance methods better.

The fundamental questions to be asked when examining any maintenance method with respect to assessing the need to conduct research into a particular maintenance method include the following:
- Can it be done with less equipment?
- Can it be done with fewer workers?
- Can it be done with less materials?
- Can it be done with less interference to traffic through the work zone?
- Can it contribute to increased productivity?
- Can it increase the quality of the result of the maintenance activity?
- Can it make the work operation safer for workers, the traveling public, or both?
- Can it reduce the cost of conducting the maintenance activity?

Obviously, no single research effort to improve a maintenance method can be expected to generate improvements in all areas. Maintenance engineering judgment has to determine the suitability of any method change that has been studied and analyzed.

5.3 MATERIALS

Maintenance costs and the performance of maintenance activities often heavily depend on the character and quality of the materials used in the various processes. There is a growing understanding that if materials were available with improved characteristics, such as durability and flexibility, a lower life-cycle cost to maintenance could be achieved even if these materials were more costly than the traditional materials used. Any agency interested in evaluating improved materials should consider interacting with the materials science and engineering profession to investigate which "designer materials" may meet the specific characteristics needed to increase the quality of materials performance in maintenance activities. A necessary first step in this direction is for maintenance personnel to study and assess what characteristics are necessary for the material to make a major difference in the performance of a maintenance activity.

5.4 HUMAN RESOURCES AND PERSONNEL

Maintenance management is an activity intended to optimize the allocation of resources (i.e., personnel, equipment, materials) for the total maintenance program to provide the best service within the budgetary constraints in effect at the time. The search for research and innovation that will improve the process requires study of equipment and materials. Too often, innovation and development of new approaches to maintenance human resources and personnel is largely an attempt to borrow private sector profit-oriented organization innovation. While there is much merit to examining what has been done in private industry and evaluating the applicability of private sector concepts to the maintenance agency, public agencies and their constituents might be better served if pilot research projects were a part of all technology transfer initiatives from private sector personnel research to maintenance agencies.

5.5 EQUIPMENT

As it should be, equipment research is active and ongoing among both manufacturers (and suppliers) and agencies. Innovation in equipment systems is probably the easiest area to evaluate and implement. Seeking innovation in equipment systems is just beginning to include some of the following areas of study, all of which may yield significant improvements in the performance of maintenance:

- Ergonomics or Human Factors: Efforts to improve the person-machine interface in maintenance equipment have the potential to increase safety, productivity, and employee satisfaction in the same ways as has been achieved in the manufacturing industry.
- Robotics: Study of the applicability of robotic methods, machine vision, and computer-controlled processes to maintenance equipment is an intriguing field of research. It is difficult to foresee if these processes will have a positive or negative effect on maintenance. So much worker skill and judgment is necessary to use equipment for a maintenance process in such a way to achieve a high-quality final product. Using robotics for maintenance may depend upon the ability of computer engineers and computer scientists to apply expert systems to maintenance processes. Highway maintenance and support is one area that has been identified for potential application of knowledge-based expert systems [3].

5.6 INTERNATIONAL ASPECTS AND TECHNOLOGY TRANSFER

Through multinational corporations, global business and commerce is increasing. Higher education is also increasingly requiring students to be aware of global connections in their field of study. While roadway and bridge systems differ on other continents, while labor and safety regulations in Europe and Asia differ from those in North America, and while the availability of materials on other continents may vary considerably from those available in North America, all nations and continents share a common purpose and intent in terms of roadway and bridge maintenance. Much can be learned from organizations outside of North America, and their practices need to be evaluated for possible technology transfer. Two documented efforts at observing practices outside North America, one in winter maintenance [4] and one in bridges [5], have already stimulated significant research and development activity. The Minnesota DOT has developed a formal personnel and information exchange with Scandinavian countries for winter maintenance.

5.7 REFERENCES

1. Minnesota Department of Transportation, Maintenance Operations Research Unit. Statewide maintenance operations research annual 1995 report. St. Paul, MN, December 1995.

2. Office of the Legislative Auditor, Program Evaluation Division, State of Minnesota. Snow and ice control: A best practices review. St. Paul, MN, May 1995.

3. Cohn, Louis F., and Roswell A. Harris. *Knowledge-based expert systems in transportation.* Synthesis of Highway Practice 183. Washington, D.C.: National Cooperative Highway Research Program, Transportation Research Board, 1992.

4. National Cooperative Highway Research Program. *Winter maintenance technology and practices—Learning from abroad.* Research Results Digest Number 204. Washington, D.C.: Transportation Research Board, January 1995.

5. National Cooperative Highway Research Program. *Report on the 1995 scanning review of European bridge structures.* Report 381. Washington, D.C.: Transportation Research Board, 1996.

6.0 TORT LITIGATION

6.1 TORT CLAIMS

Lawsuits seem to be inevitable in U.S. public agencies. U.S. society is well known as being prone to sue rather than settle disagreements through mediation and reconciliation. Thus, maintenance engineers and maintenance managers should be prepared to participate in a process intended to "manage" litigation [1]. The basic processes for managing litigation, besides having capable, aggressive legal assistance, include the following activities:

- Establishing a risk management group.
- Coordinating an approach that deals with litigation across the entire governmental unit (including all state agencies, all departments of the city, etc.).
- In a comprehensive manner, monitoring and reviewing litigation across the transportation agency's entire range of responsibilities and activities.
- Creating a claims management process that is consistent in its approach to processing claims, settling claims versus pursuing a trial, and investigating and analyzing activities.
- Establishing an analytical process to forecast and allocate expected tort claim costs, minimizing the disruption one year's claims will cause to the transportation agency's functional activities.

To the degree possible, it is advisable to have policies and mechanisms in place that facilitate resolution of claims rather than adversarial court action. This strategy is especially true for contract disputes and will become more important as some agencies increasingly adopt contract maintenance for selected activities [2]. A program to assist maintenance engineers and maintenance managers in the effective administration of contract maintenance in ways that avoid contract claims should include (1) dispute avoidance techniques, (2) early dispute recognition techniques, and (3) dispute resolution techniques.

Efforts to prevent disputes include developing partnering relationships in contract efforts so that all parties have a vested interest in cooperating. Open, direct communication activities with interested, affected parties, such as adjacent property owners, will help to avert future disputes.

Early dispute resolution requires bringing all parties together to seek a team solution to the conflict. If the agency, any contractor involved, roadway users, suppliers, etc. who are sources of the dispute can be brought together to "find an answer we can all live with," frequently persons can find a common ground without litigation erupting. The key is to get conflicting parties talking before a large emotional investment has been made in any particular position. All complaints and concerns need to be treated seriously and dealt with promptly.

Once disputes are firmly set with opposing positions, resolution without court litigation requires some form of arbitration or negotiation be available. It is usually better if the negotiator or arbitrator is a neutral third party not seen as having any real interest in either side. Both sides of a dispute have to agree to be bound by the arbitrator's finding. Otherwise, this is only a later-stage effort to recognize and avoid disputes.

An extensive survey has been conducted of the efforts of various state departments of transportation to initiate risk management [3]. The survey results were analyzed to provide typical profiles of approaches and methods that have been deemed successful. A key element in the implementation of good risk management is always good data on the system's condition, what maintenance was done when and by whom, and a paper trail of the engineering decision process in determining what maintenance was to be done when. Thus, having effective maintenance management systems and bridge maintenance management systems in place are an integral part of good risk management.

The successful defense of lawsuits arising from claims of maintenance negligence are significantly aided by good documentation of the actual conditions at the time of an accident. Field forces equipped with photographic equipment and instructed in its proper use can document the conditions existing at the time of the accident. Videotaped evidence of the conditions has also been found to be effective [4].

Maintenance organizations should establish liaisons with persons in their agency who monitor research and development in transportation safety. Maintenance engineers and maintenance managers should attempt to keep abreast of the evolving safety literature. By maintaining contact with engineers and managers who are involved in tracking safety research, therefore, they can more easily keep current with new developments that may apply to maintenance operations. For example, a research article on tort liability associated with utility poles in the right of way may interest maintenance engineers with responsibility for urban roadways [5]. Another recent research article that relates a roadway surface's potential to produce hydroplaning and roadway tort liability might interest a maintenance engineer who is responsible for the maintenance of high-speed roadways in regions with high-rainfall climates [6]. The roadway and bridge safety literature is quite extensive; therefore, no attempt has been made to cover it here. Because safety research affects the decisions of maintenance engineers and maintenance managers in planning for and conducting maintenance, however, there is some merit in keeping current with developments in this area.

Maintenance engineers and maintenance managers with little experience in responding to litigation alleging maintenance defects in a roadway or bridge network should consider reviewing some literature that will provide an understanding of liability issues. One excellent source is the National Cooperative Highway Research Program Legal Research Digest series. For example, some recent issues covering topics relevant to a maintenance engineer's or maintenance manager's scope of responsibilities include the following:

- Number 40: *Liability of Highway Departments for Damages Caused by Stormwater Runoff*, March 1998. This digest discusses how routine maintenance of drainage facilities may contribute to a reduction in liability lawsuits.
- Number 38: *Risk Management for Transportation Programs Employing Written Guidelines as Design and Performance Standards*, August 1997. The principles discussed here also apply to written and publicly available maintenance standards.
- Number 35: *Continuing Project on Legal Problems Arising Out of Highway Programs*, 1995. The contents of these periodic summaries sometimes contain information directly applicable to maintenance.
- Number 34: *Transportation Agencies as Potentially Responsible Parties at Hazardous Waste Sites*, 1995. While hazardous waste sites may be found in new route location rights of way, maintenance garage yards are also frequent hazardous waste sites.
- Number 31: *Federal Air Quality Laws Governing State and Regional Transportation Planning*, 1994. Typically, state environmental quality regulations provide guidance on how to meet these requirements. However, reading summaries such as these provides a better understanding of the degree to which maintenance operations may be exempted or required to comply.

- Number 29: *Highways and the Environment: Resource Protection and the Federal Highway Program*, October 1994. Maintenance engineers and maintenance managers can better appreciate the concerns of environmental groups with respect to maintenance methods and materials if they have reviewed an analysis of how environmental law applies to highway programs.
- Number 27: *Supplement to Liability of the State for Injury or Damage Occurring in Motor Vehicle Accidents Caused by Trees, Shrubbery, or Other Vegetative Obstruction Located in Right-of-Way or Growing on Adjacent Private Property*, 1993.
- Number 26: *Supplement to Legal Implications of Highway Department's Failure to Comply With Design, Safety, or Maintenance Guidelines*, 1992.

6.2 REFERENCES

1. Lewis, Russell M. *Managing highway tort liability*. Synthesis of Highway Practice 206. Washington, D.C.: National Cooperative Highway Research Program, Transportation Research Board, 1994.

2. Bramble, Barry B., and Mark D. Cipollini. *Resolution of disputes to avoid construction claims*. Synthesis of Highway Practice 214. Washington, D.C.: National Cooperative Highway Research Program, Transportation Research Board, 1995.

3. Demetsky, Michael J., and Kathy Yu. *Assessment of risk management and objectives on state departments of transportation*. Transportation Research Record 1401. Washington, D.C.: Transportation Research Board, 1993, pp. 1–8.

4. Turner, Daniel S. *Video evidence for highway tort trials*. Transportation Research Record 1464. Washington, D.C.: Transportation Research Board, 1994, pp.86–91.

5. Najafi, Fazil T., Fadi Emil Nassar, and Paul Kaczorowski. *Tort liability related to utility pole accidents in Florida*. Transportation Research Record 1401. Washington, D.C.: Transportation Research Board, 1993, pp. 111–116.

6. Mounce, John M., and Richard T. Bartoskewitz. *Hydroplaning and roadway tort liability*. Transportation Research Record 1401. Washington, D.C.: Transportation Research Board, 1993, pp. 117–124.

7.0 INDEX

Bridge inspection 3-155
   As performed by supervisors 3-158
   Identifying fatigue-critical bridges 3-159
   Load tests 3-165
   National Bridge Inspection Program 3-155
   Performance of the inspection 3-155
   Process 3-155
   Reports 3-158
   Resource estimating 3-158
   Steel member tests 3-163
   Testing existing bridge components 3-161
   Timber member tests 3-164

Bridge maintenance (general) 3-1
   Load-carrying capacity of bridge 3-1
   Maintenance concepts 3-11
   References 3-135
   Structural systems 3-3
      Arch bridges 3-5
      Beam bridges 3-3
      Cable-supported bridges 3-6
      Materials 3-9
      Truss bridges 3-7

Bridge maintenance management 3-138
   Contract maintenance 3-148
      Contract vs. in-house maintenance 3-152
      Cost reimbursement 3-150
      Lump sum 3-149
      Negotiated 3-151
      Unit price 3-150
   General 3-138
      PONTIS 3-140
      Process 3-138
      System interfaces 3-141
      System-level vs. project-level
        decisions 3-140
      System requirements 3-140
   Performing work 3-145
      Equipment 3-146
      Job execution 3-145
      Materials 3-147
      Personnel resources 3-146

Planning and scheduling 3-142
   Graphical scheduling methods 3-145
   Job layout 3-142
   Long-range plans 3-142
   Short-term scheduling 3-144
   Work orders 3-142
References 3-170
Reporting accomplishments 3-147
   Procedures 3-148
   Requirements 3-147
   Why report? 3-147

Bridge maintenance quality control and quality assurance 3-152
   Budget monitoring 3-154
   Methods of quality assurance 3-154
   Quality control at work site 3-152
   Rigging and climbing site review 3-153
   Schedule monitoring 3-154
   Technical site review 3-153
   Tools and equipment use site review 3-153
   Traffic control site review 3-153

Bridge structural decks 3-17
   Concrete decks 3-18
      Asphaltic concrete patching 3-21
      Crack sealing 3-22
      Emergency full-depth patching 3-21
      Epoxy patching 3-21
      Overlays 3-22
      Patching 3-19
      Preventive maintenance 3-18
      Replacement 3-24
      Sealing 3-19
   Curbs 3-34
      Asphalt 3-35
      Concrete 3-35
      Metal 3-34
      Stone 3-35
      Timber 3-35
   Deck drainage systems 3-33
      Gratings 3-34
      Open joints with troughs 3-34
      Scuppers 3-33
   General 3-17

Maintaining bridge deck joints 3-26
    Joint classifications 3-27
        Closed joints 3-29
            Compression seal 3-29
            Cushion seal 3-30
            Filled butt joint 3-29
            Membrane seal 3-30
            Modular dam joint 3-31
        Open joints 3-27
            Butt joint 3-27
            Finger joint 3-28
            Plate joint 3-28
    Joint functions 3-27
    Joint types 3-26
    Preventive maintenance 3-32
Railings 3-36
    Aluminum 3-37
    Concrete 3-36
    Steel 3-36
Sidewalks 3-36
Steel grid decks 3-25
Timber decks 3-24

Bridge substructure 3-58
    Bent and pile repair 3-70
        End bent repair 3-76
            Pile splice 3-76
        Installing deadman anchorage 3-79
        Installing helper bent 3-76
        Intermediate bent repair 3-74
            Adding a section of timber pile 3-75
            Splicing timber piles, steel columns 3-75
            Steel pile bents 3-74
        Jackets for protection and repair 3-70
        Pile repair 3-72
            Casting subfooting to cap piles 3-74
            Patching concrete pile 3-73
            Pile shell repair 3-74
            Splicing steel H-piles under footings 3-74
            Steel pile 3-72
        Pile replacement 3-72
        Preventive maintenance 3-70
        Timber cap repair 3-78
            Cap rotating 3-78
            Timber cap replacement 3-79
            Timber cap strengthening 3-79
    Maintenance (general) 3-59
    Preventive maintenance 3-59
    Repair process (general) 3-60
    Repair substructure above water line 3-61
        Abutment face 3-62
        Crack in substructure 3-63
        Deterioration at water line 3-64
        Spread footing 3-63
        Surface deterioration 3-64
        Wing walls 3-61

    Substructure cap problems 3-59
    Underwater repair of substructure 3-65
        Concrete repair 3-67
            Bagged concrete 3-68
            Concrete mix 3-68
            Concrete removal 3-67
            Forms 3-67
            Free-dump concrete 3-69
            Hand-placed concrete 3-69
            Prepackaged aggregate concrete 3-68
            Pumped concrete 3-69
            Tremie concrete 3-69
            Underwater placement 3-68
        Engineering aspects 3-65
        Pressure injection of cracks 3-66
        Protecting underwater bridge elements 3-66
        Regulatory control 3-66

Bridge superstructure 3-37
    Beam repair 3-46
        Heat straightening of steel beams 3-50
        Post-reinforcement of concrete beams 3-47
        Prestressed-concrete beams 3-48
        Reinforced-concrete beams 3-46
        Steel beams 3-49
        Timber beams 3-51
    Bearing maintenance and repair 3-40
        Elastomeric bearing 3-44
        Elevating bearings 3-46
        Pin-and-hanger bearing 3-44
        Pot bearing 3-45
        Resetting or rehabilitating bearings 3-46
        Rocker bearing 3-43
        Roller bearing 3-42
        Sliding plate bearing 3-41
    Jacking and supporting the superstructure 3-37
        Carrier beam 3-38
    Truss repair 3-54
        Heat straightening truss members 3-58
        Maintenance and repair of steel trusses 3-54
            Localized damage repair 3-57
            Repair tension members 3-55
            Replace compression members 3-56
        Modify portal bracing to increase vertical clearance 3-57

Bridge traveled surfaces 3-12
    Asphalt 3-15
        Cracking 3-15
        Disintegration 3-16
        Distortion 3-16
    Concrete 3-13
        Cracking 3-14
        Scaling 3-13
        Spalling 3-13
    Steel grate 3-17
    Stone and brick 3-16
    Wood 3-17

Buildings and facilities 2-56
   District and field offices 2-57
   Equipment repair buildings 2-57
   Equipment sheds 2-57
   Locker rooms 2-57
   Paint and sign shops 2-57
   Stockpiles 2-58
   Storerooms 2-57
   Tool rooms 2-57
   Wood and metalworking shops 2-58

Contract maintenance
   See "bridge maintenance management."

Developing issues
   Roadway maintenance 2-92
      Designing synthetic materials 2-93
      Evolving equipment systems 2-94
      Traffic demand vs. work zone capacity 2-95
      Using indirect waste products 2-92
   Roadway maintenance management 2-102

Developing issues in bridge maintenance 3-135
   Nonmetallic bridges 3-135
   Nonmetallic repair of steel bridge components 3-135

Environmental aspects of bridge maintenance 3-98
   Creosote-treated timber 3-102
   Confined spaces 3-106
   Fall protection, rigging, scaffolding, and hoisting 3-109
      Fall protection 3-109
      Hoisting 3-117
      Ladders 3-111
      Rigging 3-112
      Scaffolding 3-116
   Hauling and disposal regulations 3-101
   Hazardous wastes 3-102
   Historic structures 3-102
   Lead-based paint removal 3-98
   Maintenance management 1-26
   Noise control 3-102
   Other concerns 3-99
      Disturbing streambeds 3-100
      Stream classifications 3-101
      Water quality 3-101
   Safety review 3-120
   Toxic materials 3-106
   Work site safety review 3-105
   Worker safety 3-103
      Shoring and falsework 3-105
      Work site 3-103

Environmental aspects of roadway maintenance 2-84
   Issues 2-84
   Issues in roadway maintenance management 2-102
   Maintenance actions 2-87
   Maintenance management 1-26

Equipment systems 4-1
   Acquisition and replacement 4-5
   Application 4-2
   General 4-1
   Maintenance of equipment 4-4
   References 4-8

Human resources 1-22
   Research and development 5-3
   See "personnel."

Illumination for roadways and tunnels 2-74
   Bridge lighting 2-79
   Circuits and controls 2-78
   Cleaning 2-76
   General considerations 2-75
   Lamp replacement 2-77
   Poles and brackets 2-78
   Tunnel illumination 2-78

Litigation 6-1
   Maintenance management 1-27
   Shoulder maintenance 2-38
   Snow and ice control liability 2-61
   References 6-3
   Tort claims 6-1

Maintenance management 1-1
   Concepts 1-1
   Emergency response 1-20
   Environmental maintenance 1-26
   Functional aspects 1-8
   Incident management 1-20
   Integrated management systems 1-3, 1-25
   Litigation in maintenance 1-27
   Maintenance budgeting 1-17, 2-5
   Maintenance data systems 1-20
   Maintenance equipment 1-24
   Maintenance management 1-27
   Maintenance materials 1-24
   Maintenance planning 1-17, 2-5
   Motivation of personnel 1-22
   Organization structure for maintenance 1-13
   Planning and budgeting 1-17, 2-5
   Preventive vs. reactive maintenance 1-5
   Recruitment and development 1-24
   References 1-28

Safety of personnel 1-21, 1-23
Scheduling maintenance 1-18, 2-6
Total quality management 1-3, 1-27
Training personnel 1-23
Work zone safety 1-21

Movable bridges 3-121
    Inspection and maintenance of specific parts and components 3-123
        Bearings 3-123
        Ropes 3-123
    Inspection items 3-126
    Machinery and equipment maintenance 3-122
    Repairs by bridge type 3-123
        Bascule bridges 3-123
        Lift bridges 3-124
        Other 3-125
        Swing bridges 3-125
    Seating 3-123
    Structural maintenance 3-122
    Types 3-121

Pathways 2-59

Personnel 1-22
    Motivation 1-22
    Safety 1-23
    Training 1-23

Preliminary site visit (planning bridge repair) 3-165
    Approach damage 3-166
    Basic information required 3-165
    Channel damage 3-166
    Deck damage 3-167
    Emergency damage 3-169
    Substructure damage 3-169
    Superstructure damage 3-168

Preventive maintenance
    Bridge bents and piles 3-70
    Bridge deck joints 3-32
    Bridge substructure 3-59
    Concrete bridge decks 3-18
    Definition 2-1
    Maintenance management 1-5, 2-1
    Roadway surface 2-24 through 2-27
    Traffic control
        Electrical systems 2-72
        Electronic systems 2-72

Protective systems for bridges 3-90
    Protecting substructure 3-90
    Spot-painting superstructure 3-91
        Care and storage of paint 3-95
        Defects in paint 3-96
        Guidelines for spot-painting 3-95
        Inspection of painting 3-96
        Overcoating 3-94
        Paint systems 3-91
        Surface preparation 3-92
        Weathering steel 3-97

References
    Bridge maintenance 3-135
    Bridge maintenance management 3-170
    Equipment systems 4-8
    Litigation 6-3
    Maintenance management
        Bridges 3-170
        General 1-28
        Roadways 2-106
    Research and development 5-4
    Roadway maintenance 2-95
    Roadway maintenance management 2-106

Research and development 5-1
    Administration 5-1
    Equipment 5-3
    Human resources and personnel 5-3
    International aspects and technology transfer 5-3
    Materials 5-2
    Methods 5-2
    References 5-4

Roadside maintenance (general) 2-46
    Aesthetic priorities 2-50
    Functional priorities 2-47
    Planning 2-51
    Vegetation management 2-51
    References 2-95

Roadway drainage maintenance (general) 2-39
    Surface drainage 2-40
        Check dams and berms 2-41
        Chutes, flumes, spillways, slope drains 2-42
        Curbs and gutters 2-40
        Interceptor ditches, diversion ditches, bench-cut slope ditches 2-41
        Natural watercourses and bank protection 2-42
        Roadside ditches 2-41
        Shoulder inlets and side drains 2-41
    Subsurface drainage 2-43
        Catch basins, drain openings with grates 2-44
        Culverts 2-43
        Earth slides 2-45
        Storm sewers 2-44
        Underdrains 2-44

Roadway maintenance management 2-102
  Developing issues 2-104
  Environmental issues 2-104
  General 2-102
  Inspection and condition inventory 2-103
  Interface with other systems 2-102
  References 2-106

Roadway surface maintenance 2-4
  Aggregate (granular) surface roads 2-7, 2-17
    Compaction 2-30
    Corrugations 2-18
    Dust 2-18
    Potholes 2-18
    Ruts 2-18
    Scarifying and blading 2-30
    Soft spots 2-18
  Flexible pavements 2-7
    Bleeding (flushing surfaces) 2-11
    Compaction 2-30
    Corrugations 2-10
    Cracks (general) 2-9
      Corrective measures 2-9
      Shrinkage cracks 2-9
      Structural weakness cracks 2-9
    Flushing surfaces (bleeding) 2-11
    Polishing aggregates 2-11
    Potholes 2-8, 2-28
    Raveling surfaces 2-11
    Settlements and upheavals 2-11
    Wheel ruts 2-7
  Materials 1-24, 2-19
    Aggregates 2-22
    Asphalt materials 2-19
    Asphalts
      Asphalt-aggregate mixtures 2-20
      Asphalt-aggregate surface treatments 2-20
      Asphalt-only surface treatments 2-20
    Crack and joint fillers 2-23
    Epoxies, resins, and polymers (general) 2-21
    Premixed materials 2-23
      Hot plant mix 2-23
      Stockpile mixes 2-24
    Portland cement types 2-21
  Preventive maintenance definition 2-1, 2-24 through 2-27
  References 2-95
  Rigid pavements 2-10, 2-12
    Blowups 2-16
    Cracks and defective joints 2-14
    Faulting and settlements 2-16
    Frost heaves 2-16
    Pumping 2-16, 2-31
    Surface texture failures 2-13
    Spalling 2-15, 2-29, 2-32
  Surface maintenance methods 2-24
    Cross section 2-34
    Pavement deflection testing 2-34
    Polymer concrete patches 2-21, 2-33
    Preventive maintenance 2-24
      Asphalt rejuvenation 2-27
      Chip seals 2-25
      Crack and joint sealing 2-25
      Dust palliatives 2-25
      Fog seal 2-25
      Hot-weather sanding 2-27
      Sealing concrete surfaces 2-27
      Slurry seal 2-24
      Sweeping and cleaning 2-27
    Repairs 2-28
      Compaction 2-30
      Fills and overlays 2-32
      Patching 2-28
        Asphalt pavements 2-28
        Portland cement concrete pavements 2-29
      Pavement burning 2-33
      Scarifying and blading 2-30
      Spall repairs 2-32
      Subsurface repairs 2-31
      Tack coats and prime coats 2-28
    Soil stabilization 2-34
  System interactions 1-3 through 1-8

Roadways in tunnels 2-83

Roadways on bridges 2-80
  Inspection 2-80, 2-82
  Marine navigation lights 2-82
  Movable bridges 2-82

Safety feature maintenance 2-54
  Fences 2-55
  Guardrails and barriers 2-54
  Impact attenuators 2-55
  Traffic islands and curbs 2-55
  Truck escape lanes or ramps 2-55

Safety related to maintenance
  Fall protection, rigging and hoisting 3-109
  Hazardous wastes 3-102
  Maintenance management 1-21, 1-23
  Maintenance of safety features 2-54
  Navigation lighting on movable bridges 2-82
  Safety review for environmental aspects in bridge maintenance 3-120
  Shoulder maintenance 2-38
  Snow and ice control 2-59
  Toxic materials 3-106
  Work zone traffic control 2-88, 3-128
  Worker safety 3-103
  Worker safety notes 2-92

Shoulder maintenance (general)   2-35
   Bituminous shoulders   2-37
   Earth (turf) shoulders   2-36
   Gravel (crushed stone) shoulders   2-37
   Portland cement concrete shoulders   2-37
   Scheduling maintenance (general)   2-38
      Functional need schedule   2-38
      Litigation-driven schedule   2-38

Snow and ice control   2-59
   Bridges and overpasses   2-64
   Chemicals and abrasives   2-64
   Cooperative agreements   2-61
   Equipment   2-65
   Interurban areas   2-64
   Liability   2-61
   Mountain passes   2-64
   Planning and budgeting   2-60
   Related services   2-62
   Scheduling   2-62
      Equipment   2-63
      Personnel   2-63
      Materials   2-63
   Snow fencing and guide marking   2-66
   Training   2-61
   Urban areas   2-63

Traffic control devices and roadway instrumentation   2-67
   Electrical and electronic equipment   2-70
      Electrical systems preventive maintenance   2-72
   Highway signs   2-67
   Markings   2-69
   Traffic signals   2-73

Watercourse and embankment maintenance   3-80
   Removing debris from channel   3-80
      Bridge damage from debris   3-80
      Debris countermeasures   3-82
      Debris removal   3-81

Scour protection and repair   3-83
   Flow control measures   3-87
      Check dams   3-88
      Dikes   3-87
      Jack fields   3-88
      Retards   3-87
      Spur dikes   3-87
      Spurs   3-87
   Identifying need for countermeasures   3-83
   Material replacement for scour or undermining   3-88
      Riprap   3-88
      Concrete   3-89
   Rebuilding scoured streambeds   3-89
   Revetments   3-84
      Bulkheads   3-87
      Concrete block matting   3-85
      Concrete-filled fabric mats   3-87
      Concrete pavement   3-86
      Dumped riprap   3-84
      Grout-filled bags   3-85
      Grouted rock riprap   3-86
      Hand-placed riprap   3-85
      Sacked concrete   3-86
      Tetrapods   3-85
      Vegetation   3-85
      Wire-enclosed rock riprap   3-85
   Role of maintenance forces   3-83
   Settled and tilted pier repair   3-90
   Structure or channel modifications   3-88

Worker safety notes   2-92

Work zone traffic control   2-88
   Principles of good practice   2-89
   Some practice notes   2-91

Work zone traffic control for bridge maintenance   3-128
   Control of work zones (general)   3-128
   Temporary structures   3-133